Human Induced Environmental Threats

Human Induced Environmental Threats

Edited by **Rosemary Charles**

SYRAWOOD
PUBLISHING HOUSE

New York

Published by Syrawood Publishing House,
750 Third Avenue, 9th Floor,
New York, NY 10017, USA
www.syrawoodpublishinghouse.com

Human Induced Environmental Threats
Edited by Rosemary Charles

International Standard Book Number: 978-1-68286-147-9 (Hardback)

Contents

Preface

This book was inspired by the evolution of our times; to answer the curiosity of inquisitive minds. Many developments have occurred across the globe in the recent past which has transformed the progress in the field.

Humans have relied primarily on their surrounding environment for survival and sustenance. However, the interactions between humans and environment have resulted in massive deterioration of environment. This book explores the diverse implications of human activities that are destroying the ecosystems of our planet and endangering all the species on earth. It is a compilation of chapters that cover some of the significant aspects in this field of study such as agriculture and food security, protection and preservation of biodiversity, pollution, environment and human interaction, environmental education, etc. The extensive content of this book provides the readers with a thorough understanding of this important subject area.

This book was developed from a mere concept to drafts to chapters and finally compiled together as a complete text to benefit the readers across all nations. To ensure the quality of the content we instilled two significant steps in our procedure. The first was to appoint an editorial team that would verify the data and statistics provided in the book and also select the most appropriate and valuable contributions from the plentiful contributions we received from authors worldwide. The next step was to appoint an expert of the topic as the Editor-in-Chief, who would head the project and finally make the necessary amendments and modifications to make the text reader-friendly. I was then commissioned to examine all the material to present the topics in the most comprehensible and productive format.

I would like to take this opportunity to thank all the contributing authors who were supportive enough to contribute their time and knowledge to this project. I also wish to convey my regards to my family who have been extremely supportive during the entire project.

Editor

What are the effects of agricultural management on soil organic carbon (SOC) stocks?

Bo Söderström[1*], Katarina Hedlund[2], Louise E Jackson[3], Thomas Kätterer[4], Emanuele Lugato[5], Ingrid K Thomsen[6] and Helene Bracht Jørgensen[2]

Abstract

Background: Changes in soil organic carbon (SOC) stocks significantly influence the atmospheric C concentration. Agricultural management practices that increase SOC stocks thus may have profound effects on climate mitigation. Additional benefits include higher soil fertility since increased SOC stocks improve the physical and biological properties of the soil. Intensification of agriculture and land-use change from grasslands to croplands are generally known to deplete SOC stocks. The depletion is exacerbated through agricultural practices with low return of organic material and various mechanisms, such as oxidation/mineralization, leaching and erosion. However, a systematic review comparing the efficacy of different agricultural management practices to increase SOC stocks has not yet been produced. Since there are diverging views on this matter, a systematic review would be timely for framing policies not only nationally in Sweden, but also internationally, for promoting long-term sustainable management of soils and mitigating climate change.

Methods: The systematic review will examine how changes in SOC are affected by a range of soil-management practices relating to tillage management, addition of crop residues, manure or other organic "wastes", and different crop rotation schemes. Within the warm temperate and the snow climate zones, agricultural management systems in which wheat, barley, rye, oats, silage maize or oilseed rape can grow in the crop rotation will be selected. The review will exclusively focus on studies conducted over at least 10 years. Searches will be made in 15 publication databases as well as in specialist databases. Articles found will be screened using inclusion/exclusion criteria at title, abstract and full-text levels, and screening consistency will be evaluated using Kappa tests. Data from articles that remain after critical appraisal will be extracted using a predefined spreadsheet. Subgroup analyses will be undertaken to elucidate statistical relationships that are specific to particular type of management interventions. Meta-regression within subgroups will be performed as well as sensitivity analysis to investigate the impact of removing groups of studies with low or unclear quality.

Keywords: Soil organic carbon, Agricultural practices, Long-term, Tillage, Fertilization, Crop rotation, Cover crop, Sequestration

Background

The largest global stock of organic carbon (C) on land is contained in soils (2500 Pg of C to 2-m depth) and is about twice as large as the atmospheric C stock [1-3]. Changes in soil C stocks may thus significantly influence the atmospheric carbon dioxide (CO_2) concentration. Since approximately 12% of the soil C stock is present in cultivated soil [3] and agricultural soils occupy about 35% of the global land surface [4], soil management is potentially a powerful tool for climate change mitigation through C sequestration [5,6]. Additional benefits from increasing C stocks in agricultural soils are increased soil fertility [7,8] and improved physical and biological properties of the soil [9] by reduced bulk density, increased water-holding capacity, improved soil structure and enhanced microbial activity [10].

It is important to acknowledge that an increase in the soil C stock does not imply a decrease in the atmospheric C stock by the same amount, since the management employed to achieve increased stocks of soil organic carbon (hereafter denoted as SOC) may themselves be using

* Correspondence: bo.soderstrom@eviem.se
[1]Mistra Council for Evidence-Based Environmental Management, Royal Swedish Academy of Sciences, P.O. Box 50005, SE-104 05 Stockholm, Sweden
Full list of author information is available at the end of the article

non-renewable energy and cause changes in the atmospheric C stock [3,8]. To feed a growing world population, converting land from annual cropping to, for example, forest or grassland may require conversion of land in the opposite direction elsewhere [11]. The net effect of a certain land-use change or soil management practice on atmospheric CO_2 needs thus to be considered in a broader context [12].

Guo and Gifford [13] performed a meta-analysis of data from 74 publications indicating that soil C stocks decline after land-use changes from pasture to plantation (-10%), native forest to plantation forest (-13%), native forest to cropland (-42%), and pasture to cropland (-59%). They also found that soil C stocks increase after land-use changes from native forest to pasture (+8%), cropland to pasture (+19%), cropland to plantation forest (+18%), and cropland to secondary forest (+53%). The results varied, however, depending on factors such as annual precipitation, plant species and, not least, length of study periods.

It is quite evident that pastures and forests, whether native or plantation, compared to cropland, are more efficient in storing C in the soil. In Sweden, it has been calculated that nationwide the 270 Tg C stock in agricultural surface soil (0–25 cm) is actually decreasing at a rate of 1 Tg year^{-1} [14]. The loss of C from agricultural soils on a global scale has been a matter of considerable debate, but according to Lal [15] the C flux from soil to the atmosphere is estimated to be 0.8–1.2 Pg C year^{-1}, whereas C flux from soil to the ocean is 0.6 Pg C year^{-1}.

Even though organic C in many agricultural soils is being depleted through various mechanisms (oxidation/mineralization, leaching and erosion), there are measures other than land-use changes that potentially can slow down or reverse this trend. Such measures include: i) diverse crop rotations including, for example, leys and cover crops, ii) organic amendments such as manure or crop residues, and iii) tillage modifications such as minimum or no tillage.

Previous studies

The literature on carbon sequestration in agricultural soils is extensive. However, SOC responds slowly to changes in agricultural management [16]. Most SOC changes require many years to be detectable by present analytical methods [17], and therefore long-term experiments are required. Nevertheless, a substantial number of studies have been performed and few reviews have been published recently [6,13,15,18,19].

Gonzalez-Sanchez et al. [6] concluded from a meta-analysis of data of 29 publications (from Spain) that some forms of conservation agriculture (i.e., no tillage and implementing cover crops) can have positive effects on SOC. Govaerts et al. [18] reviewed three aspects of

conservation agriculture: reduction in tillage, retention of crop residues and use of crop rotations. The data (mainly from the Americas) indicated that the largest contribution of conservation agriculture to reducing emissions from farming activities is from the reduction of tillage operations.

Soilservice [19] reviewed the soil organic matter (SOM) content, which is closely linked to SOC, in conventional and organic farming, respectively. The conventional farming areas included management regimes with mineral fertilizer and/or pesticide application, whereas organic fields included management types with organic fertilizer and no pesticides. For the period 1945–2009 they found 29 studies meeting their screening criteria for meta-analysis. The results indicated a positive effect of organic fertilizers and/or no pesticides on SOM content (Figure 1).

Identification of topic and stakeholders

The environmental benefits of increasing SOC stocks in cropland are mainly related to climate change mitigation, agricultural sustainability and land-use issues (by enhancing the productivity of the soil less land is needed to produce a certain amount of food and fiber). The topic was suggested by Karin Hjerpe at the Swedish Board of Agriculture (May 4, 2012 and September 20, 2012). At the general stakeholder meeting arranged by Mistra EviEM (September 24, 2012), the suggestion was put forward again by Olof Johansson at the Swedish

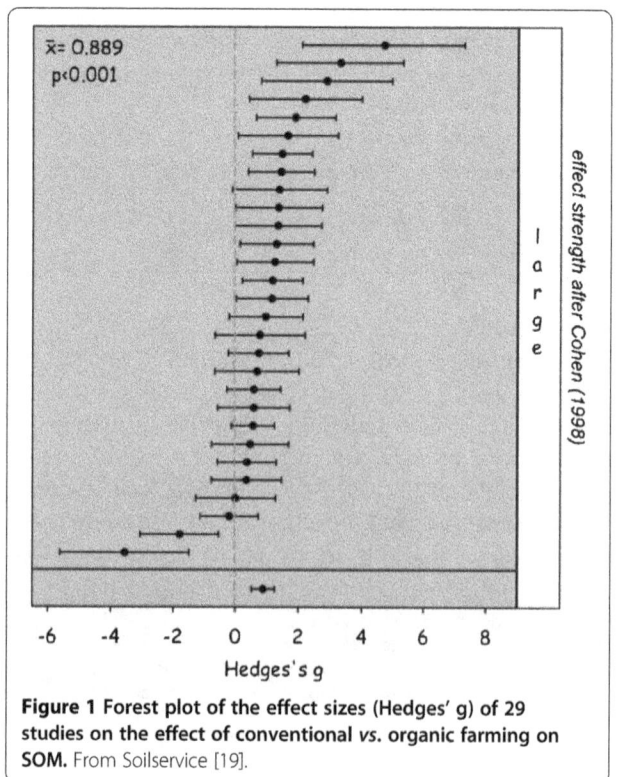

Figure 1 Forest plot of the effect sizes (Hedges' g) of 29 studies on the effect of conventional *vs.* organic farming on SOM. From Soilservice [19].

Board of Agriculture. The Swedish Board of Agriculture is responsible for the national environmental quality objective "A varied agricultural landscape". One expected outcome within this goal is that arable land will have a well-balanced nutrient status, good soil structure and humus content. Another expected outcome is that the land will be cultivated in such a way as to sustain the long-term productivity of the soil. These outcomes are closely related to SOC. The Swedish Board of Agriculture is also involved in issues regarding climate change. The agency has been commissioned by the government to work out an action plan aimed at reducing greenhouse gas emissions from Swedish agriculture. In their reports [20,21] it was concluded that while there is a large potential for C sequestration in soils globally, it is not clear how significant it is for measures that can be applied in Sweden and in Swedish climatic conditions.

The Swedish Environmental Protection Agency (EPA) is another user of the suggested review. The Swedish EPA is responsible for the environmental quality objective "Reduced Climate Impact". In this context, the Swedish Parliament has adopted a vision of zero net emissions of greenhouse gases to the atmosphere in Sweden by 2050.

The review is also of interest for the Federation of Swedish Farmers (LRF), which is interested in both the environmental issues and the productivity aspect. In their Climate Policy it is stated that increased SOM content in cropland potentially can reduce concentrations of greenhouse gases in the atmosphere and that such opportunities should be seized. The Federation of Swedish Farmers is also taking part in Focus on Nutrients ("Greppa Näringen" in Swedish), which is a joint venture between LRF, the Swedish Board of Agriculture, the County Administrative Boards and a number of companies in the farming sector. Focus on Nutrients offers advice to farmers on, e.g., climatic issues and SOC management.

The systematic review question is scientifically relevant and has received considerable research interest. Although several meta-analyses and literature reviews have been published for example [6,13,15], a systematic review comparing the efficacy of different management techniques to increase SOC stocks in agricultural areas has not yet been produced. Since there are diverging views on this matter, a systematic review would be timely. Thus, both the primary user of the review (the Swedish Board of Agriculture) and scientists from the Swedish University of Agricultural Sciences endorsed the idea of a systematic review on this topic should be conducted.

During a stakeholder meeting at the EviEM secretariat (June 4, 2013), representatives from the Swedish Board of Agriculture, Swedish Environmental Protection Agency, Federation of Swedish Farmers, and Swedish University of Agricultural Sciences discussed the formulation of the review question and exclusion/inclusion criteria. It was suggested that the focus should be on long-term studies of how agricultural management affect SOC stocks within the temperate climate zone (humid and summer dry) as well as the snow climate zone (northern Sweden). The stakeholders put forward that cereal grains such as wheat and barley were of particular interest, but also other crops that could become more important in Sweden in a changing climate (such as maize). All agricultural management types and soil types within these agricultural regions were of interest. Greenhouse gases other than CO_2, such as methane (CH_4) and nitrous oxide (N_2O), and studies solely focusing on soil phosphorus and nitrogen were considered to be outside the systematic review's scope. There is a lack of data on CH_4 since it is not often measured in upland soils. Similarly, there are few data on long-term changes in N_2O in which contrasting treatments have reached a new equilibrium. It is therefore difficult to integrate short-term N_2O processes with long-term trends in SOC changes. Stakeholders also underlined, that although the review question by definition must be fairly narrow, the narrative synthesis should attempt to have a systemic approach. For example, SOC may increase under bioenergy crops, but if the total cropped area is the same, less food will be produced. Certain interventions may also require increased use of non-renewable energy leading to a reduced net effect on carbon emissions.

Objective of the review

The effect of land-use change on SOC stocks has been documented in many parts of the world. However, more pertinent to the systematic review suggested here is that there also are a fair number of studies on the effects of various soil management practices within a given type of land-use, e.g., cropland, on SOC stocks. In order to enable a quantitative evaluation, or a meta-analysis, the various soil management practices should be well defined and, if possible, treated separately. These include i) diverse crop rotations with winter cover crops and leys, ii) organic amendments such as manure or crop residues, iii) tillage modifications such as minimum or no tillage.

Primary question: What are the effects of agricultural management on soil organic carbon (SOC) stocks?

Components of the primary question:

Population: Arable soils in agricultural regions from the warm temperate climate zone and the snow climate zone (according to the Köppen-Geiger climate classification; see *Relevant subjects* below).

Within these climate zones, agricultural management systems in which wheat, barley, rye, oats, silage maize or oilseed rape *can* grow in the crop rotation will be selected.

Intervention: A range of soil management practices relating to tillage management, addition of crop residues, manure or other organic "wastes", and different crop rotation schemes.

Comparator: Different or no intervention.

Outcome: Changes of SOC stocks, measured as a relative rate of change per year.

Methods

Searches

A review scoping exercise was conducted to test alternative search strings. The exercise resulted in the selection of the following search terms:

Population: soil*

Population: arable, agricult*, farm*, crop*, cultivat*

Intervention: till*, "direct drill*", fertili*, bio*solid*, organic, manur*, sewage, compost*, amendment*, biochar*, digestate*, crop residue*, crop straw*, mulch*, crop rotat*, break crop*, (grass OR clover) ley*, legume*, bioenergy crop*, cover crop*, "grass clover", "crop* system*", winter crop* , spring crop*, summer fallow*, "catch-crop*", intercrop*, conservation

Outcome: "soil organic carbon", "soil carbon", "soil C", "soil organic C", SOC, "carbon pool", "carbon stock", "carbon storage", "soil organic matter", SOM, "carbon sequestrat*", "C sequestrat*"

The terms within each of the categories 'population', 'intervention', and 'outcome' will be combined using the Boolean operator 'OR'. The four categories will then be combined using the Boolean operator 'AND'. An asterisk (*) is a 'wildcard' that represents any group of characters, including no character. The use of Boolean operators and truncation will be modified according to the idiosyncrasies of each publication database and how this is done will be documented.

The following search strings will be used:

English: soil* AND (arable OR agricult* OR farm* OR crop* OR cultivat*) AND (till* OR "direct drill*" OR fertili* OR bio*solid* OR organic OR manur* OR sewage OR compost* OR amendment* OR biochar* OR digestate* OR crop residue* OR crop straw* OR mulch* OR crop rotat* OR break crop* OR (grass OR clover) ley* OR legume* OR bioenergy crop* OR cover crop* OR "grass clover" OR "crop* system*" OR winter crop* OR spring crop* OR summer fallow* OR "catch-crop*" OR intercrop* OR conservation) AND ("soil organic carbon" OR "soil carbon" OR "soil C" OR "soil organic C" OR SOC OR "carbon pool" OR "carbon stock" OR "carbon storage" OR "soil organic matter" OR SOM OR "carbon sequestrat*" OR "C sequestrat*")

In addition to data in the scientific literature it is anticipated that data will be found also in the grey literature. Such data will be searched for using search engines and specialist websites using the simplified search terms given below. In each case, the first 100 hits based on relevance will be examined for appropriate data. No particular time or document type constraints will be applied. In addition, a search in Google Scholar based on title words only (advanced search) will also be made since partly different articles may be found.

English: (carbon AND sequestration AND soil AND agriculture)

Swedish: (kol AND lagring AND mark AND jordbruk)

Danish: (kulstof AND indhold AND jord AND landbrug)

French: (carbone AND stockage AND terre AND agriculture)

German: (kohlenstoff AND lagerung AND boden AND landwirt)

Italian: (carbonio AND stoccaggio AND suolo AND agricoltura)

Number of hits using the above search strings in Google Scholar on August 29, 2013 (Google Scholar based on title words only on December 9, 2013): English 65 400 (52), Swedish 1050 (0), Danish 1770 (0), French 15 190 (0), German 3550 (0), Italian 1630(0)).

Estimating the comprehensiveness of the search

The final search string resulted in 10 328 hits in Web of Knowledge and found 22 of 23 "reference articles" selected a priori as highly relevant. The only remaining reference article was a narrative review on the value of long-term experiments [17].

Bibliographies in review articles will be searched for relevant primary studies as a measure of the comprehensiveness of the search strategy. We will include relevant references in review articles previously missed by our search strategy. By using a large number of generic intervention terms and possible variations of the outcome term, our search strategy will be of a high-sensitivity and low-specificity type. This was demonstrated by the relatively small reduction in the number of articles after excluding 'particulate organic matter' and 'POM' as well as 'nitrogen' and 'N'. The specificity was judged to be too low when including 'carbon', and this outcome term was thus removed from the search string (leading to reduction from 15 649 to 9364 articles). The final number of articles after including all publication databases is expected to increase by a factor of two compared to the Web of Knowledge search.

Publication databases

The search aims to include the following online databases:

- Academic Search Premier
- Agricola
- AGRIS: Agricultural database (FAO)
- Biological Abstracts
- BioOne
- Directory of Open Access Journals

- Food Science and Technology Abstracts
- Georef and Geobase
- IngentaConnect
- JSTOR
- PubMed Central
- Scopus
- SwePub
- Web of Science
- Wiley Online Library

Internet searches

- Google (www.google.com)
- Google Scholar (scholar.google.com)
- Dogpile (www.dogpile.com)
- Scirus (www.scirus.com)

Specialist searches for grey literature

- *Aarhus University, Department of Agroecology* (http://www.au.dk/en/, http://agro.au.dk/en/)
- *African Network for Soil Biology and Fertility* (http:// agra.ciat.cgiar.org/)
- *Columbia Basin Agricultural Research Center* (http://cbarc.aes.oregonstate.edu/long_term_pubs)
- *European Environment Agency* (http://www.eea. europa.eu/)
- *European Soil Portal* (http://eusoils.jrc.ec.europa.eu)
- *Eusomnet* (http://www.ufz.de/somnet)
- *GCTE SOMNET* (http://gcmd.nasa.gov/ KeywordSearch/Keywords.do? Portal=GCMD_legacy&KeywordPath=Parameters| AGRICULTURE|SOILS|CARBON& MetadataType=0&lbnode=mdlb2)
- *GRACEnet, USDA Agricultural Research Service* http://www.ars.usda.gov/research/programs/ programs.htm?np_code=212&docid=21223
- *Indian Agricultural Statistics Research Institute* (http://iasri.res.in/)
- *National Soil Carbon Network (NSCN) of the US Forest Service* (http://www.nrs.fs.fed.us/niacs/ carbon/nscn/)
- *Rapid Assessment of US Soil Carbon (RaCA), USDA Natural Resource Conservation Service* (http://www. nrcs.usda.gov/wps/portal/nrcs/detail/soils/survey/? cid=nrcs142p2_054164)
- *Rothamsted Research* (http://www.rothamsted.ac.uk/)
- *Soil Carbon Center at Kansas State University* (http://soilcarboncenter.k-state.edu/)
- *Soilservice* (http://www4.lu.se/o.o.i.s/26761)
- *Swedish Board of Agriculture* (http://www. jordbruksverket.se)
- *Swedish Environmental Protection Agency* (http:// www.naturvardsverket.se)

- *Swedish University of Agricultural Sciences* (http:// www.slu.se)
- *UC Davis, Agricultural Sustainability Institute* (http://ltras.ucdavis.edu/)
- *University of Copenhagen* http://www.ku.dk/english
- *University of Illinois, Department of Crop Sciences* (http://cropsci.illinois.edu/research/morrow)
- *USDA Agricultural Research Service* (http://www.ars. usda.gov/research/programs/programs.htm? np_code=211&docid=22480)
- *Victorian Long Term Agro-ecological Experiments* (http://vro.dpi.vic.gov.au/dpi/vro/vrosite.nsf/pages/ lwm_ltae)
- *Videncentret for Landbrug* (http://www.vfl.dk/ English/NyEnglishsite.htm)
- *Working Group for Long-term Experiments (LTE)* (http://www.isofar.org/sections/wg-long-term-experiments.html)
- *World Bank* (www.worldbank.org/reference/)

Supplementary searches

It is anticipated that there will be a number of unpublished data sets containing information from long-term experiments. Several of the authors in the review team will search for such data sets within their respective organizations (including some of the specialist websites mentioned above).

Study inclusion/exclusion criteria

Articles found by searches in databases will be evaluated for inclusion at three successive levels. First they will be assessed by title, then by abstract, and finally by studying the full text. In cases of uncertainty, the reviewer will tend towards inclusion at all levels. One reviewer will perform the screening of all retrieved articles at the title and abstract level. To check that the screening is consistent and complies with the agreed inclusion/exclusion criteria, a subset of at least 200 articles will be screened by two reviewers at both the title and abstract levels. Kappa tests will be used to evaluate screening consistency. If Kappa tests indicate that the reviewers are inconsistent in their assessment (K < 0.6), discrepancies will be discussed and the inclusion/exclusion criteria will be clarified or modified. Next, each article found to be relevant on the basis of title and abstract will be judged for inclusion by reviewers studying the full text. Each reviewer will receive an approximately equal number of articles. Before screening full text, a subset of at least 100 articles will be double-screened and Kappa tests will be used to test consistency between reviewers.

Each study must pass each of the following criteria in order to be included at any of the three screening stages:
Relevant subject(s): Arable soils in agricultural regions. Regional factors are likely to be of much less importance

than the type of management. Following the Köppen-Geiger climate classification [22] (Figure 2), we will include agricultural regions from the warm temperate climate zone (fully humid and summer dry, i.e., Cfa, Cfb, Cfc, Csa, Csb, Csc) and the snow climate zone (fully humid, i.e., Dfa, Dfb, Dfc). Within these climate zones, agricultural management systems in which wheat, barley, rye, oats, silage maize or oilseed rape *can* grow in the crop rotation will be selected. Leys and bioenergy crops may occur as part of the crop rotation, but permanent grasslands, paddy rice systems, agroforestry systems and orchards will not be included. There will be one restriction on soil type, organic soils, since management rather leads to subsidence than to changes in SOC concentration [23]. Given the global scope of this systematic review we anticipate that a very large number of articles will be included after screening for relevance. At the abstract screening stage, we will therefore categorize studies as either belonging to the northern hemisphere or the southern hemisphere. Studies from the northern hemisphere will be prioritized if time and resource constraints prevent us from including all articles.

Relevant types of study design: The practices or systems under study must have been in operation for 10 years or more, as it is almost impossible to detect significant changes in SOC in shorter time periods (see [24]). The changes from one year to another are so small so that the change in SOC will be less than spatial variation within a field. Relevant types of studies include not only comparisons between specific agricultural practices at individual sites but also comparisons between multiple types of management factors alone or combined. Studies of how agricultural practices have affected SOC stocks at a single site over time will only be included if there is more than one treatment, as local variation in SOC can be high and information from single treatments (e.g. between farms or regions) are not possible to evaluate.

Relevant intervention(s): Any type of agricultural management that could change SOC stocks, including cropping systems that have ley, legumes or bioenergy plants in the rotation. Experimental treatments may include a range of different soil management practices introduced at the onset of the experiment, for example, different tillage practices, fertilization schemes, and cover/catch crops. Studies of biochar will also be included in the systematic review. As long as relevant data are found in the articles they will be included regardless of study purpose (e.g. C sequestration to counteract climate change or management intended to increase soil fertility). Multiple interventions without information on specific management/

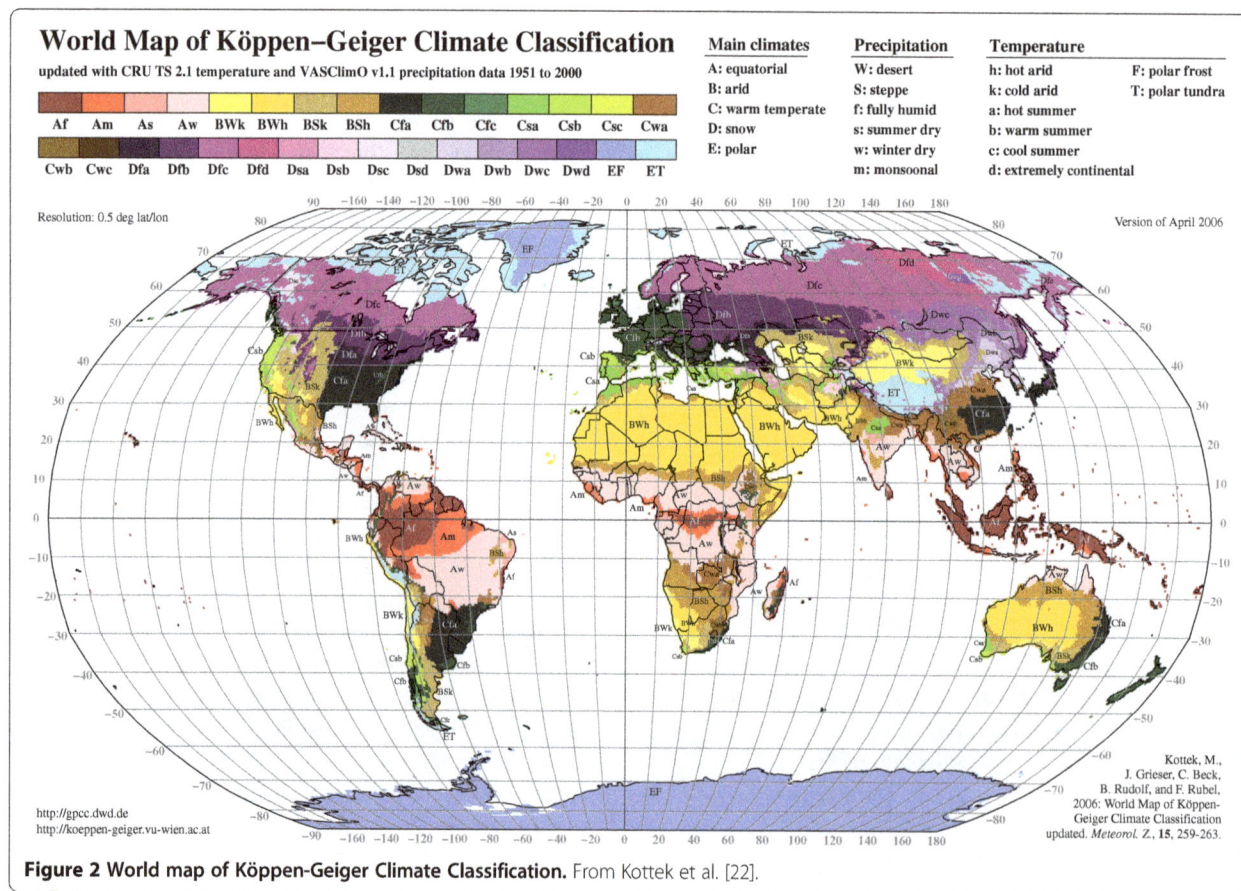

Figure 2 World map of Köppen-Geiger Climate Classification. From Kottek et al. [22].

treatments made in the same crop field preclude the opportunity to assess the effect of each intervention separately. For example, comparisons of organic and conventional farming may not always separate between different crop field treatments and such studies will then be rejected on the basis of study quality.

Relevant comparator(s): In studies of specific types of interventions, the relevant comparator is a treatment where no such intervention has occurred, or where it has been applied at a different level. In studies of entire agricultural systems, 'current' or 'conventional' farming practices that encompass a suite of management practices may be a relevant comparator as long as differences in all management factors are provided.

Relevant outcome(s): Relative changes of SOC will be the main focus. The C stock can be given as Soil Organic Carbon (SOC), Total Organic Carbon (TOC), Total Carbon (TC) or Soil Organic Matter (SOM), and it can be measured as mass and/or concentration.

Potential effect modifiers and reasons for heterogeneity

The following potential effect modifiers (non-intervention variables that might influence the outcome) will be considered and recorded in the review:

- Type of crop
- Type of crop rotation
- Soil type, mineral soil texture class
- Amount/type of fertilizers
- Latitude and longitude
- Climate (average annual precipitation and average annual temperature)
- Topography (altitude, slope)
- Previous land-use

The above list was compiled by the review team after consultation with stakeholders and external experts. Further modifiers and causes of heterogeneity will be identified and defined in an iterative process.

Study quality assessment

Studies still included after full text screening will be subject to quality assessment. During this process the studies will be categorized into one of the categories: a) do not meet quality criteria, b) acceptable study quality, and c) high study quality. Studies that do not meet the quality criteria will be excluded from data synthesis, whereas studies of acceptable and high quality will be retained. Assessment of study quality will be based on:

- Level of replication
- Method of sample selection (randomization)
- Paired, blocked or nested designs to avoid spatial effects

- Experimental duration
- Sampling frequency
- Soil sampling method (surface soil versus subsoil)
- To what extent potential effect modifiers have been assessed

Ideally, studies should sample both surface soil and subsoil. Studies sampling only surface soil may be biased and lead to misinterpretation of intervention effects, since the SOC concentration may increase with soil depth depending on the treatment applied [25]. Changes in SOC stock may go along with changes in bulk density [26]. Ideally, SOC would thus be measured not only by volume but also by soil mass. In most cases, however, SOC is measured as concentration rather than mass.

When assessing study quality the articles will be evenly distributed among the reviewers. A subset of at least 25% of studies will be appraised by a second reviewer. Conclusions will be compared, and where reviewers differ, discrepancies will be discussed and reconciled individually. A study may be included even if not all criteria have been fulfilled. Detailed reasoning will be recorded in a transparent manner. A list of studies rejected on the basis of quality assessment (i.e., do not meet quality criteria) will be provided in an appendix to the review together with the reasons for exclusion.

Data extraction strategy

All authors in the review team will participate in extracting metadata (effect modifiers such as types of crop, crop rotation, soil etc.). To make data extraction as consistent as possible, metadata will be entered into a spreadsheet with predefined categories. In case it is not possible to assign metadata to a specific category, it will be assigned to 'Other, please specify' (to allow the use of further categories if needed). One reviewer will be solely responsible for extracting numerical data from main text, tables and graphs. Data extraction will be double-screened for a subset of articles to check for consistency. The image analysis software WebPlotDigitizer will be used to extract data from graphs. To enable comparison between different interventions when measured at different sites, change in SOC will be recorded as the relative rate of change per year.

Data synthesis and presentation

A narrative synthesis of data from all studies with weighting as 'acceptable' or of 'high quality' will describe the quality of the results along with the findings. Tables will be produced to summarize the results. Precise details of the quantitative analysis will only be known when full texts have been assessed for their contents and quality.

Subgroup analyses will be undertaken to elucidate statistical relationships that are specific to particular type of management interventions. Overall effects of different management effects on SOC will be presented visually in forest plots. Separate analyses of surface soil and sub-soil rates of SOC change will be undertaken for studies reporting both measures. Meta-regression within sub-groups will be performed using rates of SOC change as a response variable and the effect modifiers as explanatory variables. Finally, we will perform sensitivity analysis to investigate the impact of removing studies with acceptable study quality.

Competing interests
The authors declare that they have no competing interests.

Authors' contributions
This review protocol is based on a draft written by BS. The draft was discussed with all authors at a meeting on 15-16 August, 2013. All authors participated in the drafting, revision and approval of the manuscript.

Acknowledgements
This protocol and the forthcoming review are financed by the Mistra Council for Evidence-Based Environmental Management (EviEM). The authors thank two anonymous reviewers whose advice improved this protocol considerably.

Author details
[1]Mistra Council for Evidence-Based Environmental Management, Royal Swedish Academy of Sciences, P.O. Box 50005, SE-104 05 Stockholm, Sweden. [2]Department of Biology, Lund University, SE-223 62 Lund, Sweden. [3]Department of Land, Air and Water Resources, University of California Davis, One Shields Avenue, Davis, CA 95616, USA. [4]Department Ecology, SLU, P.O. Box 7044, 750 07 Uppsala, Sweden. [5]Joint Research Centre, Land Resource Management – Soil action, Institute for Environment & Sustainability (IES), European Commission, Ispra, VA, Italy. [6]Department of Agroecology, Aarhus University, P.O. Box 50, DK-8830 Tjele, Denmark.

References
1. Batjes NH: **Total carbon and nitrogen in the soils of the world.** *Eur J Soil Sci* 1996, **47:**151–163.
2. Batjes NH, Dijkshoorn JA: **Carbon and nitrogen stocks in the soils of the Amazon region.** *Geoderma* 1999, **89:**273–286.
3. Schlesinger WH: *Biogeochemistry: an analysis of global change.* San Diego, Calif: Academic Press; 1997.
4. Betts RA, Falloon PD, Goldewijk KK, Ramankutty N: **Biogeophysical effects of land use on climate: model simulations of radiative forcing and large-scale temperature change.** *Agr Forest Meteorol* 2007, **142:**216–233.
5. Lal R, Delgado JA, Groffman PM, Millar N, Dell C, Rotz A: **Management to mitigate and adapt to climate change.** *J Soil Water Conserv* 2011, **66:**276–285.
6. Gonzalez-Sanchez EJ, Ordonez-Fernandez R, Carbonell-Bojollo R, Veroz-Gonzalez O, Gil-Ribes JA: **Meta-analysis on atmospheric carbon capture in Spain through the use of conservation agriculture.** *Soil & Tillage Research* 2012, **122:**52–60.
7. Lal R, Follett F: **Soils and climate change.** In *Soil Carbon Sequestration and the Greenhouse Effect.* 2nd edition. Edited by Lal R, Follett F. Madison, Wisconsin, USA: SSSA Special Publication; 2009:57.
8. Bolinder MA, Katterer T, Andren O, Ericson L, Parent LE, Kirchmann H: **Long-term soil organic carbon and nitrogen dynamics in forage-based crop rotations in Northern Sweden (63-64 degrees N).** *Agr Ecosyst Environ* 2010, **138:**335–342.
9. Hati KA, Swarup A, Dwivedi AK, Misra AK, Bandyopadhyay KK: **Changes in soil physical properties and organic carbon status at the topsoil horizon of a vertisol of central India after 28 years of continuous cropping, fertilization and manuring.** *Agr Ecosyst Environ* 2007, **119:**127–134.
10. Yang XY, Li PR, Zhang SL, Sun BH, Chen XP: **Long-term-fertilization effects on soil organic carbon, physical properties, and wheat yield of a loess soil.** *J Plant Nutr Soil Sci* 2011, **174:**775–784.
11. Powlson DS, Whitmore AP, Goulding KWT: **Soil carbon sequestration to mitigate climate change: a critical re-examination to identify the true and the false.** *Eur J Soil Sci* 2011, **62:**42–55.
12. Katterer T, Bolinder MA, Berglund K, Kirchmann H: **Strategies for carbon sequestration in agricultural soils in northern Europe.** *Acta Agriculturae Scandinavica Section a-Animal Science* 2012, **62:**181–198.
13. Guo LB, Gifford RM: **Soil carbon stocks and land use change: a meta analysis.** *Glob Chang Biol* 2002, **8:**345–360.
14. Andren O, Katterer T, Karlsson T, Eriksson J: **Soil C balances in Swedish agricultural soils 1990-2004, with preliminary projections.** *Nutr Cycl Agroecosyst* 2008, **81:**129–144.
15. Lal R: **Carbon sequestration.** *Philosophical Transactions of the Royal Society B-Biological Sciences* 2008, **363:**815–830.
16. Ludwig B, Geisseler D, Michel K, Joergensen RG, Schulz E, Merbach I, Raupp J, Rauber R, Hu K, Niu L, Liu X: **Effects of fertilization and soil management on crop yields and carbon stabilization in soils. A review.** *Agron Sustain Dev* 2011, **31:**361–372.
17. Rasmussen PE, Goulding KWT, Brown JR, Grace PR, Janzen HH, Korschens M: **Agroecosystem - long-term agroecosystem experiments: assessing agricultural sustainability and global change.** *Science* 1998, **282:**893–896.
18. Govaerts B, Verhulst N, Castellanos-Navarrete A, Sayre KD, Dixon J, Dendooven L: **Conservation agriculture and soil carbon sequestration: between myth and farmer reality.** *Crit Rev Plant Sci* 2009, **28:**97–122.
19. SOILSERVICE: *Conflicting demands of land use, soil biodiversity and the sustainable delivery of ecosystem goods and services in Europe.* Final publishable report; 2012 [http://www4.lu.se/upload/Ekologi/soilservice/FinalPubl.pdf].
20. Andersson R, Bång M, Frid G, Paulsson R: *Minskade växtnäringsförluster och växthusgasutsläpp till 2016 – förslag till handlingsprogram för jordbruket.* Swedish: Swedish Board of Agriculture; 2010. Report 10.
21. Stenberg M: *Reducerad jordbearbetning på rätt sätt – en vinst för miljön!.* Swedish: Swedish Board of Agriculture; 2010. Report 36.
22. Kottek M, Grieser J, Beck C, Rudolf B, Rubel F: **World map of the Koppen-Geiger climate classification updated.** *Meteorol Z* 2006, **15:**259–263.
23. Schipper LA, McLeod M: **Subsidence rates and carbon loss in peat soils following conversion to pasture in the Waikoto Region, New Zealand.** *Soil Use Manag* 2002, **18:**91–93.
24. Smith P: **How long before a change in soil organic carbon can be detected?** *Glob Chang Biol* 2004, **10:**1878–1883.
25. Baker JM, Ochsner TE, Venterea RT, Griffis TJ: **Tillage and soil carbon sequestration - what do we really know?** *Agr Ecosyst Environ* 2007, **118:**1–5.
26. Ellert BH, Bettany JR: **Calculation of organic matter and nutrients stored in soils under contrasting management regimes.** *Can J Soil Sci* 1995, **75:**529–538.

Evaluating effects of land management on greenhouse gas fluxes and carbon balances in boreo-temperate lowland peatland systems

Neal R Haddaway[1*†], Annette Burden[2†], Chris D Evans[2], John R Healey[3], Davey L Jones[3], Sarah E Dalrymple[4] and Andrew S Pullin[1]

Abstract

Background: Peatlands cover 2 to 5 percent of the global land area, while storing between 30 and 50 percent of all global soil carbon (C). Peatlands constitute a substantial sink of atmospheric carbon dioxide (CO_2) via photosynthesis and organic matter accumulation, but also release methane (CH_4), nitrous oxide (N_2O), and CO_2 through respiration, all of which are powerful greenhouse gases (GHGs). Lowland peats in boreo-temperate regions may store substantial amounts of C and are subject to disproportionately high land-use pressure. Whilst evidence on the impacts of different land management practices on C cycling and GHG fluxes in lowland peats does exist, these data have yet to be synthesised. Here we report on the results of a Collaboration for Environmental Evidence (CEE) systematic review of this evidence.

Methods: Evidence was collated through searches of literature databases, search engines, and organisational websites using tested search strings. Screening was performed on titles, abstracts and full texts using established inclusion criteria for population, intervention/exposure, comparator, and outcome key elements. Remaining relevant full texts were critically appraised and data extracted according to pre-defined strategies. Meta-analysis was performed where sufficient data were reported.

Results: Over 26,000 articles were identified from searches, and screening of obtainable full texts resulted in the inclusion of 93 relevant articles (110 independent studies). Critical appraisal excluded 39 studies, leaving 71 to proceed to synthesis. Results indicate that drainage increases N_2O emission and the ecosystem respiration of CO_2, but decreases CH_4 emission. Secondly, naturally drier peats release more N_2O than wetter soils. Finally, restoration increases CH_4 release. Insufficient studies reported C cycling, preventing quantitative synthesis. No significant effect was identified in meta-analyses of the impact of drainage and restoration on DOC concentration.

Conclusions: Consistent patterns in C concentration and GHG release across the evidence-base may exist for certain land management practices: drainage increases N_2O production and CO_2 from respiration; drier peats release more N_2O than wetter counterparts; and restoration increases CH_4 emission. We identify several problems with the evidence-base; experimental design is often inconsistent between intervention and control samples, pseudoreplication is common, and variability measures are often unreported.

Keywords: Lowland, Boreo-temperate, Peat, Greenhouse gas, Carbon, Carbon dioxide, Methane, Nitrous oxide, Land management, Drainage, Restoration, Agriculture, Extraction

* Correspondence: neal_haddaway@hotmail.com
†Equal contributors
[1]Centre for Evidence-Based Conservation, School of Environment, Natural Resources and Geography, Bangor University, Bangor, Gwynedd LL57 2UW, UK
Full list of author information is available at the end of the article

Background

Peat and peatlands are composed of partly decomposed plant material deposited under saturated soil conditions. Peatland is a generic term including all types of peat-covered terrain and as a land cover class peatlands are a complex of swamps, bogs, and fens, sometimes called a "mire complex" [1]. 'Peatland' is sometimes taken to apply only to peat soils which retain their natural (peat-forming) vegetation cover, but is applied here to any land-use on peat soils.

Peatlands are estimated to cover between 2 and 5 percent of the global land surface area while storing between 30 and 50 percent of all global soil carbon [2-4]. Whilst rates of C accumulation in peats are relatively low [5], they can continue to accumulate C over millennia, and exceed mineral soils in the length of time C is stored [6].

C is sequestered into peatlands via photosynthetic uptake of carbon dioxide (CO_2). It then accumulates due to the slow decomposition rates of organic matter (OM) in anaerobic conditions associated with high water tables. C loss can occur in a range of forms, of which the most important are gaseous CO_2 and methane (CH_4), and dissolved C, primarily dissolved organic carbon (DOC) and dissolved inorganic carbon (DIC). Low water tables associated with artificial drainage or natural dry periods are generally associated with net emission of CO_2, whilst high water tables can result in significant emission of CH_4 [7-11]. CO_2 and CH_4 are the two most important greenhouse gases (GHGs) [12,13] and peatlands may, therefore, contribute significantly to global GHG production under a changing climate [6]. In some peatlands, emission of nitrous oxide (N_2O), which is a powerful GHG, may also be significant; this is typically associated with fertilisation, drainage and dry-rewet cycles e.g. [14].

This review was undertaken with the primary aim of evaluating land-management impacts on 'lowland peats' in England and Wales, in the United Kingdom (UK). In contrast to many boreo-temperate regions, the majority of the UK's peatland area is occupied by blanket bog, a peat type that forms in oceanic regions, and which therefore has a restricted global extent [15]. In the UK, blanket bogs occur extensively in upland high-rainfall areas, and comprise over 90% of the total peatland area of the country. Whilst large areas of blanket bog have indeed been modified by anthropogenic pressures, including drainage, managed burning, afforestation and grazing pressure, a high proportion of the total area nevertheless remains under natural or semi-natural vegetation. In contrast, although lowland peatlands (comprising lowland fen and lowland raised bog) occupy less than 10% of the overall UK peatland area, they are subject to disproportionately high levels of land-use pressure. Worrall et al. [16] estimated that over 70% of the total GHG emissions from peatlands were the result of land-

use activities (arable cultivation, improved grassland and peat extraction) that are almost entirely associated with lowland peat, with an even higher proportion of peat GHG emissions within England associated with these activities. Despite this, relatively little research has been undertaken on lowland peats in the UK, and this review was therefore undertaken in order to assess the current evidence base relating to this peatland type. As the basis for this, it was considered that research undertaken on continental bogs and fens would be more analogous to UK lowland peats (in terms of both their functioning and the associated land-management activities) than UK blanket bogs, and studies from the wider boreal and temperate zones were therefore included in the review. As a consequence, the review may be considered applicable to boreo-temperate lowland peats in general, whilst excluding studies undertaken within blanket bogs and other peat types occurring in upland regions.

Lowland peats differ in many respects from the blanket mires of the uplands. Whereas blanket peat forms under conditions of high rainfall and low temperatures, lowland fens and raised bogs can form under drier and warmer conditions where natural drainage is poor. Consequently they tend to form only in flat areas, whereas blanket bogs can develop on areas of moderately undulating topography. Fens are characterised by the lateral input of water from groundwater or rivers, and have moderate to high levels of nutrients and low acidity. Raised bogs form low 'domes' of peat which are fed by rainwater, and which are therefore low in nutrients and are acidic. Natural vegetation in lowland peats may be quite different to that occurring in blanket bogs, particularly within fens where tall herb species and brown mosses may predominate, rather than the dwarf shrubs, low-growing sedges and Sphagnum mosses associated with bogs. Given these major differences in both the natural properties of lowland peats, and the different land-uses to which they have been subjected it is doubtful whether data obtained from studies of upland blanket bogs can be extrapolated to lowland systems. The difficulty of quantifying C cycling and GHG fluxes for lowland peats is increased by their greater heterogeneity in terms of both typology and management, as well as their fragmented nature. Because of their importance for a wide range of ecosystem services (notably provisioning services such as food and water, but also cultural services such as access to natural landscapes in otherwise often highly developed regions, and regulating services such as flood control in some areas; Bonn et al. [17]), the role of lowland peats in climate regulation must be weighed against these other ecosystem services to enable appropriate management decisions.

Recent reviews of measurements of peatland C stocks and GHG fluxes for the Department for the Environment,

Food and Rural Affairs (Defra, UK) and the Joint Nature Conservation Committee (JNCC, UK), [16,18-20] have highlighted both the high degree of uncertainty in GHG flux estimates for lowland peats (in particular fens) and their high relative importance in terms of overall C and GHG emissions from peatlands. Worrall et al. [16] estimated that around 54% of total GHG emissions from peatlands in the UK originate from lowland peats. On this basis, Evans et al. [20] suggested that the (few) existing studies on sites in lowland areas should be augmented in order to provide full C/GHG budgets, and that new measurement sites should be established in fens, particularly in areas under intensive agriculture, for which no existing flux measurement sites could be identified. A clear distinction was made in this assessment between emissions from sites undergoing a transition in land-use/management (which may lead, for example, to a short-term pulse of CH_4 emission), and emissions from sites under stable long-term management.

Land use change over the past century has greatly altered many temperate and boreal lowland peatland ecosystems, favouring agriculture, forestry and peat extraction sites with artificially lowered water tables [21,22]. The Kyoto Protocol aims to mitigate GHG emissions [23], and restoration of peatlands via re-wetting has increasingly been identified as a potential means of doing so [24]. However, whilst a body of primary literature on the impacts of different land management practices on C cycling and GHG fluxes on lowland peats does exist, these data have yet to be synthesised. A previous systematic review has assessed the impact of drainage and rewetting on peat soils [25], but this review was undertaken in 2008, and focused on drainage and rewetting and included upland and lowland peats on a global scale. Here we report on the results of a Collaboration for Environmental Evidence (CEE) systematic review of the available evidence with regard to GHG emissions and C cycling on lowland peatlands under land management activities.

Objective of the review
Primary question
The primary question of this review was:

What is the impact of land management on GHG and C fluxes of boreo-temperate lowland peatland systems? The question components were defined as follows:

Population: Boreo-temperate lowland peat systems.

Exposure: Areas with different long-term hydrological regimes.

or

Intervention: Draining and re-wetting/cessation of draining, extraction, conversion to agricultural production, agricultural or forestry practice.

Comparator: Control (with no intervention) or before-after studies or comparisons of areas with different

management regimes over long periods of time, i.e. not short-term or seasonal changes.

Outcome: Net change (sequestration or release) in C or GHG balance, or net change of individual components of the C/GHG balance.

Methods
The systematic review question in hand was selected in order to provide an underpinning evidence base for a UK Department of Environment, Food and Rural Affairs (Defra) funded project to evaluate GHG flux and C balances of lowland peatland ecosystems in England and Wales. The review has been conducted alongside the establishment of a set of intensive C and GHG flux measurement sites under a range of land-management in England and Wales, with the joint aim of providing greater understanding and quantification of C/GHG emissions associated with different forms of land-use and management. A systematic review protocol was developed according to this research question, peer reviewed and posted in the CEE Library [26].

Search strategy
The search aimed to capture an unbiased and comprehensive sample of the literature relevant to the question, both published and unpublished. A number of different sources of information were searched in order to maximise coverage.

Search terms
Combinations of the following search terms (where * denotes a wild card that may represent zero or more characters and $ represents zero or one character only) were applied to databases (Table 1).

As all databases and websites vary in the way they handle complex search strings and the use of Boolean operators, the exact search strings used differed between databases. Details of the different search terms used in each of the search facilities employed are provided in Additional file 1.

Databases
The following databases were searched using the terms detailed in Search terms:

1) ISI Web of Knowledge (inc. ISI Web of Science and ISI Proceedings)
2) Science Direct
3) Directory of Open Access Journals
4) Copac
5) Index to Theses Online
6) Agricola
7) CAB Abstracts
8) CSA Illumina/Proquest
9) Scopus

Table 1 Habitat and outcome search terms

Habitat search terms	Outcome search terms	Intervention search terms
Aapa*	Accret*	Afforest*
Bog*	Accumulation	Arable
Carr	Carbon	Cut$over
Fen$	CH_4	Cutt*
Fenland	CO_2	Ditch*
Histosol*	Depth	Drain*
Hochmoortorf	DOC	Drought
Mire	DOM	Extract*
Mor	Erosion	Fertili*
Muck	"GHG*"	Flood*
Muskeg	"Green$house gas*"	Forest*
Niedermoortorf	Methane	Graz*
Palsa	N_2O	"Grip block*"
Peat*	Nitrous Oxide	Pastor*
Pocosin*	"Organic content"	Pastur*
Quag*	"Organic matter"	Plough*
Sedge	Shrink*	Plow*
Slough	SOM	Re$veg*
Suo	Subside*	Re$wet*
Swamp$	Wastage	Restor*
Torfmoor		Till*
Tourbe		Turf$strip*
Tourbièr*		
Turvesuo		

The characters* and $ denote wildcards for multiple or single characters respectively.

No time, language or document type restrictions were applied. References retrieved from the computerised databases were exported into a bibliographic software package (Endnote X3) and duplicates removed prior to assessment of relevance using inclusion criteria (Study inclusion).

Search engines

Searches of the following internet databases were performed:

http://scholar.google.com

http://www.Scirus.org (All journal and web sources)

Additional file 2 details the terms employed in these searches and the number and relevance of the hits returned. Assessments were limited to the first 50 hits returned for each search.

Specialist sources

Websites of the following relevant specialist organisations, identified during the planning stages of the review, were also searched for relevant material (see Table 2).

Websites were searched manually, by navigating through the site 'Publications' section, if available, and also by using any provided automated search with a number of key search terms. Details of the search terms employed for each website and the number and relevance of hits returned are provided in Additional file 3.

Search comprehensiveness assessment

The search results were tested for comprehensiveness in two ways. Firstly, a test library of eleven articles of known relevance was compared with search results to identify any missing sections of literature (see Additional file 1). Secondly, the bibliographies of seven reviews on the subject that were identified during searching were screened to identify potentially missed literature. Additional file 4 details the reviews that were screened and the results of the screening.

Unobtainable/non-English language articles

Articles that could not be obtained through subscription or open-access journals were obtained where possible by contacting authors with a request for proofs. Articles in foreign languages were screened at abstract using in-house translators and online translation software, where possible. Articles that could not be obtained or translated at full text are detailed in Additional file 5.

Study inclusion
Study screening

Studies retained in the Endnote database from the search were screened for relevance in a three stage process. This process systematically removed studies that were not relevant or did not contain relevant information or data. At each stage, if there was insufficient information to exclude a study it was retained until the next stage.

In the first instance, the inclusion criteria identified below were applied to titles only in order to remove spurious citations. Articles remaining after this stage were subsequently screened by viewing the abstract and those remaining were also screened at full text.

To assess and limit the effects of between-reviewer differences in determining relevance, two reviewers (AB and SD) applied the inclusion criteria to a random sample of 90 articles at the abstract level. The kappa statistic [27,28] was calculated, which measures the level of agreement between reviewers. After an initially low kappa score was obtained (kappa=0.356), the reviewers discussed the discrepancies and clarified the interpretation of the inclusion criteria. Following this discussion, the inclusion criteria were applied by the two reviewers to the rest of the citations.

Table 2 List of specialist organisations, the websites of which were searched for evidence

Organisation Name	
Agriculture And Agri Foods Canada	Ministry Of Natural Resources Of The Russian Federation
Agri-Food And Biosciences Institute	Moorland Association
Alterra	Moors For The Future
British Association For Shooting And Conservation	National Council For Forest Research And Development (COFORD)
Centre For Ecology And Hydrology	National Parks
Countryside Council For Wales	National Soil Resources Institute
Department Of Energy And Climate Change	National Trust
Department For The Environment, Food And Rural Affairs	Natural England
Dŵr Cymru/Welsh Water	Natural Resources Canada
Environment Agency	Peat-Portal.Net
Environment Canada	Plantlife UK
Environmental Protection Agency	RAMSAR
Environment Protection Agency Ireland	Royal Society For The Protection Of Birds
EHS –Northern Ireland Environment Agency	Russian Guild Of Ecologists
European Commission Joint Research Centre	Russian Regional Environmental Centre
European Environment Agency	Severn Trent Water
Finnish Peatland Society	Scottish Agricultural College
Farmers Unions - UK	Scottish Executive
Finland's Environmental Administration	Scottish Environment Protection Agency
Finnish Environment Institute SYKE	Scottish Natural Heritage
Food And Agriculture Organization Of The United Nations	Society For Ecological Restoration
Forest Research	Society For Wetlands Scientists
Forestry Commission	Tyndall Centre For Climate Change Research
Global Environment Centre	UK Climate Impacts Programme
Greenpeace	United Nations Environment Programme
Intergovernmental Panel For Climate Change	United States Environment Protection Agency
International Association For The Study Of The Commons	United Utilities
International Mire Conservation Group	Welsh Assembly Government
International Union For Conservation Of Nature	Wetlands International
International Peat Society	Wildfowl And Wetlands Trust
Irish Agriculture And Food Development Authority (Teagasc)	Wildlife Trusts UK
Irish Peatland Conservation Council	World Wildlife Fund (organised by country)
Joint Nature Conservation Committee	Yorkshire Water
Macaulay Land Use Research Institute	

See Additional file 3 for details of the websites searched and search results.

Inclusion criteria

Each article was required to contain the following criteria in order to be included after each filter. However, in cases of uncertainty (including the absence of an abstract), the reviewer tended towards inclusion.

- Relevant population(s): Lowland peatland systems in temperate and boreal regions. In general, studies from outside the British Isles were considered to be from lowland-type peatlands (such as continental

mires) unless specifically characterised as 'blanket bog' or associated with upland topographic locations.

- Exposure/intervention: Long-term re-wetting or draining of peat or peat related soils. Agricultural conversion of peat or peat-related soils. Afforestation of peat or peat-related soils. Agricultural or forestry management on peat or peat-related soils. Natural experiments comparing areas of peat or peat-related soils in the same region with different long term (not seasonal or sporadic) hydrology.

- Types of comparator: Control or no intervention or before-after comparisons.
- Types of outcome: Amount of C or GHG stored in, sequestered or released from soils. Other physical measures of C loss or gain from peat or peat-related soils, e.g. erosion, efflux of DOC, subsidence and/or accumulation.
- Types of study: Any primary study including measures of C or GHG storage, sequestration or release from peat or peat-related soils. Studies that measured C or GHG storage, sequestration or release under laboratory conditions were excluded. Additionally, studies were only included if an intervention was administered for a period in excess of one year, so as to retain relevance and generalisability to the overall research aims of the review.

Critical appraisal

Individual studies that met the required inclusion criteria were subjected to critical appraisal. This process involved a detailed assessment of internal and external validity. Internal (the appropriateness of the experimental design in measuring the desired outcomes) and external (the generalisability of the study findings to the review question in hand) validity assessment included the extraction of a number of variables from included studies (see Table 3).

Once this information had been assessed for each article, a final judgment on study *susceptibility to bias* was made using all collated information in concert (see Additional file 6 for details of the information upon which this critical appraisal was based). Each study was classified as either 'low' or 'high' susceptibility to bias. Where insufficient detail was provided in study methodology, the category 'unclear' was given. Where design or methodology was considered to be significantly flawed, or where detail was lacking and flawed design implied, studies were excluded from further consideration. Studies with low external validity (i.e. generalisability to the review question in hand) were also excluded at this stage. Details of these excluded studies are provided in Additional file 7. Information extraction for critical appraisal was divided and carried out independently by two reviewers (NRH and AB), and all excluded articles were discussed with additional advice from the Review Team (ASP, DLJ, CDE and JRH). All included studies were given *susceptibility to bias* categories by two reviewers (NRH and AB), and agreement was assessed using a kappa test, which revealed substantial agreement between reviewers (Kappa statistic=0.61). Disagreements, particularly those between the 'low' and 'high' judgement groups, were discussed at length and a consensus agreed upon with the inclusion of advice from the Review Team where necessary.

Data extraction
Data extraction strategy

Data were extracted from included studies using a predefined form (Additional file 8), recording the mean, standard deviation, sample size, units, and notes for the study comparator and intervention. Pooled means and standard deviations were themselves calculated using a series of individual spreadsheets, which were populated with data from tables and figures of results from each article. The software DataThief [29] was used to extract data from figures.

Given the varied nature of data presentation in the included literature, a detailed protocol for data extraction was established in order to maintain consistency and transparency (see Additional file 8). Most importantly, whilst the use of pseudoreplicates (i.e. within sample replication as opposed to true replication of samples) is not ideal, the limited use of true replication within the evidence-base meant that spatial pseudoreplication (subsamples taken from within one sample where study sample size was very low) was used to estimate the variability in effect sizes (see Additional file 8 for details).

Potential reasons for heterogeneity were also extracted from studies (see Reasons for heterogeneity (subgroup analysis and meta-regression)):

Data handling

Extracted data were prepared for synthesis by producing effect sizes (raw mean difference). Raw mean difference $\bar{x}_{intervention} - \bar{x}_{control}$ was chosen since the effect size units are inherently meaningful. In order to perform meta-analysis with data measured in a variety of units, data for each measured outcome were converted to a common unit; mg l^{-1} for DOC, mg m^{-2} h^{-1} for GHG flux (where mg represents the whole molecule) and other units specified for individual outcome measures. Details of data extracted and how standard deviations were calculated can be found in Additional file 8.

Where GHG flux was reported as CO_2-C, CH_4-C or N_2O-N, data were converted to CO_2, CH_4 and N_2O by applying a multiplication factor of 3.664, 1.336 and 1.571 respectively.

CO_2 data were extracted in their commonly reported forms; ecosystem respiration (R_{eco}) and net ecosystem exchange (NEE).

Summary effects from meta-analyses identifying significant patterns across studies were used to generate estimated summary effects in the units of kg or × 10^3 kg (tonnes) of GHG flux per hectare per year. For CH_4 and N_2O these numbers were then used to calculate 100-year global warming potentials (as adopted in the Kyoto Protocol; UNFCCC/CP/1997/7/Add.1/Decision 2/CP.3) by conversion into CO_2 equivalents according to Forster et al. [30], using multiplication factors of 1 for CO_2, 25 for CH_4 and 298 for N_2O.

Table 3 Variables assessed and criteria used for critical appraisal of included studies

		Low Susceptibility to Bias	High Susceptibility to Bias
Study design	Study season, length	Long (>3 months) study period, multiple seasons	Short (<3 months) study period, winter measurement
	Intervention timescale	Long-term intervention maintained for multiple years	Intervention in place for <2 years
	Replication, randomisation	Replication at level of intervention, large sample size (>3), some degree of randomisation in sample selection	Pseudoreplicated, low sample size (<3), no randomisation
	Control matching	Control and treatment/exposure samples well-matched (i.e. close in proximity but low chance of spillover effects)	No evidence of matching, potentially influential differences between treatment and control
	Clarity and detail of methods	General study design very clear and repeatable	Some missing information
Specific methodology	Eddy covariance measurement details	Full description of methodology, accounting for wind direction/speed	Some missing methodological detail, no accounting for wind direction/speed
	Flux chamber measurement details	Full description of methodology, measurement disturbance mitigation measures	Some missing methodological detail, no mitigation for measurement disturbance
	Soil porewater/air measurement details	Full description of methodology, representative sampling	Some missing methodological detail, sampling unlikely to be representative of variability in environment
	Surface water measurement details	Full description of methodology, representative sampling	Some missing methodological detail, sampling unlikely to be representative of variability in environment
Bias	Potential measurement bias	Measurement bias unlikely or evidently not present	Bias likely as a result of methodology
	Presence of confounders	No obvious confounders stated or evident, or stated but adequately accounted for	Confounders stated and unaccounted for or likely to be present

Unclear classification given to any study where substantial details within the methods are either unclear of missing.

Data synthesis

Interventions

Groups of interventions and exposures were identified in an iterative process, with comparable studies retrospectively classed into one of 16 intervention/exposure groups, as detailed in Table 4.

Narrative synthesis

Tables of studies for each of the 16 intervention/exposure groups were arranged by measured outcome to display effect sizes and standard deviation along with summary descriptive data.

Meta-analysis

Where sufficient data existed, studies that reported means and variability were included in meta-analyses for the most common intervention/exposure groups by outcome. Meta-analysis was performed in the statistics package R [31] using the 'metafor' library [32] using the rma function. This function is equivalent to a random effects meta-analysis; a mixed effects model when moderators are included. Marginally significant test results ($0.075 > p > 0.050$) are reported for clarity, since thresholds for statistical significance are inherently arbitrary [33]. The presence of significant heterogeneity (i.e. differences between significant studies' results) was assessed using the test for residual heterogeneity (QE) [32]. In addition, forest plots [34] for each intervention/exposure group were produced for each outcome to display these data visually. Diagnostic plots to identify studies with strong influence on the inter study variability measure, T^2, and the summary effect were plotted using the R library 'metafor' [32] and can be found in Additional file 9.

Subgroup analysis

Where sufficient data existed, subgroup analysis was performed by including susceptibility to bias judgement category (see Critical appraisal) in the meta-analysis using tests for moderators (QM, p-values and AIC) in the *rma* function of the 'metafor' library [32].

Publication bias (funnel plots)

The presence of potential publication bias was assessed for each intervention/exposure group by plotting study effect size against standard error. Asymmetry in both axes of the plot may suggest publication bias and was tested for with Egger's z-test. Funnel plots and egger's test results can be found in Additional file 9.

Table 4 Intervention/exposure groups generated iteratively during the review process with brief descriptions of definitions

Intervention/exposure group	Definition
Cropped-vs-bare	Actively cropped fields compared to those with no vegetation (extracted or de-vegetated)
Drained and restored-vs-undrained	Drained and then restored peatlands compared to pristine fen
Drained-vs-undrained	Drained peatlands compared to pristine/undrained peats, or heavily drained peatlands compared to lightly drained peats
Dry-vs-wet	Dry peat soils compared to wetter soils
Extracted and restored-vs-natural	Both extracted and then restored peats compared to unextracted/pristine peats
High intensity farmed-vs-low intensity farmed	Intensively compared with extensively farmed peats (in terms of a wide variety of unspecified farming activities)
Fertilised and grazed-vs-unfertilised and mown	Both fertilised and grazed compared to unfertilised and ungrazed but mown peats
Fertilised-vs-less fertilised	High fertilisation rates compared to low fertilisation rates
Grass-vs-forest	Grass fields compared to forested fields
Grazed-vs-mown	Grazed peatlands compared to ungrazed but mown fields
Irrigated-vs-non-irrigated	Rain fed irrigated peatlands compared to non-irrigated peats
Mineral soil dressed-vs-undressed	Peats with added mineral soil compared to soils lacking mineral addition
Old abandoned-vs-recently abandoned	Peats abandoned for a longer period compared to more recently abandoned peats
Old afforested-vs-recently afforested	Peats planted with trees for a longer period compared to more recently planted forest peatlands
Poor-vs-rich	Peats described by authors as 'poor' compared to those described as 'rich' (corresponding to natural gradients in nutrient status and acidity)
Restored-vs-unrestored	Previously extracted peats with deliberately raised water tables compared to extracted and unaltered peatlands

Reasons for heterogeneity (subgroup analysis and meta-regression)

Potential sources of heterogeneity that were identified *a priori* during protocol development were; peat or peat-related soil type (e.g. bog versus fen peats), depth of drainage, vegetation, annual mean temperature/rainfall, timescale of study, timescale of hydrological intervention, peat condition.

Studies that reported sufficient data to allow meta-analysis of measured outcomes within intervention/exposure groups were not sufficient in number nor sufficiently consistent in the reporting of sources of potential heterogeneity. This precluded meaningful subgroup analysis or meta-regression using these reasons for heterogeneity.

Results

Review descriptive statistics

Searches of literature databases identified 25,665 potentially relevant titles. Of these, 1,794 were assessed as relevant at the title-level and 229 at abstract-level screening. Eleven articles were unable to be translated and 79 articles could not be obtained at full text (see Additional file 5). Whilst efforts were made to contact the authors to request copies, 62 of these articles were published between 1973 and 2003, making successful contact with authors unlikely. After the addition of articles from other sources of searching, 93 articles were identified as relevant at full text screening. The 89 articles excluded as a result of full text screening, along with exclusion reasons, are identified in Additional file 7.

Figure 1 displays the different stages of the systematic review process and the number of articles and studies retained at each stage. Following full text assessment, data sources are referred to as studies rather than articles. This definition is important, since many articles report multiple studies. Here we define the unit of study as a geographically distinct study system. Hence, where different peatland systems were studied within the same article their results were treated as independent sample points. One hundred and ten individual studies were identified from the 93 included articles, and these studies were subjected to critical appraisal.

Critical appraisal

Critical appraisal of study internal and external validity resulted in the exclusion of 39 studies that, upon closer inspection, did not meet the inclusion criteria precisely and/or provided insufficient detail in their description of methodologies or results to warrant inclusion in any synthesis (see Additional file 7 for these excluded studies).

Of the 71 studies that were suitable for inclusion, 11 studies reported 19 outcome measures that were judged to be of high susceptibility to bias. Thirty-four studies reported 71 outcome measures that were judged to be of

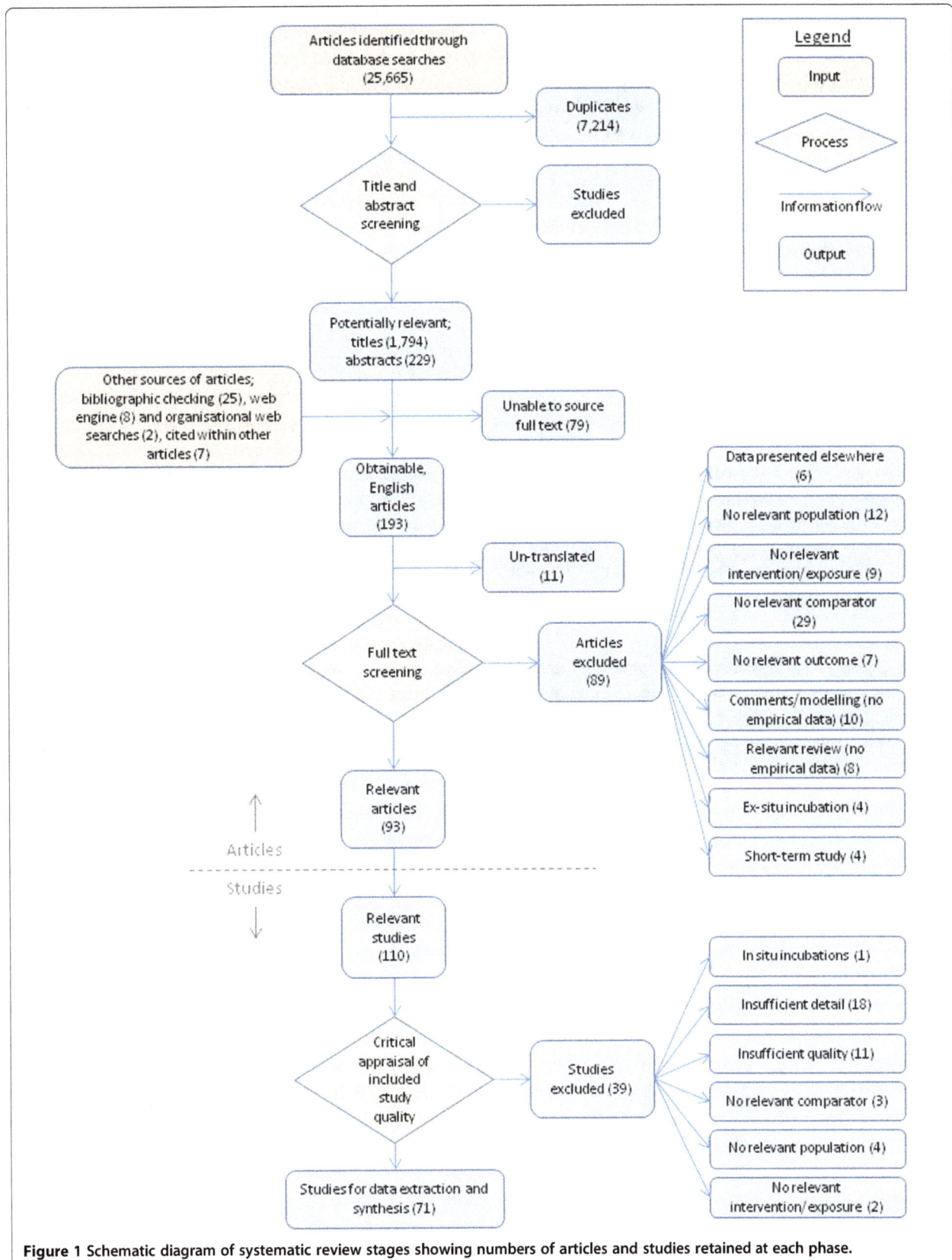

Figure 1 Schematic diagram of systematic review stages showing numbers of articles and studies retained at each phase.

low susceptibility to bias. Twenty-seven studies reported 48 outcome measures that were judged to be of unclear susceptibility to bias (n.b. some studies reported multiple outcomes of different susceptibility to bias). Supporting information extracted from each study relating to the judgment for critical appraisal can be found in Additional file 6.

Some countries were very frequently studied. Finland and Canada were overwhelmingly the most studied countries, followed by the Netherlands and Germany with no other country exceeding three studies (Figure 2). Eleven regions were studied more than once (Figure 3), with Lakkasuo (in Finland) and Zegveld (in the Netherlands) being the most common.

Eleven studies employed full BACI (before-after-control-impacts) study designs, with one study BA (before-after), and the remaining studies all CI (control-impacts). The majority of studies used flux chambers to measure reported outcomes, with a very small number using eddy covariance and other less frequently employed methods. This is also reflected in the less frequently measured outcomes, e.g. subsidence measured via long-term peat depth records Figure 4. Figure 5 displays the period of included studies, showing that the majority of studies lasted less than 2 years.

Synthesis

Table 5 displays the numbers of outcome measures reported for different studies across the 16 intervention/exposure groups. A number of intervention/exposure groups were represented by individual studies within each measured outcome, and quantitative synthesis of

these studies' results is not feasible as a result of low commonality. These studies are described below in narrative syntheses, and where sufficient studies within outcomes allow, the impact of interventions/exposures is assessed quantitatively using meta-analysis. Effect sizes are all in mg m^{-2} h^{-1} for CO_2, CH_4 and N_2O, and mg l^{-1} for DOC unless otherwise stated.

Table 6 displays the number of studies with sufficient data for outcomes and groups with sufficient studies to allow meta-analysis (i.e. both mean and variability measure reported within the study and a total of more than three studies in one outcome per group). Drained versus undrained was the most frequent comparison, followed by restored versus unrestored and then high intensity versus low intensity farmed. Table 7 summarises the results of the meta-analyses carried out on these studies. The four outcomes with a greater significance than the conventional p <0.05 threshold were; greater R_{eco} in drained compared to undrained, greater CH_4 emission in restored compared to unrestored, and greater N_2O emission in drained compared to undrained and in dry compared to wet peatlands (Table 5). There was also a marginally significant (p=0.055) effect of drainage on CH_4 emission, and a marginally significant (p=0.051) effect of drainage on DOC.

Studies reporting combined GHG outcomes

Nine studies measured CH_4, N_2O and CO_2 emissions and fluxes together (see Table 8). Only one study, however, measured both NEE and R_{eco}. This precluded a combined analysis of net 100-year global warming potential, since photosynthetic uptake of CO_2 by plants

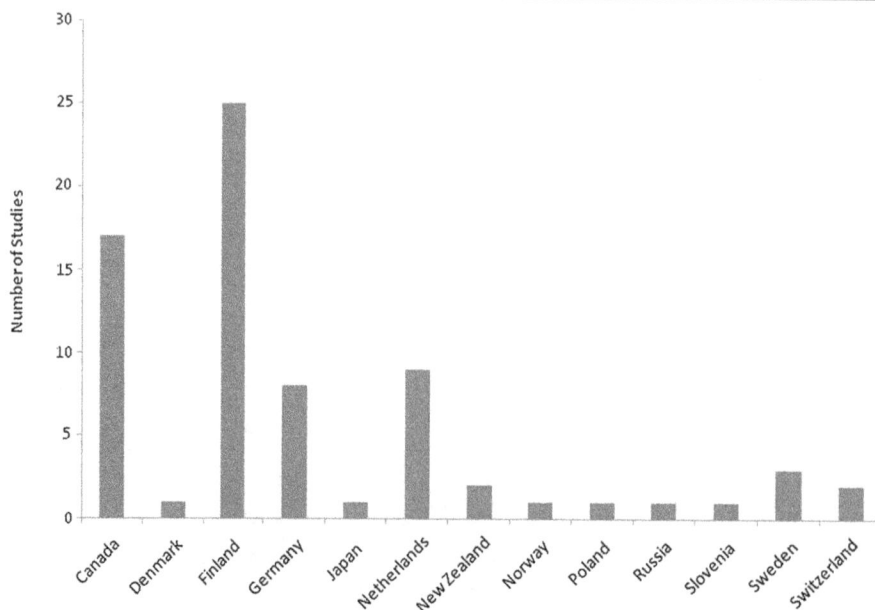

Figure 2 Frequency plot showing numbers of included studies undertaken in each country for all studies in the review.

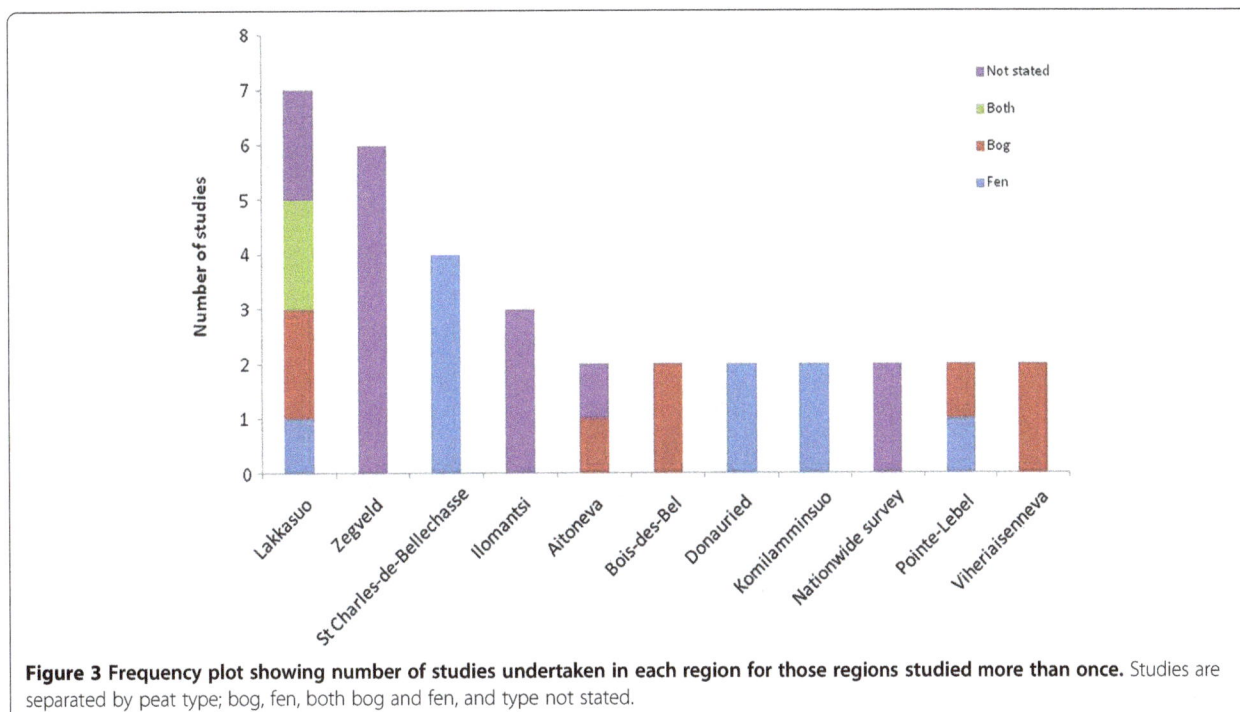

Figure 3 Frequency plot showing number of studies undertaken in each region for those regions studied more than once. Studies are separated by peat type; bog, fen, both bog and fen, and type not stated.

was unaccounted for in the remaining 8 studies. Furthermore, these studies' outcomes were not combined in a multivariate meta-analysis since the outcomes can be viewed as independent of one another.

Peat extraction and restoration interventions, and ground water level exposure

Drained-vs-undrained

Twenty-six studies reported a total of 48 outcomes dealing with drainage interventions. Meta-analyses were possible for NEE, R_{eco}, CH_4, N_2O and DOC. Only one study measured concentration of particulate organic carbon

(POC) and one study C stores. Both of these studies indicated lower levels following drainage, although POC effect size had a large variance (see narrative synthesis tables in Additional file 10).

Net ecosystem exchange (NEE) No significant patterns in NEE CO_2 release were identified across the three synthesised studies (all of which were from Canada) for drained versus undrained peatlands (Table 7 and Figure 6). Significant heterogeneity was found, however, as suggested by the substantially higher flux figure for the Point-Lebel study of Strack et al. [42].

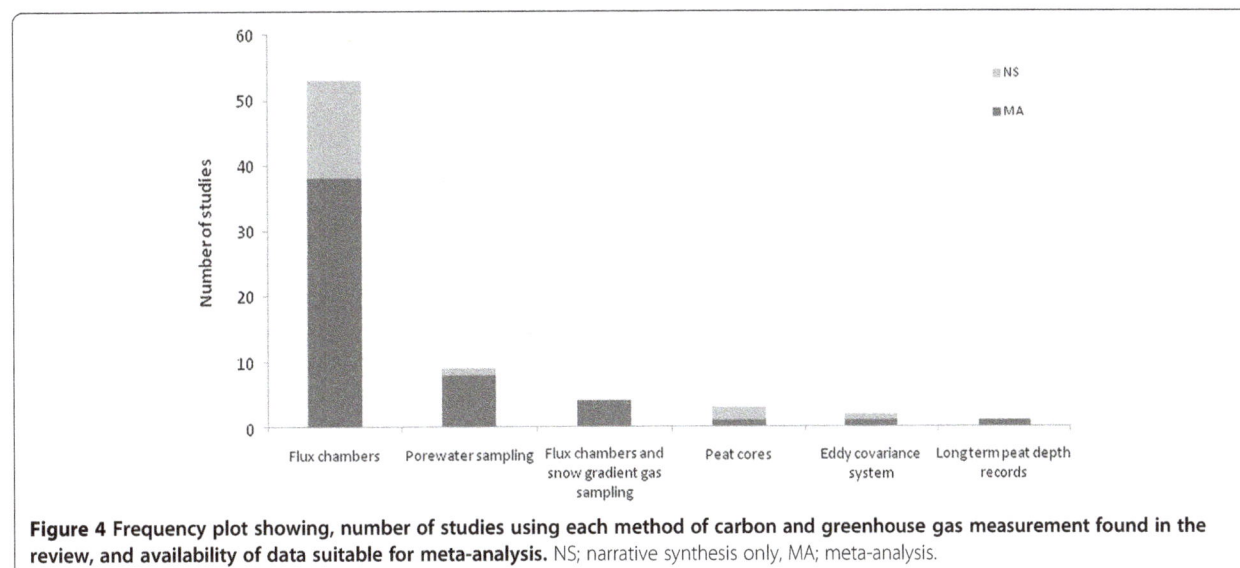

Figure 4 Frequency plot showing, number of studies using each method of carbon and greenhouse gas measurement found in the review, and availability of data suitable for meta-analysis. NS; narrative synthesis only, MA; meta-analysis.

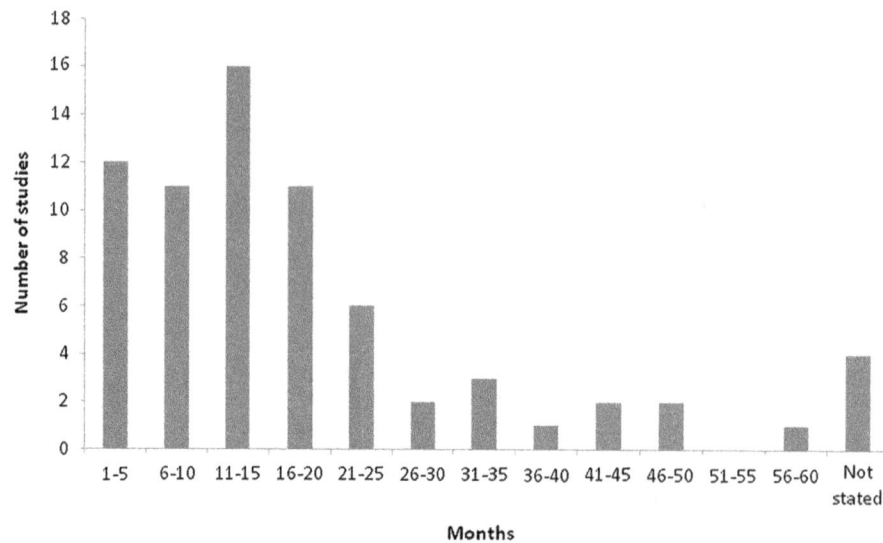

Figure 5 Frequency plot showing study length (total study months) for included studies.

Ecosystem respiration (R_{eco}) Drained peatlands released, on average, 125 mg CO_2 m^{-2} h^{-1} (95% CI=6.71 to 243) more CO_2 through respiration than undrained peatlands, approximating to an annual difference of 11,000 kg CO_2 ha^{-1} (95% CI=0.588 to 21.3) resulting from drainage (p=0.038) (Table 7 and Figure 7). Studies varied considerably in their findings, shown clearly in the forest plot and by the presence of significant heterogeneity. Studies were undertaken over 4 [43], 6 [44], 12 [45], 15 [35], 16 [36,42], and 32 [46] months (study period not stated in [47]).

Five other studies reported mean R_{eco} values but no measure of variability. Effect sizes ranged from -6.88 to 123 mg m^{-2} h^{-1} (see narrative synthesis tables in Additional file 10).

CH_4 Drained peatlands may release 0.126 mg CH_4 less than undrained peatlands per square metre per hour; equating to an approximate annual additional release of 11.0 kg CH_4 ha^{-1} (95% CI = 22.3 to 0.263) and a 100-year global warming potential difference of 276 kg CO_2 equivalents ha^{-1}, although this result is only marginally significant (p=0.055) (Table 7 and Figure 8). No significant heterogeneity between study results was identified. The results indicate an effect of -0.126 mg CH_4 m^{-2} h^{-1} (95% CI=-0.254 to -0.003). Studies were undertaken over 12 [45], 14 [48], 15 [10,35], 16 [36], 19 [49], and 32 [46] months (study period not stated in [47]).

Four studies could not be included in this meta-analysis due to missing measures of variability, but these studies all showed a negative effect of drainage (if only marginally); their mean effect sizes ranged from -0.047 to -4.72 mg m^{-2} h^{-1} (see narrative synthesis tables in Additional file 10).

N_2O Drained peatlands were found to release, on average, 0.008 mg N_2O m^{-2} h^{-1} (95% CI=0.001 to 0.016) more on drained peatlands relative to undrained peatlands (p=0.033) (Table 7 and Figure 9), corresponding to an approximate annual excess of 0.701 kg N_2O ha^{-1} (95% CI=0.088 to 1.40), and a 100-year global warming potential difference of 209 kg CO_2 equivalents ha^{-1}. No significant heterogeneity in study results was identified. Only Nykanen et al. [10] studied peatlands under active agricultural management, possibly accounting for the high variability in N_2O emission measured. Studies were undertaken over 9 [41], 15 [10,35], 16 [36], and 32 [46] months.

Five studies reported mean N_2O emission but no measure of variability, but these effects were positive (4 studies, effect sizes ranged from 0.004 to 0.227 mg m^{-2} h^{-1}) or undetectable (1 study) (see narrative synthesis tables in Additional file 10).

DOC Drained peatlands possessed, on average, a greater concentration of porewater DOC of 27.3 mg l^{-1} (95% CI=-0.099 to 54.7) relative to undrained peatlands, although this result was only marginally significant (p=0.051) (Table 7 and Figure 10). Significant heterogeneity existed between studies, indicating that a source of variation in study results was present and unaccounted for. Studies were undertaken over 2 [50], 6 [51,52], and 25 [53] months (study period not stated in [54]).

Restoration (Drained and restored-vs-undrained, restored-vs-unrestored, extracted and restored-vs-natural)
Seventeen studies reported a total of 26 outcomes for restoration interventions. Only CH_4 and R_{eco} emission,

Table 5 Measured outcomes within studies falling into the 16 intervention/exposure groups tabulated across 11 outcomes

Intervention/exposure-vs-comparator	NEE CO_2	R_{eco} CO_2	CH_4	CH_4 pore water	N_2O	Soil air N_2O	Soil surface N_2O	C stores	DOC	SOC	POC (TOC-DOC)	Long-term subsidence
Cropped-vs-bare		2	3		1							
Drained and restored-vs-undrained	1	1										
Drained-vs-undrained	3	15	13		10			1	5		1	
Dry-vs-wet	2	2	2		3							
Extracted and restored-vs-natural	1	1										
High intensity farmed-vs-low intensity farmed		3	5		6			1	1	1		1
Fertilised and grazed-vs-unfertilised and mown					1							
Fertilised-vs-less fertilised	2	3	3		6							
Grass-vs-forest		1	1		1	1						
Grazed-vs-mown			3		1							
Irrigated-vs-non-irrigated					1		1					
Mineral soil dressed-vs-undressed					1							
Old abandoned-vs-recently abandoned		1										
Old afforested-vs-recently afforested		1	1		1							
Poor-vs-rich		1	1						2			
Restored-vs-unrestored	6	4		1	1				8		1	

NEE=net ecosystem exchange CO_2, R_{eco}=ecosystem respiration CO_2, DOC=dissolved organic carbon, SOC=soil organic carbon, POC=particulate organic carbon (total organic carbon (TOC) – DOC).

and DOC concentration outcomes had sufficient data reported to permit meaningful meta-analysis.

Ecosystem respiration (R_{eco}) Restoration was found to have no significant effect on R_{eco} (Table 7 and Figure 11). No heterogeneity between studies was found, although this result was marginal.

One other study provided means but no variability [51]. This study compared restored with unrestored peatlands using three different comparisons; restored-vs-actively

Table 6 Study results across outcomes for key intervention/exposure groups

Intervention/exposure-vs-comparator	CH_4	N_2O	NEE	R_{eco}	DOC
Drained-vs-undrained	9	5	3	10	5
Dry-vs-wet		3			
High intensity farmed-vs-low intensity farmed	5	6			
Grazed-vs-mown	3				
Restored-vs-unrestored	4		3		8

Numbers refer to those studies with sufficient data (i.e. mean and variability measure) and quantities of studies to allow meta-analysis (i.e. >3). NEE=net ecosystem exchange CO_2, R_{eco}=ecosystem respiration CO_2.

harvested, closed-ditch restored-vs-abandoned, and inundated restored-vs-abandoned. The former restored-vs-harvested comparison yielded a positive effect, whilst the latter two restored-vs-abandoned had large negative effects.

Two studies compared drained and restored peatlands with 'natural' or undrained peatlands. Both studies found negative effect sizes of -37.5 ±85.4 (SD) mg m^{-2} h^{-1} [55] and -101 ±110 (SD) mg m^{-2} h^{-1}, indicating that the drained and restored peatlands released less CO_2 from respiration than undrained/'natural' peatlands.

CH_4 Restoration resulted in an increase in emission of 0.248 mg CH_4 m^{-2} h^{-1} (95% CI=0.052 to 0.446) (p=0.014) (Table 7 and Figure 12), corresponding to 21.7 kg ha^{-1} annually (95% CI=4.56 to 39.1) and 543 kg CO_2 equivalents ha^{-1}. Studies showed no identifiable heterogeneity. Studies were undertaken over 16 [56,57] and 43 months [58].

This finding was supported by higher concentrations of pore water CH_4 in restored peatlands reported by Waddington and Day [58].

Table 7 Test results for random effects meta-analyses for five intervention/exposure categories with sufficient data

Intervention/exposure-vs-comparator	Measured outcome	No. studies	SEE[1]	SE	Lower 95% CI	Upper 95% CI	Description	p-value*	T^2 (±SE)	Test for Heterogeneity; $Q_{(df)}$, p-value
Drained-vs-undrained	NEE	3	116	161	–200	431	Greater in drained	0.473	7.40×10^3 (±77.8×10^3)	$85.3_{(2)}$, <0.001
	R_{eco}	10	125	60.2	6.71	243	Greater in drained	0.038*	31.1×10^3 (±16.9×10^3)	$150_{(9)}$, <0.001
	CH_4	9	–0.126	0.066	–0.254	0.003	Greater in undrained	0.055	0.000 (±0.016)	$12.8_{(8)}$, 0.119
	N_2O	5	0.008	0.004	0.001	0.016	Greater in drained	0.033*	0.000 (±0.0001)	$2.33_{(4)}$, 0.676
	DOC	5	27.3	14.0	–0.099	54.7	Greater in drained	0.051	839 (±677)	$134_{(4)}$, <0.001
Dry-vs-wet	N_2O	3	0.221	0.103	0.019	0.424	Greater in dry	0.032*	0.018 (±0.032)	$5.46_{(2)}$, 0.065
Grazed-vs-mown	CH_4	3	–5.39	4.40	–14.0	3.23	Greater in grazed	0.220	44.4 (±58.8)	$8.71_{(2)}$, 0.013
Restored-vs-unrestored	R_{eco}	3	48.7	104	–155	253	Greater in restored	0.640	20.0×10^3 (±33.7×10^3)	$5.89_{(2)}$, 0.053
	CH_4	4	0.248	0.101	0.052	0.446	Greater in restored	0.014*	0.000 (±0.053)	$1.18_{(3)}$, 0.759
	DOC	6	15.6	11.6	–7.20	38.3	Greater in restored	0.180	622 (±497)	$26.1_{(5)}$, <0.001
High intensity farmed-vs-low intensity farmed	CH_4	4	–0.004	0.004	–0.011	0.004	Greater in low intensity farmed	0.349	0.000 (±0.0003)	$3.51_{(3)}$, 0.314
	N_2O	5	0.143	0.100	–0.053	0.340	Greater in high intensity farmed	0.153	0.043 (±0.035)	$18.0_{(4)}$, 0.001

[1]Units: CH_4=mg CH_4 m^{-2} h^{-1}, DOC=mg l^{-1}, N_2O=mg N_2O m^{-2} h^{-1}, NEE and R_{eco}=mg CO_2 m^{-2} h^{-1}.
Negative 'summary effect estimates' indicate a greater value for the control than the intervention and vice versa. SEE=summary effect estimate. SE=standard error, CI=confidence interval, T^2=estimate of tau-squared, inter-study variability. '*' indicates marginal significance. '*' indicates significance at α=0.05, '.' indicates marginal significance. NEE=net ecosystem exchange CO_2, R_{eco}=ecosystem respiration CO_2, DOC=dissolved organic carbon.

Table 8 Studies measuring emissions/fluxes of all three GHGs together

Study	Region	Measured outcomes
Danevcic et al. [35]	Ljubljana	CH_4, N_2O, CO_2 (R_{eco})
Klove et al. [36]	Bodin	CH_4, N_2O, CO_2 (R_{eco})
Lund et al. [37]	Fajemyr	CH_4, N_2O, CO_2 (R_{eco} and NEE)
Maljanen et al. [38]	Western Finland	CH_4, N_2O, CO_2 (R_{eco})
Maljanen et al. [39]	Kannus	CH_4, N_2O (air and soil), CO_2 (R_{eco})
Martikainen et al. [10]	Lakkasuo	CH_4, N_2O, CO_2 (R_{eco})
Nykanen et al. [10]	Ilomantsi	CH_4, N_2O, CO_2 (R_{eco})
Petersen et al. [40]	Nationwide survey	CH_4, N_2O, CO_2 (R_{eco})
Von Arnold et al. [41]	Asa	CH_4, N_2O, CO_2 (R_{eco})

DOC No significant effect on DOC concentration of restored versus unrestored peatlands was apparent across the included studies (Table 7 and Figure 13). Significant heterogeneity was identified, however, with the three studies in River-de-Loup having particularly large variability around their effect sizes.

Two studies did not report variability about their means for DOC; they showed both positive [59, 22.8 mg l⁻¹] and negative [60, -28.8 mg l⁻¹] effects. The former study fails to describe the restoration time scale, whilst the latter investigated a peatland that had been drained in the 19th Century and rewetted in 1984, 20 years prior to the study.

Other measured outcomes Two studies reported net ecosystem exchange (CO_2) for drained and restored peatlands relative to undrained/'natural' peatlands. These studies found positive effect sizes of 42.2 ±145 (SD) mg m⁻² h⁻¹ [55] and 289 ±161 (SD) mg m⁻² h⁻¹ [61]. One study found a small negative effect of restoration on drained peatlands of -0.006 mg N_2O m⁻² h⁻¹ [62]. Finally, one study reported a negative effect size of -5.67 mg l⁻¹ in concentration of POC as a result of restoration of drained peatlands [53].

Dry-vs-wet

Three studies reported mean and variability measures for N_2O emission on dry and wet peatlands; whilst a range of other measured outcomes were reported in a small number of studies.

N_2O Drier peat released 0.221 mg N_2O m⁻² h⁻¹ (95% CI=0.019 to 0.424) more than wet peat (p=0.032) (Table 7 and Figure 14), corresponding to an approximate annual difference of 19.4 kg N_2O ha⁻¹ (95% CI=1.66 to 37.1) and 5.77 tonnes of CO_2 equivalents ha⁻¹. However, whilst the results are from three different studies, the data are collected from the same peatland area. Furthermore, during the critical appraisal process, these studies were classified as unclear [63,64] and high [65] in their susceptibility to bias. The results should therefore be treated with caution. No heterogeneity was evident, although the result is marginal, a possible concern given the small number of studies in the analysis. However, the consistency between this result and the comparison of N_2O emissions from drained and

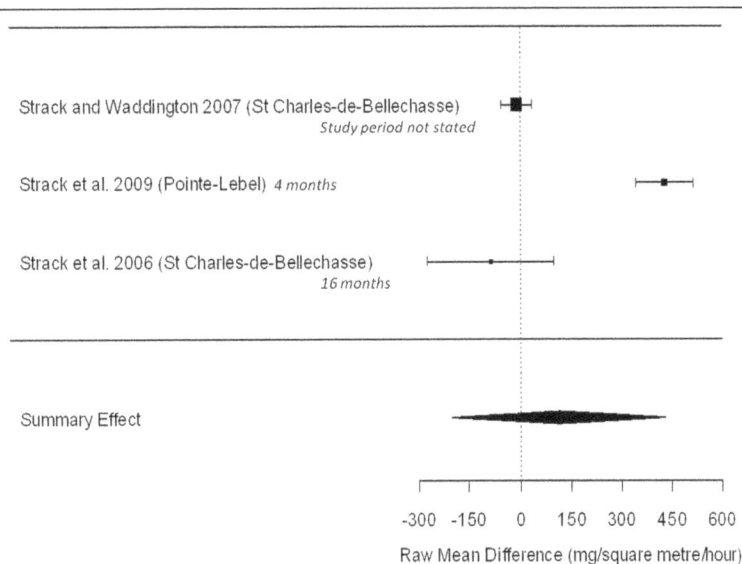

Figure 6 Forest plot of meta-analysis of effect sizes (raw mean difference) comparing net ecosystem exchange (NEE CO_2) in drained versus undrained peatlands. The dashed line represents no effect. Size of boxes represents the weight of the study in the analysis. Horizontal lines represent 95% confidence intervals. Positive effect sizes indicate greater flux of CO_2 in drained peatlands than undrained controls. The position of the centre of the diamond represents the overall summary effect (a weighted average) and its horizontal extent represents the positive and negative 95% confidence intervals.

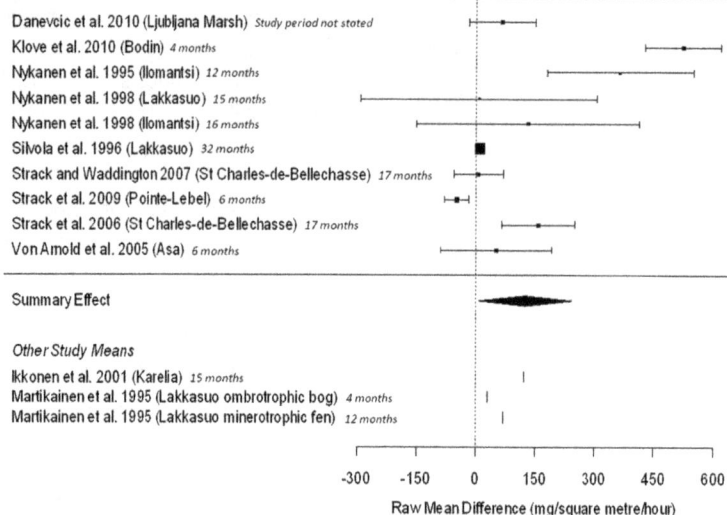

Figure 7 Forest plot of meta-analysis of effect sizes (raw mean difference) comparing total ecosystem respiration (R_eco) in drained versus undrained peatlands. The dashed line represents no effect. Size of boxes represents the weight of the study in the analysis. Horizontal lines represent 95% confidence intervals. Positive effect sizes indicate greater flux of CO_2 in drained peatlands than undrained controls. The position of the centre of the diamond represents the overall summary effect (a weighted average) and its horizontal extent represents the positive and negative 95% confidence intervals. Other Study Means are provided for studies where a measure of variability could not be extracted.

undrained peatlands provides some confidence that the results may reflect a genuine pattern. Studies were undertaken over 9 [65], 19 [64], and 25 [63] months.

Other measured outcomes Four studies reported comparisons of dry versus wet peatlands across three other outcomes. Methane was found to be released in greater quantities from wet peats; effect size=-0.003 ±0.005 (SD) mg m^{-2} h^{-1} [66], Hedges d=-2.081 ±0.412 (SD) [67]. Two studies found lower NEE in dry peats than wet peats; one reporting only a mean [68] and the other reporting wide variability about a small effect size [61]. Finally, two studies reported

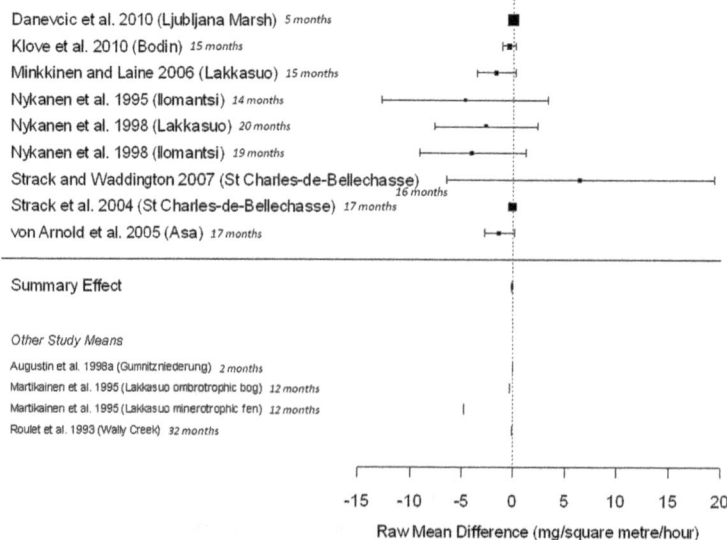

Figure 8 Forest plot of meta-analysis of effect sizes (raw mean difference) comparing CH_4 from drained versus undrained peatlands. The dashed line represents no effect. Size of boxes represents the weight of the study in the analysis. Horizontal lines represent 95% confidence intervals. Positive effect sizes indicate greater flux from drained peatlands than undrained controls. In the Summary Effect row, the position of the centre of the diamond represents the overall summary effect (a weighted average) and its horizontal extent represents the positive and negative 95% confidence intervals. Other Study Means are provided for studies where a measure of variability could not be extracted.

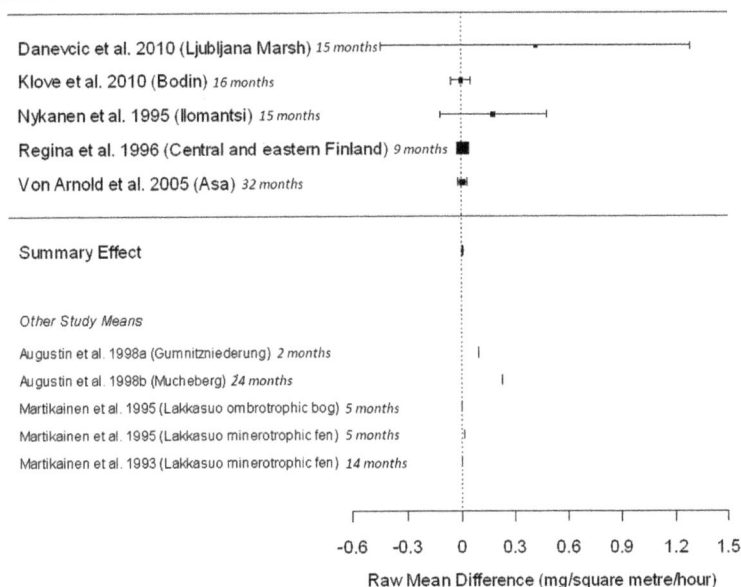

Figure 9 Forest plot of meta-analysis of effect sizes (raw mean difference) comparing N₂O flux from drained versus undrained peatlands. The dashed line represents no effect. Size of boxes represents the weight of the study in the analysis. Horizontal lines represent 95% confidence intervals. Positive effect sizes indicate greater flux in drained peatlands than undrained controls. The position of the centre of the diamond represents the overall summary effect (a weighted average) and its horizontal extent represents the positive and negative 95% confidence intervals. Other Study Means are provided for studies where a measure of variability could not be extracted.

positive R_{eco} effect sizes of dry versus wet peatlands; 72.2 ±110 (SD) mg m^{-2} h^{-1} [61] and 269 mg m^{-2} h^{-1} [68].

Poor-vs-rich

Two studies classified peatlands as 'poor' or 'rich'. Poor fens are generally characterised by lower pH and nutrient concentrations, and 'rich' sites by higher pH and nutrients. The

authors also note that their 'poor' sites tended to have lower water tables than their 'rich' sites [69]. Although not explicitly stated, this was interpreted as a natural contrast between sites, rather than an effect of management. In the two studies, which were undertaken at the same sites, CH₄ was found to be produced at higher rates in richer peats [69], DOC concentration to be higher in poorer peats [69,70],

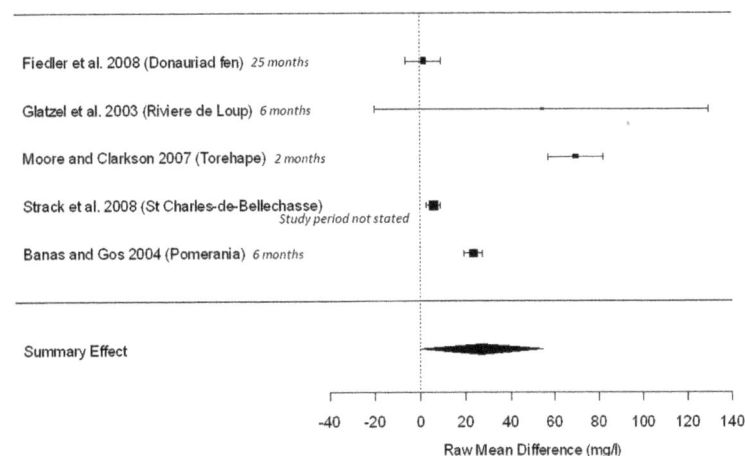

Figure 10 Forest plot of meta-analysis of effect sizes (raw mean difference) comparing dissolved organic carbon (DOC) concentration in drained versus undrained peatlands. The dashed line represents no effect. Size of boxes represents the weight of the study in the analysis. Horizontal lines represent 95% confidence intervals. Positive effect sizes indicate greater concentration in drained peatlands than undrained controls. The position of the centre of the diamond represents the overall summary effect (a weighted average) and its horizontal extent represents the positive and negative 95% confidence intervals.

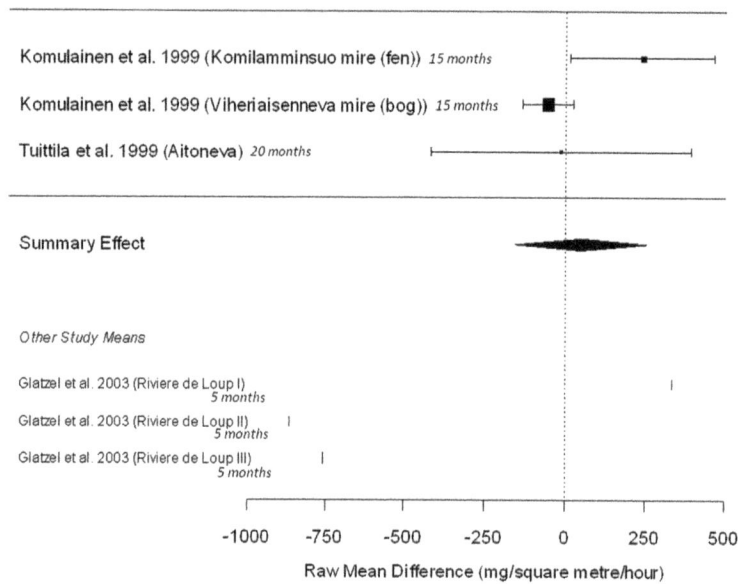

Figure 11 Forest plot of meta-analysis of effect sizes (raw mean difference) comparing total ecosystem respiration (R_{eco}) in restored versus unrestored peatlands. The dashed line represents no effect. Size of boxes represents the weight of the study in the analysis. Horizontal lines represent 95% confidence intervals. Positive effect sizes indicate greater flux in restored peatlands than unrestored peatlands. The position of the centre of the diamond represents the overall summary effect (a weighted average) and its horizontal extent represents the positive and negative 95% confidence intervals. Other Study Means are provided for studies where a measure of variability could not be extracted.

and ecosystem respiration rate to be higher in poorer peats [69] (see narrative synthesis tables in Additional file 10). These results are consistent with the comparison of drained vs undrained sites, and thus with water table providing the primary control on emissions and fluxes from these sites.

Irrigated-vs-non-irrigated

One study [71] compared emission and soil surface concentration of N_2O from rain-fed peat soils with non-irrigated peat soils. N_2O emission was higher in non-irrigated soils, whilst the opposite was true for soil

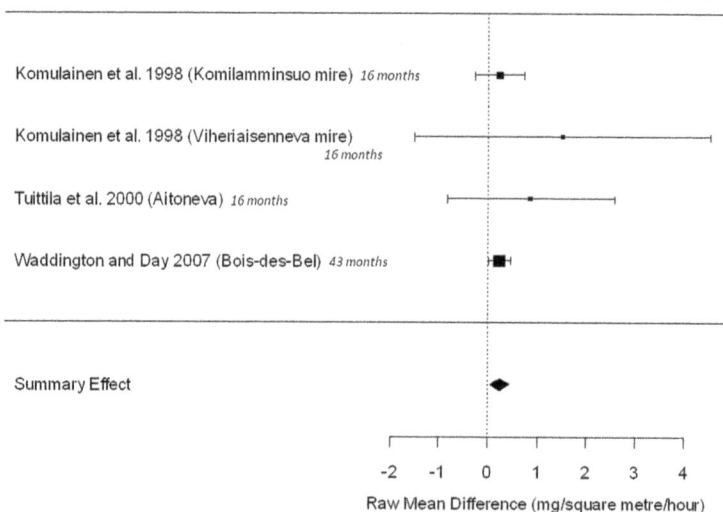

Figure 12 Forest plot of meta-analysis of effect sizes (raw mean difference) comparing CH_4 flux from restored versus unrestored peatlands. The dashed line represents no effect. Size of boxes represents the weight of the study in the analysis. Horizontal lines represent 95% confidence intervals. Positive effect sizes indicate greater flux in restored peatlands than unrestored peatlands. The position of the centre of the diamond represents the overall summary effect (a weighted average) and its horizontal extent represents the positive and negative 95% confidence intervals.

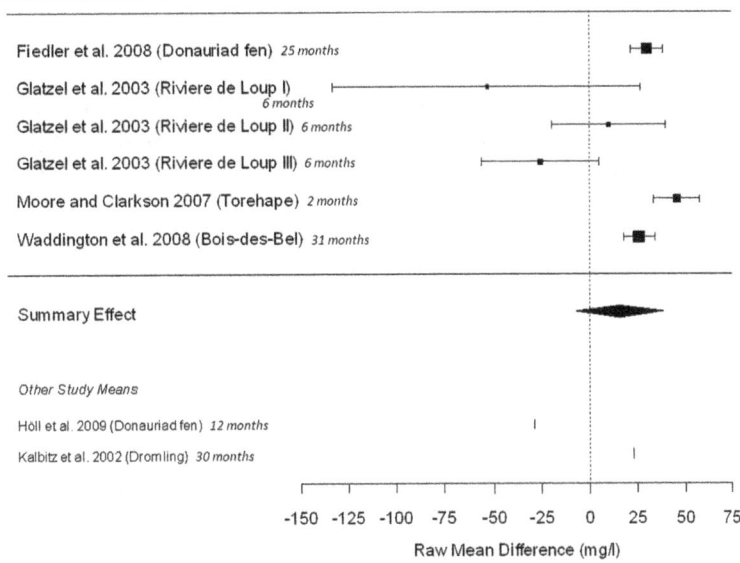

Figure 13 Forest plot of meta-analysis of effect sizes (raw mean difference) comparing dissolved organic carbon (DOC) concentration from restored versus unrestored peatlands. The dashed line represents no effect. Size of boxes represents the weight of the study in the analysis. Horizontal lines represent 95% confidence intervals. Positive effect sizes indicate greater flux in restored peatlands than unrestored peatlands. The position of the centre of the diamond represents the overall summary effect (a weighted average) and its horizontal extent represents the positive and negative 95% confidence intervals. Other Study Means are provided for studies where a measure of variability could not be extracted.

surface N_2O concentration (see narrative synthesis tables in Additional file 10).

Farmland interventions
High intensity farmed-vs-low intensity farmed
A total of nine studies report seven different measured outcomes, but only two of these outcomes (CH_4 and N_2O) had sufficient studies to allow meta-analysis.

CH_4 Farming intensity had no significant effect on CH_4 emission across the four included studies (Table 7 and Figure 15). Whilst no heterogeneity was found across studies, the studies by Petersen et al. [40] showed much higher variability than the other two studies, possibly because they used nationwide survey data rather than site-based experimental comparisons. The Petersen et al. [40] studies were classed as 'unclear' in their susceptibility to bias, and more information on the methodology

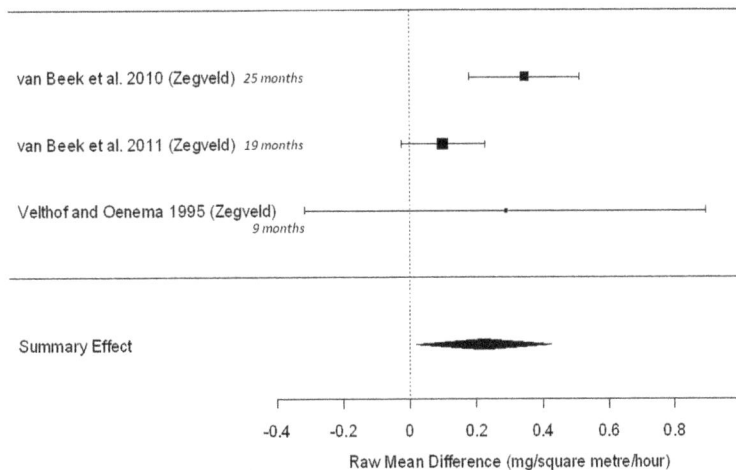

Figure 14 Forest plot of meta-analysis of effect sizes (raw mean difference) comparing N_2O flux from wet versus dry peatlands. The dashed line represents no effect. Size of boxes represents the weight of the study in the analysis. Horizontal lines represent 95% confidence intervals. Positive effect sizes indicate greater flux in wet peatlands than dry peatlands. The position of the centre of the diamond represents the overall summary effect (a weighted average) and its horizontal extent represents the positive and negative 95% confidence intervals.

involved would be needed before the low precision can be accounted for.

N2O Farming intensity was not found to have a significant effect on N_2O emission across the included studies (Table 7 and Figure 16). Significant heterogeneity was identified, however. Only two of the five included studies cross the line of no effect, indicating that in individual locations there may be an effect of farming intensity on N_2O emission and suggesting that further studies would be particularly valuable.

Other measured outcomes Five other outcomes were reported across seven studies (see narrative synthesis tables in Additional file 10); namely, C store, DOC concentration, long-term subsidence, SOC concentration and R_{eco}. The various different measures of C concentration showed a range of results; both positive and negative with respect to farming impact. R_{eco} showed varied results with effect sizes from -251 to 93.6 mg m^{-2} h^{-1} (high intensity relative to low intensity farming).

Cropped-vs-bare Three studies reported CH_4 emission on cropped versus bare peat (i.e. actively extracted) (see narrative synthesis tables in Additional file 10 for details). Two studies reported negative effect sizes [72, -0.070 ±0.053 (SD)], [73, -0.003] and one a positive effect size [74, 0.007 ±0.008 (SD)]. Two studies measured R_{eco} for cropped and bare peats [73,74], finding conflicting positive (48.8 mg m^{-2} h^{-1}) and negative (-11.4 ±8.75 (SD) mg m^{-2} h^{-1}) effect sizes respectively.

Fertilised-vs-less fertilised Ten studies contributed 14 lines of data across four different outcomes for peatlands

with experimental gradients in fertilisation rate (see narrative synthesis tables in Additional file 10). For CH_4 emission effect sizes were negative (-2.58 ±1.61 (SD) mg m^{-2} h^{-1}, and -0.117 mg m^{-2} h^{-1}) or zero (0.000 ±0.005 (SD) mg m^{-2} h^{-1}) indicating greater or negligible emission in controls than intervention soils. For N_2O emission effect sizes were predominantly positive (six values ranging from 0.004 to 17.0 mg m^{-2} h^{-1}) with one negative value (-0.092 ±0.103 (SD) mg m^{-2} h^{-1}), with positive values signifying greater emission on fertilised soils and vice versa. The two studies reporting NEE showed very different outcomes, with effect sizes of 214 ±459 (SD) mg m^{-2} h^{-1} and -272 ±70.3 (SD) mg m^{-2} h^{-1}. Similarly, the three measured total respiration (CO_2) effect sizes ranged from -16.1 to 601 ±522 (SD) mg m^{-2} h^{-1} indicating a substantial degree of variability amongst studies.

Grazed-vs-mown Four studies compared grazed and mown peatland fields. Three of these reported CH_4 emissions and are analysed below, whilst only one measured N_2O emissions.

CH4 No significant difference in CH_4 emissions was identified between grazed and mown peat soils (Table 7 and Figure 17). However, significant heterogeneity was identified, highlighting the very small variability reported in the van den Pol-van Dasselaar et al. [66] study.

Other measured outcomes One study reported N_2O emission in grazed and mown peats, finding greater emission on grazed soils [65, 0.256 ±0.220 (SD) mg m^{-2} h^{-1}].

Fertilised and grazed-vs-unfertilised and mown One study compared fertilised grazed peatland with unfertilised

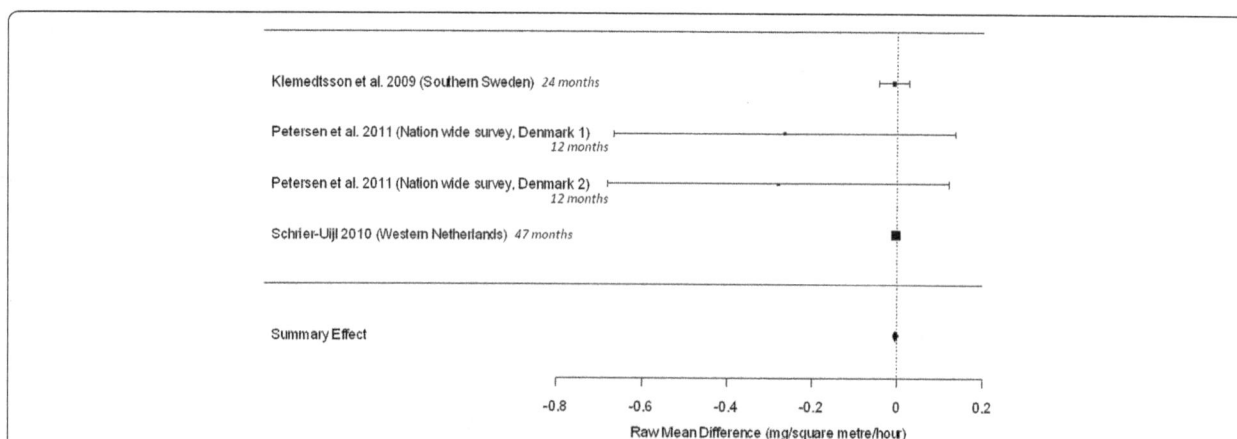

Figure 15 Forest plot of meta-analysis of effect sizes (raw mean difference) comparing CH₄ flux from high intensity versus low intensity farmed peatlands (see text for definition). The dashed line represents no effect. Size of boxes represents the weight of the study in the analysis. Horizontal lines represent 95% confidence intervals. Positive effect sizes indicate greater flux in high-intensity farmed peatlands than in low-intensity. The position of the centre of the diamond represents the overall summary effect (a weighted average) and its horizontal extent represents the positive and negative 95% confidence intervals.

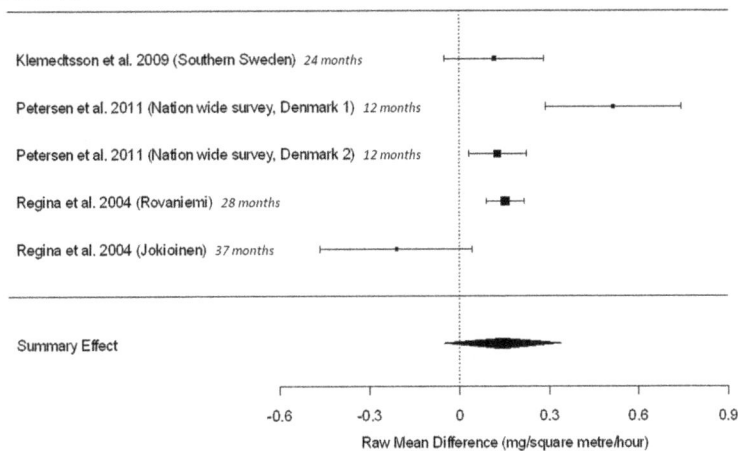

Figure 16 Forest plot of meta-analysis of effect sizes (raw mean difference) comparing N$_2$O flux from high intensity versus low intensity farmed peatlands (see text for definition). The dashed line represents no effect. Size of boxes represents the weight of the study in the analysis. Horizontal lines represent 95% confidence intervals. Positive effect sizes indicate greater flux in high-intensity farmed peatlands than low-intensity. The position of the centre of the diamond represents the overall summary effect (a weighted average) and its horizontal extent represents the positive and negative 95% confidence intervals.

mown peatland. The confounding nature of these two interventions prevented their collation with studies in Sections Fertilised-vs-less fertilised and Grazed-vs-mown. This study showed that N$_2$O emissions were greater under the fertilized grazed regime by 0.260 ±0.311 (SD) mg m^{-2} h^{-1} [64].

Other interventions
Grass-vs-forest
One study measured CH$_4$ emission, N$_2$O emission, soil air N$_2$O concentration, and total respiration

(CO$_2$) flux in a grass peatland field versus a forested peatland field [39]. All measured GHG emissions/ concentrations were greater on the grass field than the forested field (see narrative synthesis tables in Additional file 10).

Old abandoned-vs-recently abandoned One study examined the impact of time since abandonment on R$_{eco}$ [75], finding a positive effect size but not reporting a measure of variability (see narrative synthesis tables in Additional file 10).

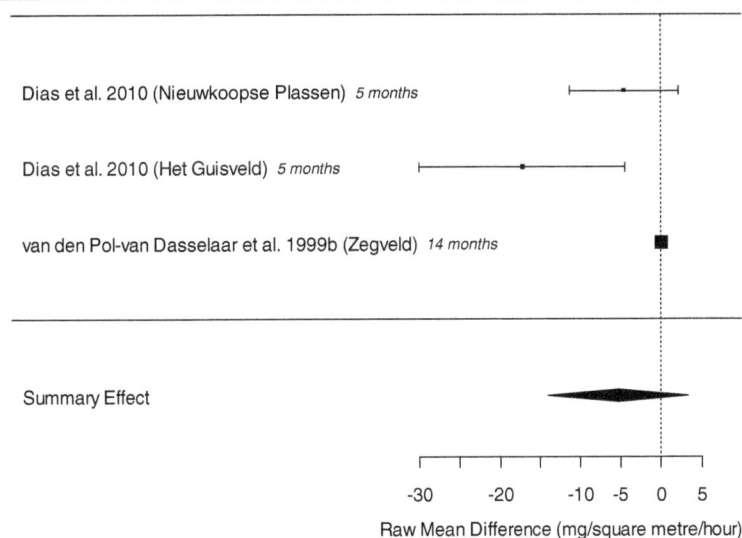

Figure 17 Forest plot of meta-analysis of effect sizes (raw mean difference) comparing CH$_4$ flux from grazed versus mown peatlands. The dashed line represents no effect. Size of boxes represents the weight of the study in the analysis. Horizontal lines represent 95% confidence intervals. Positive effect sizes indicate greater flux in grazed peatlands than mown peatlands. The position of the centre of the diamond represents the overall summary effect (a weighted average) and its horizontal extent represents the positive and negative 95% confidence intervals.

Recently-vs-long-term afforested One study reported CH_4, N_2O and R_{eco} emissions for peatlands afforested two years prior to study versus 24 years previous [38]. Both CH_4 and R_{eco} emissions were greater on more recently afforested peatlands (-0.410 ±0.634 and -125 ±90.2 mg m^{-2} h^{-1} respectively), whilst N_2O emission was greater in older forests by 0.072 ±0.197 (SD) mg m^{-2} h^{-1} [38].

Mineral soil dressed-vs-undressed drained peat Only one study measured the impact of mineral soil-dressing on drained peatland, finding N_2O to be released at a greater rate from the undressed peat [76, -0.111 ±0.247 (SD)].

Combined 'drier-vs-wetter' peat (drained-vs-undrained, dry-vs-wet, and unrestored-vs-restored) Three intervention/exposure categories (drained-vs-undrained, dry-vs-wet, and unrestored-vs-restored) were combined for CH_4 and R_{eco} into one category ('drier-vs-wetter') since this allowed the addition of one extra study in each case that was excluded from the above meta-analyses as it was a single study within its intervention/exposure group. In this larger group, restored-vs-unrestored has been reversed to reflect the same direction of effect relative to 'drier versus wetter' peatlands.

Ecosystem respiration (R_{eco}) Drier peat soils released, on average, 88.8 mg m^{-2} h^{-1} (95% CI=-7.97 to 186) more CO_2 through respiration than wetter peats, although this result is only marginally significant (p=0.072) (Figure 18). There is significant heterogeneity across studies (QE$_{13}$=158 p<0.001), as found with the individual group analyses. This was investigated using meta-regression, including mean site air temperature (T_{air}) and mean precipitation (PPT) reported for each site or taken from 30 year mean data within other articles. These analyses showed that neither mean air temperature nor mean precipitation were significantly responsible for the heterogeneity observed (T_{air}; QM$_1$=0.002 p=0.968. PPT; QM$_1$=0.271 p=0.602).

CH_4 Drier peats released, on average, -0.131 mg m^{-2} h^{-1} (95% CI=-2.54 to -0.009) more CH_4 than wetter peat soils (p=0.035) (Figure 19). Combining these three categories produces a significant effect size of lower CH_4 emissions from the drier peatland, whereas the reduction in emission in drained-vs-undrained alone was marginal (p=0.055, Table 7 and Figure 8), although unrestored peatland did show significantly lower emissions than restored. The study by van den Pol-van Dasselaar et al. [66] of naturally dry versus wet peatland is strongly weighted in this combined analysis, but it possesses a small effect size.

Discussion
Major findings
The evidence base concerning the impact of land management activities on GHG flux and C stores in lowland peatlands is rather limited in extent. Furthermore, the disparate nature of the different management drivers studied, variables measured and ways in which the results are reported makes quantitative synthesis difficult or in some cases impossible. Some interventions/exposures are better covered (e.g. drainage) than others (e.g. restoration). This may reflect the fact that restoration through rewetting is a fairly recent intervention [77] and that long-term studies are lacking relative to those on earlier drainage [25]. However, whilst the scopes and designs of included studies vary considerably, our review has highlighted a number of significant patterns across the evidence base (see Table 9).

Firstly, less CH_4 is emitted by drained lowland peatlands than from undrained peats. Secondly, more CH_4 is emitted by restored (i.e. rewetted) lowland peatlands than those that have been similarly previously drained but not restored. The finding that wetter peat soils produce more CH_4 is in accordance with the concept that CH_4 emission increases, as higher water tables increase anaerobic decomposition of soil organic matter [78,79]. Similar findings were reported in the global review of all peatlands by Bussell et al. [25]. Our review found 0.126 mg m^{-2} h^{-1} (95% CI=0.003 to 0.254; 9 studies) more CH_4 released from undrained than drained peats, whilst Bussell et al. [25] reported a figure of 0.335 mg m^{-2} h^{-1} (95% CI=0.200 to 0.467; 27 studies). Our review found 0.248 mg CH_4 m^{-2} h^{-1} (95% CI=0.052 to 0.446; 4 studies) more CH_4 released from restored than unrestored peats, whilst Bussell et al. [25] reported a figure of 0.667 mg m^{-2} h^{-1} (95% CI=0.017 to 1.32; 5 studies). Upland (blanket bog) peatlands accounted for a high proportion of the studies reviewed by Bussell et al. [25]. Although the comparison of the two reviews may indicate that the impacts of re-wetting drained peatlands, in terms of greater CH_4 emission, may be less in the lowlands than the uplands, this observed difference is based on a relatively few studies and may arise by chance.

Thirdly, more N_2O is emitted by drained lowland peatlands and dry peatlands than their undrained and wet counterparts. Emission of N_2O may increase as aerobic conditions that extend to greater depths in drained peats favour nitrification [80], whilst in undisturbed wetter peat soils the anaerobic conditions of elevated water tables may limit N_2O production. A similar finding was reported in the review of all peatlands by Bussell et al. [25], with drained peatlands producing on average 3.97 µg m^{-2} h^{-1} more N_2O than undrained peats (95% CI=2.63 to 5.33; 13 studies), compared with our finding for lowland peatlands of a much larger effect size of 8.28

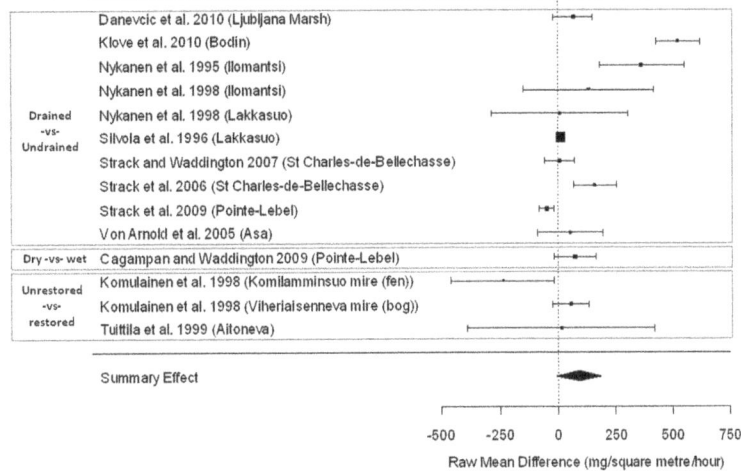

Figure 18 Forest plot of meta-analysis of effect sizes (raw mean difference) comparing total ecosystem respiration (R$_{eco}$) in 'drier versus wetter' peatlands. The dashed line represents no effect. Size of boxes represents the weight of the study in the analysis. Horizontal lines represent 95% confidence intervals. Positive effect sizes indicate greater flux in drier peatlands than wetter counterparts. The position of the centre of the diamond represents the overall summary effect (a weighted average) and its horizontal extent represents the positive and negative 95% confidence intervals. Red boxes indicate where the original grouping of intervention/exposure categories lies.

μg m^{-2} h^{-1} (95% CI=0.659 to 15.9; 5 studies). A greater N$_2$O response in lowland peats may reflect their typically higher nutrient levels, either naturally (in fens) or as a consequence of management (e.g. associated with fertilisation or higher livestock densities). Alternatively, this difference may reflect chance variability between the studies included in the respective reviews.

Finally, more CO$_2$ was produced through respiration by drained lowland peatlands than undrained peats. This finding can be related to the rapid release of CO$_2$ that results from aerobic microbial respiration as the peat profile moves away from the anoxic conditions of undrained

peat. The review of all global peatlands by Bussell et al. [25] reported the same direction of effect, but a non-significant, lower value; with drained peats respiring 59.8 mg m^{-2} h^{-1} more CO$_2$ than undrained peats (95% CI=16.7 to 100; 21 studies), compared with our finding of 125 mg CO$_2$ m^{-2} h^{-1} (95% CI=6.71 to 243; 10 studies).

Reasons for heterogeneity
Given the low number of primary studies investigating lowland-type peatlands which met the criteria for inclusion in the systematic review, few opportunities for investigation of sources of heterogeneity were available. As

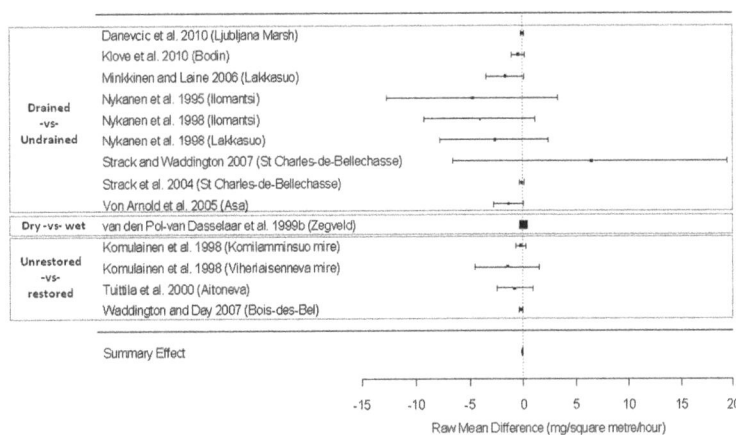

Figure 19 Forest plot of meta-analysis of effect sizes (raw mean difference) comparing CH$_4$ flux from 'drier versus wetter' peatlands. The dashed line represents no effect. Size of boxes represents the weight of the study in the analysis. Horizontal lines represent 95% confidence intervals. Positive effect sizes indicate greater flux in drier peatlands than wetter counterparts. The position of the centre of the diamond represents the overall summary effect (a weighted average) and its horizontal extent represents the positive and negative 95% confidence intervals. Red boxes indicate where the original grouping of intervention/exposure categories lies.

Table 9 Major findings from the quantitative synthesis of the evidence-base

Intervention/exposure-vs-comparator	CH$_4$	DOC	N$_2$O	NEE CO$_2$	R$_{eco}$ CO$_2$
Drained-vs-undrained	SEE=-0.126	SEE=27.3	SEE=0.008		SEE=125
	p=0.055 (.)	p=0.051 (.)	p=0.033 (*)	n.s.	p=0.038 (*)
	n=9	n=5	n=5	n=3	n=10
Dry-vs-wet			SEE=0.221		
			p=0.019 (*)		
			n=3		
Grazed-vs-mown	n.s.				
	n=3				
Restored-vs-unrestored	SEE=0.248				
	p=0.014 (*)	n.s.			n.s.
	n=4	n=8			n=3
High intensity farmed-vs-low intensity farmed	n.s.		n.s.		
	n=5		n=6		

SEE, summary effect estimate; n.s., non-significant; p, p-value; n, number of studies; (*), significant; (.), marginally significant (α=0.05).

future research is integrated into these meta-analyses, other sources of heterogeneity, such as soil type, climate, and susceptibility to bias judgements should be included in meta-regression and subgroup analysis.

It should be added that the amalgamation of studies within an intervention adds heterogeneity, since studies will differ in the intensity of their intervention and form of the comparator. This will increase the variability in the summary effect estimate. However, this will make the results herein more conservative, since the variability around the summary effect will be larger than for a meta-analysis of studies with identical intervention and comparator pairs.

Strengths and limitations of the review

This review is the first collation and synthesis of the literature on lowland peat C balances and GHG flux. As a systematic review it has included all of the available evidence on the subject and has objectively and transparently assessed the relevance and validity of all included studies. The review results, however, are more limited than similar findings for upland peats (or peats in general) because of the relative scarcity of research in lowland peatland environments. Many of the studies that did meet the criteria for inclusion were obtained from boreal 'mire' systems in Scandinavia and Canada. These systems could potentially differ considerably from the smaller, temperate lowland raised bog and fen systems of other regions, such as the UK. As research on lowland peats in other regions continues, the relevance of updates to this review's conclusions to areas outside the currently more studied regions will increase. Despite these limitations, this review has taken a first step in compiling and synthesising the results of the available literature, and will act as a strong foundation for future

updates of the review as further research on lowland peats becomes available.

Of the 282 articles identified as potentially relevant from abstract-level assessment and other sources of relevant literature, 79 could not be obtained due to restrictions on access or inability to identify the holding institutions. At 28 percent, this is rather a high proportion of unobtainable texts. Many of these articles were published in non-English language journals and the titles provided by the listing literature database may be English translations of their published counterparts, which is a possible reason we have been unable to locate them. This number is likely to be an overestimate of the relevance of the full texts, since those articles that lacked an abstract were included out of caution. Since primary literature accessibility is continually improving, future updates to this review will be able to lower the proportion of unobtainable articles as more research is published.

We have not combined those studies with multiple outcomes in a multivariate meta-analysis, since the separate outcomes can be viewed as independent of one another. For the same reason we have chosen not to correct our p-values across different outcomes for the same studies for multiplicity, which itself is a subject of controversy [81].

Limitations of the evidence base

The number of studies investigating the impacts of land management on GHG fluxes and C stores in lowland peatlands is rather low. This was also noted in the review of drainage and rewetting on GHG and C stores in peatlands worldwide [25]. In addition, studies often failed to report measures of variability within their results; also a finding of Bussell et al. [25]. Means alone cannot be included in meta-analysis, and a substantial proportion of the data from the literature could not be quantitatively synthesised as a result. Due to the lack of studies focusing on lowland peatlands in general and the lack of reporting of variability measures, the meta-regressions and subgroup analyses undertaken by Bussell et al. [25] in their review of global peatlands could not be undertaken in this review. Further investigation of the factors that may contribute to the heterogeneity seen across some of the evidence (for example total ecosystem respiration responses to drainage) will become possible as more evidence is made available in future years, and regular updates to this review are vital to capture this important information.

The included studies employed a range of different study periods, from 1 month to 60. Shorter studies may identify different net changes in outcome measures relative to longer studies: interventions or exposures may have very different short term consequences relative to the longer term. There were insufficient studies in our meta-analyses to allow study period to be included as a covariate. Future updates to this review may be able to examine the influence of study length on heterogeneity amongst studies.

It is surprising that very few studies focused on changes to C stores relative to GHG fluxes. Changes in organic C content of peatlands can, theoretically, occur without the release of GHGs, for example through DOC export, which has been found to be significant in some studies e.g. [82]. While C stock changes are difficult to measure directly, since changes are small relative to the total pool size, methods based on peat subsidence have been successfully applied in other regions, notably in tropical peatlands e.g. [83]. Similarly, few (9) studies examined CO_2 sequestration either through plant CO_2 photosynthesis (gross primary production) or through whole ecosystem net CO_2 exchange (NEE) relative to the number of studies that examined CO_2 respiration (R_{eco}) alone (37 studies). The inclusion of measurement of fluxes of all GHG gases and all forms of C store would allow a more thorough investigation of the impact of land management on peatlands, and future studies should include such C measurements. Where the hydrological regime permits, measurement of fluxes in DOC and POC would provide a valuable addition to the completeness of the picture of ecosystem C cycling.

Furthermore, as elaborated upon below, studies often reported temporal variability in measurements as a proxy for spatial replication, or combined spatial and temporal replication together in their presentation of results. Primary research is often not designed with forethought of meta-analysis, and the reporting of temporal variability is perfectly acceptable. However it should not be treated as providing replication that is comparable in rigour to that achieved through true spatial replication. In some instances the statistics undertaken within the primary research papers were incorrect through the use of temporal replication instead of spatial (true) replication.

Review conclusions
Implication for policy/management
The results of this review provide some empirical underpinning for policy on the management of lowland peatlands in relation to GHG emissions. The data show a consistent pattern of lower CH_4 emissions from drained sites, and that this is reversed by re-wetting over the short to medium term (12 to 32 months). These results are consistent with previous findings e.g. [19], although they do not take account of potential modifying factors, such as the role of drainage ditches as a CH_4 source in drained peatlands [24] or the influence of different vegetation types on emissions from re-wetted sites. For N_2O, the significant pattern of increased emissions from drained sites is also consistent with previous assessments e.g. [14,25]. It is interesting to note that, in terms of 100 year Global Warming Potential, the average increase in N_2O emissions following drainage approximately equates to the decrease in CH_4, suggesting a small net effect. However, this is unlikely to hold true on an individual site basis, since the emissions of both GHGs will potentially be affected by a range of site factors. For example, the review showed a general tendency towards higher N_2O emissions from intensively farmed or fertilised sites, but neither effect was statistically significant given the small number of available studies. Management factors are therefore likely to be influential in moderating the effects of lowland peat management on GHG fluxes, but additional primary data are required to quantify these effects.

For CO_2, the scarcity of comprehensive, replicated experimental studies makes it surprisingly difficult to draw clear conclusions about the effects of peatland management on the overall C balance. The number of studies measuring NEE was insufficient to draw clear conclusions, and similarly there were insufficient studies to elucidate the effects of other land-management activities on CO_2 fluxes. While peat subsidence records e.g. [84] and individual flux measurement studies e.g. [85] in drained agricultural lowland peats strongly indicate that they are large net CO_2 sources, there remains a lack of rigorous experimental data from which to derive robust effect sizes, or emissions factors, as a function of management. For DOC, again based on limited data, there appears to be a tendency towards higher DOC concentrations from drained lowland peats, but this effect was not significant based on the currently available studies. Re-wetting studies do not currently demonstrate a corresponding DOC reduction following the raising of water tables, perhaps due to the confounding effects of time since restoration and the short duration of many studies.

In summary, the results of the systematic review are broadly consistent, or at least do not generally conflict, with current understanding that lowland peat drainage is likely to lead to greater C loss (as CO_2 and/or DOC), increased N_2O emissions, and reduced CH_4 emissions, and that these responses may, to varying degrees, be reversed by re-wetting and restoration. However, only in the case of CH_4 and N_2O can these inferences be clearly demonstrated by a statistical meta-analysis of current experimental studies. Furthermore, few inferences can currently be drawn from the literature regarding the size/ magnitude of effects of peatland agricultural activities including farming intensity, fertiliser use or irrigation, or of other lowland peat land-uses such as peat extraction or forestry. In consequence there is a need for additional well-designed primary measurement studies of lowland peats under a range of management regimes (see below). In the meantime, policy on lowland peat management should take account of the level of knowledge and uncertainty in the current evidence base.

Implication for research
Priorities for primary research
In the UK, which formed the initial focus for this review, lowland peat ecosystems are less commonly studied than upland peatlands. In many other areas of the boreotemperate zone, where continental type bogs and fens predominate, a distinction between 'upland' and 'lowland' peats is rarely made, with most peatlands (and hence most peat-related research) falling within the latter category. Nevertheless, this systematic review has demonstrated that there are relatively few studies in the evidence base that provide robust comparisons of C and GHG fluxes in relation to management, and that more studies are required on the impacts of land management on lowland peatland systems. We have identified a range of commonly investigated land management practices and a list of commonly recorded outcome measures. These findings demonstrate the key knowledge gaps within this topic area. They also highlight areas for which some evidence currently exists but where additional data are required to strengthen current findings. Table 10 demonstrates the major gaps in the evidence-base in regards to meta-analysable groups of studies. These knowledge gaps appear to lie in the following areas: the effect of restoration on N_2O emissions; the effect of fertilizer on fluxes of all GHGs, especially N_2O; the specific effects of ploughing/ cultivation; studies of any treatment on NEE; and the effect of farming activities on DOC.

Recommended study design
Some recommendations for the basic design of peat GHG measurement studies (including spatial scale, use of controls and pre-intervention data, replication, and duration) were previously described by Evans et al. [20]. The following highlights some specific issues affecting the use of currently published data that were highlighted by the results of this review.

Replication Pseudoreplication, where replicates are not sampled at the same level as the treatment or exposure of interest, is a major issue within the evidence base of this field. Of the 71 studies included in the synthesis, only 21 involved appropriate field- or site-level replication, and only 17 of these studies provided meta-analysable data. The remaining 49 used spatial pseudoreplication to generate measures of variability, i.e. more than one chamber per field/site. Use of pseudoreplicates as true replicates in statistical analyses is, strictly speaking, flawed. However, the limited pool of adequately replicated studies identified in this review is insufficient to permit meaningful meta-analysis, which is why we have relaxed our requirements for true replication in accordance with previous authors [25]. Pseudoreplication is only a cause for concern where authors or reviewers use the results to generalise outside the study system. In these cases independent samples must be replicated at the level at which the intervention or exposure is enacted. In experimental design, increasing replication and sample size improve the accuracy of estimated means with respect to the true population mean. With replication at the 'within-site' level the true population mean being estimated by sampling is the site mean, and the estimated mean will trend towards this value as sample size increases. When we wish to make conclusions beyond the site level, however, increasing such within-site replication (pseudoreplication) does not necessarily improve accuracy, only precision. For this reason, the external validity, or generalisability, of pseudoreplicated studies is lower than for those with true replication.

Whilst we appreciate that logistical limitations may make it difficult to maintain high levels of true (i.e. 'site-level') replication, pseudoreplication is not a like-for-like substitute. The only way to increase accuracy of these studies without sacrificing precision is to improve design which may require more resources/funding, since apportioning the same resources in terms of practical effort in the field across a larger number of site-level replicates will not only introduce undesirable variability (through additional moderating variables) but also stretch researchers too thinly.

More of a problem than spatial pseudoreplication is the treatment of repeated sampling, or providing true replication. This temporal pseudoreplication is not an appropriate means of estimating the variability in a comparison between spatially applied treatments or exposures. By repeatedly measuring from fixed chambers, for example, the variability we are measuring is across time, e.g. associated with variation in weather conditions or season. If we

Table 10 Current knowledge gaps for the 16 key intervention/exposure groups for lowland peatland

Intervention/exposure-vs-comparator	CH_4	N_2O	NEE CO_2	R_{eco} CO_2	DOC
Cropped-vs-bare	2 (1)	1	0	1 (1)	0
Drained and restored-vs-undrained	0	0	1	1	0
Drained-vs-undrained	9 (4)	5 (5)	3	10 (5)	4
Dry-vs-wet	2	3	1 (1)	1 (1)	0
Extracted and restored-vs-natural	0	0	1	1	0
High intensity farmed-vs-low intensity farmed	5	6	0	2 (1)	1
Fertilised and grazed-vs-unfertilised and mown	0	1	0	0	0
Fertilised-vs-less fertilised	2 (1)	2 (4)	2	2 (1)	0
Grass-vs-forest	1	1	0	1	0
Grazed-vs-mown	3	1	0	0	0
Irrigated-vs-non-irrigated	0	1	0	0	0
Mineral soil dressed-vs-undressed peat	0	1	0	0	0
Old abandoned-vs-recently abandoned	0	0	0	(1)	0
Old afforested-vs-recently afforested	1	1	0	1	0
Poor-vs-rich	1	0	0	1	2 (1)
Restored-vs-unrestored	4	(1)	0	3 (3)	8

Numbers indicate the number of studies presenting meta-analysable data for that outcome and intervention/exposure group. Bracketed numbers indicate the number of additional studies that present *mean only* data (i.e. no measure of variability).

wished to extrapolate forward or backwards in time and predict C stores or GHG flux, temporal replication would be appropriate. However, by confusing temporal replication with spatial replication the possibility of committing both Type I (spurious significant findings) and Type II (failure to find significance) errors is increased.

In light of these comments future research should avoid using temporal replication as a proxy for spatial replication, and should seek to increase the number of sampling sites to increase external validity.

Flux chamber equipment Many studies employed measurement-error mitigation measures as part of flux chamber equipment in order to minimise sampling impacts on measured outcomes. The reason for these efforts can be broadly summarised as attempts to maximise the signal-to-noise ratio, i.e. to reduce sampling error in an effort to increase the chance of detecting a small but significant difference between sampling populations.

Some studies used fans for this purpose, to ensure air circulation, or sealed chambers onto carefully installed collars to reduce gas emissions resulting from soil disturbance. Others used radiators to mitigate heating effects of solar radiation, or used valves to reduce suction through the soil as gas samples were extracted. Some studies were not able to use all of these elaborate equipment designs, however, because of the nature of the exposures investigated. For example, permanent collars could not be installed on actively harvested peatlands, since regular machine access was required

for peat extraction. These studies used a more basic chamber lacking many of these mitigation measures that could be more readily transported and inserted into the peat.

Such differences in experimental equipment between intervention and control sites may be practically necessary, but such experimental designs are open to a number of problems. Firstly, any differences in GHG fluxes found using these different methods cannot decisively be explained by the intervention alone as they could result from differences in the reliability of the experimental apparatus. Secondly, the efforts to reduce sampling error in one site are rendered moot if other samples are not measured to the same level of precision. Thirdly, different measurement techniques such as these are highly likely to result in heterogeneous variance in the compared sample populations, requiring complex statistical analysis to compensate, and possibly reducing the rigor of the analyses used.

In order to avoid this problem, identical methods should be used on comparator and intervention/exposure sites. Where this is absolutely impossible tests should be undertaken to demonstrate that this difference in methodology does not affect the results. Surprisingly no studies identified in this review have undertaken this latter option.

Increasingly, studies are seeking to overcome the problems associated with measurements using flux chambers on a small sample area by using eddy covariance measures, which integrate fluxes over a much larger area, measure (semi-)continuously rather than periodically, and are not subject to some of the potential artefacts from use of chambers; although they are affected by a different set of methodological issues e.g. [20]. On the other hand, replication is difficult to achieve using eddy covariance systems given their cost and complexity, and they have therefore tended to be used singly e.g. [85] or at best on paired comparison sites. This lack of replication and comparators was a key factor limiting the use of eddy covariance data in the systematic review (e.g. Figure 4). Moving eddy covariance systems between sites provides the potential to address this problem, although the sampling issues inherent to moving towers between locations in order to obtain replication, but where the sampling is carried out over different time periods, needs to be addressed. Whilst there are increasingly sophisticated modelling methods into which GHG flux measurements can be fitted, reporting only the outputs of these and not the primary measurement results together with their quantification of variability should be avoided as it will greatly hinder the potential for future meta-analyses.

Additional files

Additional file 1: Literature Database Search Strategy.

Additional file 2: Search Engine Search Strategy.

Additional file 3: Specialist Websites Search Strategy.

Additional file 4: Bibliographic Testing Strategy.

Additional file 5: Unobtainable and Un-translated Articles.

Additional file 6: Critical Appraisal Tables.

Additional file 7: Articles Excluded at Full Text Screening.

Additional file 8: Data Extraction Proforma and Strategy.

Additional file 9: Funnel Plots and Egger's Test Results for Publication Bias.

Additional file 10: Narrative Synthesis Tables (Study Details and Extracted Data).

Competing interests
The authors declare that they have no competing interests.

Authors' contributions
ASP and CDE conceived and planned the review, with input from JRH, DLJ, AB and SED. SED and NRH undertook the searches. NRH and AB screened articles, with final checking by CDE. NRH and AB extracted data and performed critical appraisal. NRH performed meta-analysis and carried out synthesis. NRH drafted the review report. AB, CDE, JRH, DLJ and ASP assisted in planning of and provided advice for review activities, and assisted in drafting the review report. All authors read and approved the final manuscript.

Acknowledgements
This review forms an initial stage of Defra project SP1210, to evaluate emissions and storage of C and GHGs. The authors wish to thank Defra for their support. We also wish to thank members of the wider research project for their comments on a draft of the manuscript.

Sources of support
This review is funded by the UK Department for Environment, Food and Rural Affairs project SP1210. The systematic review is led by CEBC with significant contributions by CEH staff, who are also responsible for the coordination of the project as a whole.

Author details
[1]Centre for Evidence-Based Conservation, School of Environment, Natural Resources and Geography, Bangor University, Bangor, Gwynedd LL57 2UW, UK. [2]NERC Centre for Ecology and Hydrology, Bangor, UK. [3]School of Environment, Natural Resources and Geography, Bangor University, Bangor, Gwynedd LL57 2UW, UK. [4]School of Natural Sciences and Psychology, Liverpool John Moores University, James Parsons Building, Byrom Street, Liverpool L3 3AF, UK.

References
1. NWWG: *Wetlands of Canada: Ecological Land Classification Series No. 24*. Ottowa: Environment Canada; 1988.
2. Lavoie M, Paré D, Bergeron Y: **Impact of global change and forest management on carbon sequestration in northern forested peatlands.** *Environ Rev* 2005, **13**:199–240.
3. Grønlund A, Sveistrup TE, Sovik AK, Rasse DP, Klove B: **Degradation of cultivated peat soils in Northern Norway based on field scale CO_2, N_2O and CH_4 emission measurements.** *Arch Agron Soil Sci* 2006, **52**:149–159.
4. Treat C, Turetsky M, Harden J, McGuire A: **Methane emissions from boreal peatlands in a changing climate: quantifying the sensitivity of methane fluxes to experimental manipulations of water table and soil temperature regimes in an Alaskan boreal fen.** *AGU Fall Meeting Abstracts* 2006, **1**:1191.
5. Vitt DH, Halsey LA, Bauer IE, Campbell C: **Spatial and temporal trends in carbon storage of peatlands of continental western Canada through the Holocene.** *Can J Earth Sci* 2000, **37**:683–693.
6. Yu Z, Campbell ID, Vitt DH, Apps MJ: **Modelling long-term peatland dynamics: I: concepts, review, and proposed design.** *Ecol Model* 2001, **145**:197–210.
7. Gorham E: **Northern peatlands: role in the carbon cycle and probable responses to climatic warming.** *Ecol Appl* 1991, **1**:182–195.
8. Martikainen PJ, Nykänen H, Crill P, Silvola J: **Effect of a lowered water table on nitrous oxide fluxes from northern peatlands.** *Nature* 1993, **366**:51–53.
9. Freeman C, Lock M, Reynolds B: **Fluxes of carbon dioxide, methane and nitrous oxide from a Welsh peatland following simulation of water table draw-down: potential feedback to climatic change.** *Biogeochemistry* 1993, **19**:51–60.
10. Nykänen H, Alm J, Lång K, Silvola J, Martikainen PJ: **Emissions of CH_4, N_2O and CO_2 from a virgin fen and a fen drained for grassland in Finland.** *J Biogeogr* 1995, **22**:351–357.
11. Hendriks DMD, van Huissteden J, Dolman AJ, van der Molen MK: **The full greenhouse gas balance of an abandoned peat meadow.** *Biogeosciences* 2007, **4**:411–424.
12. Moore T, Knowles R: **The influence of water table levels on methane and carbon dioxide emissions from peatland soils.** *Can J Soil Sci* 1989, **69**:33–38.
13. Moore T, Roulet N: **Methane flux: water table relations in northern wetlands.** *Geophys Res Lett* 1993, **20**:587–590.
14. Couwenberg J, Thiele A, Tanneberger F, Augustin J, Bärisch S, Dubovik D, Liashchynskaya N, Michaelis D, Minke M, Skuratovich A, Joosten H: **Assessing greenhouse gas emissions from peatlands using vegetation as a proxy.** *Hydrobiologia* 2011, **674**:67–89.
15. Gallego-Sala AV, Prentice IC: **Blanket peat biome endangered by climate change.** *Nat Clim Chang* 2013, **3**:152–155.
16. Worrall F, Chapman P, Holden J, Evans C, Artz R, Smith P, Grayson R: *A review of current evidence on carbon fluxes and greenhouse gas emissions from UK peatlands*. UK: Report to JNCC; 2011:86.
17. Bonn A, Holden J, Parnell M, Worrall F, Chapman P, Evans C, Termansen M, Beharry-Borg N, Acreman M, Rowe E, *et al*: *Ecosystem services of peat – Phase 1: Final report*. UK: Defra project SP0572; 2009:141.
18. JNCC: *Towards an assessment of the state of UK peatlands: JNCC Report No. 445*. Peterborough: JNCC; 2011.
19. Baird A, Holden J, Chapman P: *A literature review of evidence on emissions of methane in peatlands*. University of Leeds; UK: Defra Project SP0574; 2009:54.
20. Evans C, Worrall F, Holden J, Chapman P, Smith P, Artz R: *A programme to address evidence gaps in greenhouse gas and carbon fluxes from UK peatlands*. UK: Report to JNCC; 2011:52.
21. Gustavsen HG, Heinonen R, Paavilainen E, Reinikainen A: **Growth and yield models for forest stands on drained peatland sites in southern Finland.** *For Ecol Manage* 1998, **107**:1–17.

22. Charman DJ, Warner BG: *Peatlands and environmental change.* Chichester: John Wiley & Sons Ltd; 2002.

23. Dumanski J: **Carbon sequestration, soil conservation, and the Kyoto Protocol: summary of implications.** *Clim Change* 2004, **65**:255–261.

24. IPCC: *Supplement to the 2006 IPCC Guidelines for National Greenhouse Gas Inventories: Wetlands.* in press.

25. Bussell J, Jones D, Healey J, Pullin A: In *How do draining and re-wetting affect carbon stores and greenhouse gas fluxes in peatland soils,* Review, CEE review 08-012 (SR49), Collaboration for Environmental Evidence. http://www.environmentalevidence.org/SR49.html 2010.

26. Dalrymple S, Burden A, Evans C, Healey J, Jones D, Pullin AS: In *Evaluating effects of management on greenhouse gas fluxes and carbon balances in boreo-temperate lowland peatland systems: CEE protocol 11-010,* Collaboration for Environmental Evidence. www.environmentalevidence.org/SR11010.html 2012.

27. Cohen J: **A coefficient of agreement for nominal scales.** *Educ Psychol Meas* 1960, **20**:37–46.

28. Edwards P, Clark M, DiGuiseppi C, Pratap S, Roberts I, Wentz R: **Identification of randomized controlled trials in systematic reviews: accuracy and reliability of screening records.** *Stat Med* 2002, **21**:1635–1640.

29. Tummers B: *DataThief III.* ; 2006. http://datathief.org/.

30. Forster P, Ramaswamy V, Artaxo P, Berntsen T, Betts R, Fahey DW, Haywood J, Lean J, Lowe DC, Myhre G: **Changes in atmospheric constituents and in radiative forcing.** In *Climate Change 2007: The Physical Science Basis Contribution of Working Group I to the Fourth Assessment Report of the Intergovernmental Panel on Climate Change.* Edited by Solomon S, Qin D, Manning M, Chen Z, Marquis M, Averyt K, Tignor M, HL H. Cambridge, UK and New York, USA: Cambridge University Press; 2007.

31. R Development Core Team: *R version 2.9. 2.* Vienna, Austria: R Project for Statistical Computing; 2009.

32. Viechtbauer W: **Conducting meta-analyses in R with the metafor package.** *J Stat Softw* 2010, **36**:1–48.

33. Sterne JAC, Smith GD: **Sifting the evidence—what's wrong with significance tests?** *Phys Ther* 2001, **81**:1464–1469.

34. Lewis S, Clarke M: **Forest plots: trying to see the wood and the trees.** *BMJ: Bri Med J* 2001, **322**:1479.

35. Danevčit T, Mandic-Mulec I, Stres B, Stopar D, Hacin J: **Emissions of CO_2, CH_4 and N_2O from Southern European peatlands.** *Soil Biol Biochem* 2010, **42**:1437–1446.

36. Kløve B, Sveistrup TE, Hauge A: **Leaching of nutrients and emission of greenhouse gases from peatland cultivation at Bodin, Northern Norway.** *Geoderma* 2010, **154**:219–232.

37. Lund M, Christensen TR, Mastepanov M, Lindroth A, Ström L: **Effects of N and P fertilization on the greenhouse gas exchange in two northern peatlands with contrasting N deposition rates.** *Biogeosciences* 2009, **6**:2135–2144.

38. Maljanen M, Hytönen J, Martikainen PJ: **Fluxes of N_2O, CH_4 and CO_2 on afforested boreal agricultural soils.** *Plant Soil* 2001, **231**:113–121.

39. Maljanen M, Hytönen J, Martikainen PJ: **Cold-season nitrous oxide dynamics in a drained boreal peatland differ depending on land-use practice.** *Can J Forest Res-Revue Canadienne De Recherche Forestiere* 2010, **40**:565–572.

40. Petersen SO, Hoffmann CC, Schäfer CM, Blicher-Mathiesen G, Elsgaard L, Kristensen K, Larsen SE, Torp SB, Greve MH: **Annual emissions of CH_4 and N_2O, and ecosystem respiration, from eight organic soils in Western Denmark managed by agriculture.** *Biogeosci Discuss* 2011, **8**:10017–10067.

41. Regina K, Nykänen H, Silvola J, Martikainen PJ: **Fluxes of nitrous oxide from boreal peatlands as affected by peatland type, water table level and nitrification capacity.** *Biogeochemistry* 1996, **35**:401–418.

42. Strack M, Waddington JM, Rochefort L, Tuittila ES: **Response of vegetation and net ecosystem carbon dioxide exchange at different peatland microforms following water table drawdown.** *J Geophys Res-Biogeosci* 2006, **111**:G2.

43. Strack M, Waddington JM, Lucchese MC, Cagampan JP: **Moisture controls on CO_2 exchange in a Sphagnum-dominated peatland: results from an extreme drought field experiment.** *Ecohydrology* 2009, **2**:454–461.

44. Silvola J, Alm J, Ahlholm U, Nykänen H, Martikainen PJ: **CO_2 fluxes from peat in boreal mires under varying temperature and moisture conditions.** *J Ecol* 1996, **84**:219–228.

45. Nykänen H, Alm J, Silvola J, Tolonen K, Martikainen PJ: **Methane fluxes on boreal peatlands of different fertility and the effect of long-term experimental lowering of the water table on flux rates.** *Global Biogeochem Cycles* 1998, **12**:53–69.

46. Von Arnold K, Weslien P, Nilsson M, Svensson BH, Klemedtsson L: **Fluxes of CO_2, CH_4 and N_2O from drained coniferous forests on organic soils.** *For Ecol Manage* 2005, **210**:239–254.

47. Strack M, Waddington JM: **Response of peatland carbon dioxide and methane fluxes to a water table drawdown experiment.** *Global Biogeochem Cycles* 2007, **21**:GB1007.

48. Minkkinen K, Laine J: **Vegetation heterogeneity and ditches create spatial variability in methane fluxes from peatlands drained for forestry.** *Plant Soil* 2006, **285**:289–304.

49. Strack M, Waddington JM, Tuittila ES: **Effect of water table drawdown on northern peatland methane dynamics: implications for climate change.** *Global Biogeochem Cycles* 2004, **18**:GB4003.

50. Moore TR, Clarkson BR: **Dissolved organic carbon in New Zealand peatlands.** *N Z J Mar Freshw Res* 2007, **41**:137–141.

51. Glatzel S, Kalbitz K, Dalva M, Moore T: **Dissolved organic matter properties and their relationship to carbon dioxide efflux from restored peat bogs.** *Geoderma* 2003, **113**:397–411.

52. Banaś K, Gos K: **Effect of peat-bog reclamation on the physico-chemical characteristics of the ground water in peat.** *Pol J Ecol* 2004, **52**:69–74.

53. Fiedler S, Höll BS, Freibauer A, Stahr K, Drösler M, Schloter M, Jungkunst HF: **Particulate organic carbon (POC) in relation to other pore water carbon fractions in drained and rewetted fens in Southern Germany.** *Biogeosciences* 2008, **5**:1615–1623.

54. Strack M, Waddington JM, Bourbonniere RA, Buckton EL, Shaw K, Whittington P, Price JS: **Effect of water table drawdown on peatland dissolved organic carbon export and dynamics.** *Hydrol Process* 2008, **22**:3373–3385.

55. Soini P, Riutta T, Yli-Petäys M, Vasander H: **Comparison of vegetation and CO_2 dynamics between a restored cut-away peatland and a Pristine Fen: evaluation of the restoration success.** *Restor Ecol* 2010, **18**:894–903.

56. Komulainen VM, Nykänen H, Martikainen PJ, Laine J: **Short-term effect of restoration on vegetation change and methane emissions from peatlands drained for forestry in southern Finland.** *Can J Forest Res* 1998, **28**:402–411.

57. Tuittila ES, Komulainen VM, Vasander H, Nykänen H, Martikainen PJ, Laine J: **Methane dynamics of a restored cut-away peatland.** *Glob Chang Biol* 2000, **6**:569–581.

58. Waddington JM, Day SM: **Methane emissions from a peatland following restoration.** *J Geophys Res-Biogeosciences* 2007, **112**:G3.

59. Kalbitz K, Rupp H, Meissner R: *N, P- and DOC-dynamics in soil and groundwater after restoration of intensively cultivated fens;* 2002.

60. Höll BS, Fiedler S, Jungkunst HF, Kalbitz K, Freibauer A, Drösler M, Stahr K: **Characteristics of dissolved organic matter following 20 years of peatland restoration.** *Sci Total Environ* 2009, **408**:78–83.

61. Cagampan JP, Waddington JM: **Net ecosystem CO_2 exchange of a cutover peatland rehabilitated with a transplanted acrotelm.** *Ecoscience* 2008, **15**:258–267.

62. Augustin J, Merbach W, Steffens L, Snelinski B: **Nitrous oxide fluxes of disturbed minerotrophic peatlands.** *Agribiol Res-Zeitschrift Fur Agrarbiologie Agrikulturchemie Okologie* 1998, **51**:47–57.

63. van Beek CL, Pleijter M, Jacobs CMJ, Velthof GL, van Groenigen JW, Kuikman PJ: **Emissions of N_2O from fertilized and grazed grassland on organic soil in relation to groundwater level.** *Nutr Cycl Agroecosyst* 2010, **86**:331–340.

64. van Beek CL, Pleijter M, Kuikman PJ: **Nitrous oxide emissions from fertilized and unfertilized grasslands on peat soil.** *Nutr Cycl Agroecosyst* 2011, **89**:453–461.

65. Velthof GL, Oenema O: **Nitrous oxide fluxes from grassland in the Netherlands: II. Effects of soil type, nitrogen fertilizer application and grazing.** *Eur J Soil Sci* 1995, **46**:541–549.

66. van den Pol-van Dasselaar A, van Beusichem ML, Oenema O: **Effects of nitrogen input and grazing on methane fluxes of extensively and intensively managed grasslands in the Netherlands.** *Biol Fertil Soils* 1999, **29**:24–30.

67. van den Pol-van Dasselaar A, van Beusichem ML, Oenema O: **Effects of grassland management on the emission of methane from intensively managed grasslands on peat soil.** *Plant Soil* 1997, **189**:1–9.

68. Dirks BOM, Hensen A, Goudriaan J: **Effect of drainage on CO_2 exchange patterns in an intensively managed peat pasture.** *Climate Res* 2000, **14**:57–63.

69. Godin A, McLaughlin JW, Webster KL, Packalen M, Basiliko N: **Methane and methanogen community dynamics across a boreal peatland nutrient gradient.** *Soil Biol Biochem* 2012, **48**:96–105.

70. Webster KL: *Importance of the Water Table in Controlling Dissolved Carbon along a Fen Nutrient Gradient [electronic resource];* 2010.

71. Rochette P, Tremblay N, Fallon E, Angers DA, Chantigny MH, MacDonald JD, Bertrand N, Parent LE: N₂O emissions from an irrigated and non-irrigated organic soil in eastern Canada as influenced by N fertilizer addition. *Eur J Soil Sci* 2010, **61**:186–196.

72. Hyvönen NP, Huttuneen JT, Shurpali NJ, Tavi NM, Repo ME, Martikainen PJ: Fluxes of nitrous oxide and methane on an abandoned peat extraction site: effect of reed canary grass cultivation. *Bioresour Technol* 2009, **100**:4723–4730.

73. Glenn S, Heyes A, Moore T: Carbon-dioxide and methane fluxes from drained peat soils, Southern Quebec. *Global Biogeochem Cycles* 1993, **7**:247–257.

74. Mäkiranta P, Hytönen J, Aro L, Maljanen M, Pihlatie M, Potila H, Shurpali NJ, Laine J, Lohila A, Martikainen PJ, Minkkinen K: Soil greenhouse gas emissions from afforested organic soil croplands and cutaway peatlands. *Boreal Environ Res* 2007, **12**:159–175.

75. Waddington JM, Warner KD, Kennedy GW: Cutover peatlands: a persistent source of atmospheric CO₂. *Global Biogeochem Cycles* 2002, **16**(1):GB001298.

76. Nagata O, Yazaki T, Yanai Y: Nitrous oxide emissions from drained and mineral soil-dressed peatland in central Hokkaido, Japan. *J Agric Meteorol* 2010, **66**:23–30.

77. Rupp H, Meissner R, Leinweber P: Effects of extensive land use and re-wetting on diffuse phosphorus pollution in fen areas—results from a case study in the Drömling catchment, Germany. *J Plant Nutr Soil Sci* 2004, **167**:408–416.

78. Clymo R: Peat. In *General Studies Ecosystems of the world 4A Mires: Swamp, Bog, Fen and Moor.* Edited by Gore AJP. Amsterdam: Elsevier; 1983:159–224.

79. Charman D: *Peatlands and environmental change.* Chichester: John Wiley & Sons Ltd; 2002.

80. Bremner J, Blackmer A: Nitrous oxide: emission from soils during nitrification of fertilizer nitrogen. *Science* 1978, **199**:295–296.

81. Perneger TV: What's wrong with Bonferroni adjustments. *BMJ* 1998, **316**:1236–1238.

82. Kortelainen P, Saukkonen S: Leaching of organic carbon and nitrogen from forested catchments. In *The Finnish research programme on climate change, second progress report. Volume 1/94.* Edited by Kanninen M, Heikinheimo P. Publications of the Academy of Finland; 1994:285–290.

83. Hooijer A, Page S, Jauhiainen J, Lee W, Lu X, Idris A, Anshari G: Subsidence and carbon loss in drained tropical peatlands: reducing uncertainty and implications for CO₂ emission reduction options. *Biogeosciences Discuss* 2011, **8**:9311–9356. doi: 10.5194/bgd-8-9311-2011.

84. Waltham T: Peat subsidence at the Holme post. *Mercian Geol* 2000, **15**:49–51.

85. Morrison R, Cumming A, Taft H, Kaduk J, Page S, Jones D, Harding R, Balzter H: Carbon dioxide fluxes at an intensively cultivated temperate lowland peatland in the East Anglian Fens, UK. *Biogeosci Discuss* 2013, **10**:4193–4223.

What are the effects of the cultivation of GM herbicide tolerant crops on botanical diversity? A systematic review protocol

Jeremy Sweet[1*] and Kaloyan Kostov[2]

Abstract

Background: There are concerns that the cultivation of genetically modified herbicide tolerant (GMHT) crops treated with broad spectrum herbicides will cause declines in botanical diversity and hence loss of biodiversity. Cultivation systems of these have different levels of inputs and management interventions and some incorporate the use of minimal/no tillage. Research results show a range of effects and the priority is to determine whether research studies show shifts in botanical diversity and/or declines in plant populations in GMHT compared with conventionally managed crops.

Methods: We will perform a rigorous review of studies of plant populations in fields and field margins of GMHT crops by complying with CEE requirements for Systematic Reviews (SR) and the EFSA Guidance on Systematic Reviewing. A Review Protocol (RP) is presented for the SR of data from field studies of GMHT crops, comparing the effects of GM crop, herbicide regimes and associated management applied to HT crops with conventional crops and their weed management for impacts on plant populations in fields and field margins as assessment end points or indicators of impacts on botanical diversity and associated food chain and ecosystem services effects. The literature search will include all the main GMHT crops including maize, soya, oilseed rape, sugar beet, cotton and rice. The keywords will be broad and the search strategy is developed to capture all literature relevant to the primary objective of the review. The range of data bases for the searches is described and all articles discovered in the searches will be collated by Endnote. The criteria against which studies will be included in the review and how they will be assessed are described. They include appropriate study designs, statistical power and comparators. The RP outlines the type of analyses that will be performed to assess bias of the selected studies and if covariables describing the heterogeneity of the studies introduce biases. Publications meeting the selection criteria will be filed separately and subjected to more detailed analysis and data on different plant species and types will be analysed separately, to determine outcomes.

Keywords: Herbicide tolerant, Herbicide resistant, Genetically modified (GM), Genetically engineered (GE), Weed, Plant, Management, Diversity, Populations

Background

Impact of GM Herbicide Tolerant (GMHT) crops

HT crops are widely grown in N and S America and are being developed for many other regions, including Europe. Since use of herbicides has been associated with declines in farmland biodiversity in some regions, there are concerns that GMHT crops treated with broad spectrum herbicides will also cause declines in biodiversity [1-3]. However HT systems have different levels of inputs and management interventions and some incorporate the use of minimal or no tillage. Research results show a range of effects depending on crop type, the specific herbicide tolerance, ai applied, dose, timing and number of herbicide treatments, tillage system and cropping system [4-12]. Consequently it is not clear how GMHT crop management is affecting botanical diversity or affecting diversity of other biota, either through food chain effects or directly, in agricultural land. In order to determine GMHT

* Correspondence: jeremysweet303@aol.com
[1]Sweet Environmental Consultants, 6 Green Street, Willingham, CB24 5JA Cambridge, UK
Full list of author information is available at the end of the article

cropping effects on farmland biodiversity a starting point is to determine effects on botanical diversity since this will indicate likely food chain effects [13]. There are also concerns that repeated use of the herbicides used on HT crops will promote herbicide resistance development [14] and further inappropriate use of herbicides leading to reductions in biodiversity [15].

In agroecosystems, sustainable agricultural production, integrated crop production and integrated pest management are broad protection goals associated with sustainable food production, which embrace the exploitation of a range of ecosystem services which are considered desirable to protect. These include pollination, predation, nutrient cycling, etc.... These protection goals are relevant in relation to herbicide usage in conventional and GMHT crops.

Concept

The main concern is that the cultivation of GMHT crops will reduce biodiversity and adversely affect ecosystem services in farmland regions [16]. Conceptual models and logic maps for the most significant environmental issues associated with GMHT crops were distributed to stakeholders. At the stakeholder workshop on 23–24.4.2013 presentations were made explaining the conceptual models and the rationale for the systematic review of GMHT crops. Stakeholders (Appendix listed in Table 1) were invited to comment on our proposed approach and the review topics and questions. In addition comments were received by email. The questions were modified in response to these comments and clarifications provided for some aspects of our approaches to the topics. The updated

review questions were returned to stakeholders for prioritisation by them.

Stakeholder concern

At the Stakeholder meeting of 23 & 24 April 2013 and subsequent consultation with stakeholders it was agreed that this topic was a high priority, raised public concern and was scientifically controversial because of the mixed results from field studies and cropping experience in different crops in different regions. The EC is currently considering applications to commercialise GMHT crops in Europe and this study will inform them of the current state of knowledge on this topic. The SR questions on GMHT crop effects were scored above 3.5 (out of 5) on average with the highest scores for scientific disagreement. The SR will initially focus on the question receiving the higher score for scientific importance, with the question on weed resistance to herbicides used in GMHT systems considered according to time availability.

Objective of the review

The conceptual model shows that the Intervention is the introduction of GMHT crops, the main stressors are the changes in the herbicide applications and management practices, particularly cultivation and tillage. The populations directly affected are graminaceous and broadleaved weed populations and the species diversity in these populations (Figure 1). The concerns are that these will change and/or decline leading to effects through the food chain and reductions in biodiversity of a range of species including those with important functional roles in crop

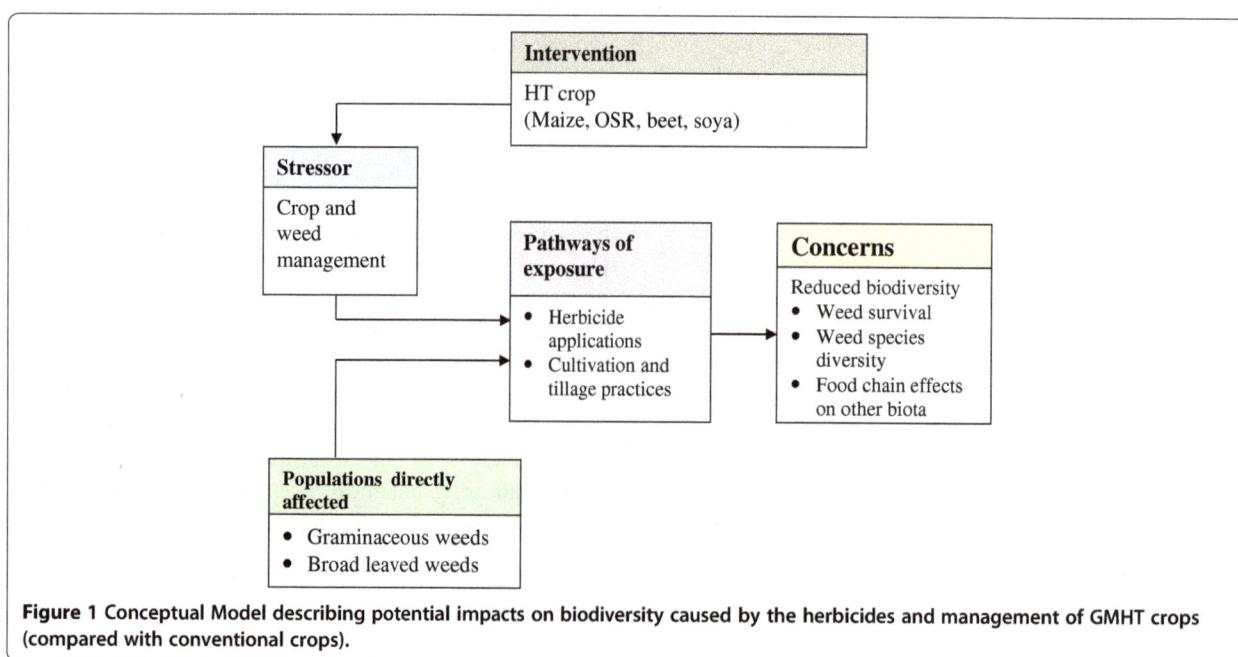

Figure 1 Conceptual Model describing potential impacts on biodiversity caused by the herbicides and management of GMHT crops (compared with conventional crops).

protection and fertilisation, and also species of conservational interest.

Several narrative reviews of GMHT crops have been conducted and it is considered that a systematic review (SR) would bring some added-value in comparison with a narrative review. The SR would determine whether research studies and field data from commercialised GMHT crops show shifts in botanical diversity or declines in plant populations in GMHT compared with conventionally managed crops.

Primary review question

Are populations of plant species changed by management regimes applied to GM HT crops compared with conventional crop management?

The SR will review data from field studies of GMHT crops and compare effects of GM crop, herbicide regimes and associated management applied to HT crops and conventional crops for impacts on plant species populations in fields, as assessment end points or indicators of impacts on botanical diversity and associated food chain and ecosystem services effects.

Secondary question

Arising from this primary question is the secondary review question of whether there is evidence that weeds are developing resistance to broad spectrum herbicides as a consequence GMHT crop management.

Methods

We will perform a rigorous review of studies of plant populations in fields and field margins of GMHT crops by complying with CEE requirements for Systematic Reviews (SR) and the EFSA Guidance on Systematic Reviewing [6].

Search strategy

The aim of the literature search is to be as comprehensive as possible in revealing all available scientific and technical publications and reports which compare the environmental effects of GMHT and conventionally managed crops, with a specific focus on the effects on the botanical diversity and weed populations in farmland receiving the two comparable managements. The search will be indiscriminate and non-selective and include all the main GMHT crops being grown and studied including maize, soya, oilseed rape, sugar beet, cotton and rice. Commercialised GMHT crops such as maize, soya and rapeseed are being extensively cultivated in N and S America [17] with more recent introductions of cotton, sugar beet and alfalfa into USA and rapeseed into Australia. Field studies have been conducted extensively in these countries and experimental studies have been conducted in many European and other countries.

From existing literature reviews [1-3,12,18-23] it can be anticipated that there is considerable literature on this topic.

The literature search strategy will capture the relevant literature for the systematic review by identifying as many relevant datasets as possible. The keywords will be broad and have a high-sensitivity and low-specificity in order to capture a high proportion of existing datasets. The search will include abstracting databases and reviews, as well as full text databases.

Search terms

Terminology for the effects of the management of genetically modified herbicide tolerant crops will consider the main herbicides used in GMHT systems (glyphosate and glufosinate). Other herbicides such as dicamba 2,4-D, and other sulphonyl ureas will also be captured in the search but not specifically searched for as these GMHT systems have only been recently been developed and it is not likely that there will be sufficient data to review separately. The search terms will be selected from the following relevant terms.

Population terms

1. Weed*
2. "Weed species"
3. "Weed control"
4. "Weed population*"
5. "weed cover"
6. "weed biomass"
7. "Weed diversity"
8. "Weed community"
9. "Weed resistan*"
10. diversity
11. Biodiversity
12. Botanic*
13. "herbicide resistan*"
14. "herbicide toleran*"
15. "Weed seed"
16. "seed bank"
17. "Seed rain"
18. HT

Intervention terms

19. "Herbicide management"
20. Roundup*
21. "Liberty Link"
22. "Glyphosate toleran*"
23. "glyphosate resistan*"
24. "Glufosinate toleran*"
25. "glufosinate resistan*"
26. Till*

27. Cultivation*
28. "Rotation"

Outcome terms

29. Sample
30. Quadrat
31. Plot
32. "weed abundance"
33. "weed densit"

Others

34. trial
35. Field
36. Farm

Search strings

The search strings for the literature databases consist of 3 parts (linked with AND). Terms in each part are linked with "OR". Asterisks are used to include terms which are plural or have additional letters. Quotation marks denote multi-word terms (e.g. "herbicide toleran*").

Part 1: ("herbicide toleran*" OR "herbicide resistan*" OR "glyphosate toleran*" OR "glufosinate toleran*" OR "glyphosate resistan*" OR "glufosinate resistan*" OR HT).

Limits the query to the HT crops for which this review is conducted.

Part 2: (field* OR plot* OR trial* OR farm* OR sampl* OR quadrat*).

This provides data from field studies as the terms refer to methods used in field studies.

Part 3: (weed* OR "weed species" OR "weed population" OR "weed abundance" OR "weed biomass" OR "seed bank" OR "seed return").

This provides the entry for the following full search string:

("herbicide toleran*" OR "herbicide resistan*" OR "glyphosate toleran*" OR "glufosinate toleran*" OR "glyphosate resistan*" OR "glufosinate resistan*" OR HT) (field* OR plot* OR location* OR trial* OR farm* OR sampl* OR quadrat*) (weed* OR "weed species" OR "weed population" OR "weed biomass" OR "weed cover" OR "seed bank" OR "seed return").

Search terms for full text literature databases (e.g. Google Scholar, Scirus), are more restricted in order to avoid large amounts of non-relevant data.

Part 1: (herbicide* OR tolerant OR resistan* OR GM OR GE).

Part 2: (field* OR plot* OR location* OR trial* OR farm*).

Part 3: (weed* OR seed OR "seed bank").

Part 4: (sample OR "weed abundance").

The primary searches will be in English. Searches in other languages (e.g. Spanish and Portugese as these are relevant for S America where there is extensive cultivation of GMHT crops) will be conducted using search engines that contain non-English information (e.g. Google Scholar). One example might be to use international terms for GMOs and products, such as.

OGM OR glyphosate OR glufosinate

The precise format of the search strings will be adapted to the requirements of each database and other search words listed above will be used to broaden the search or in case some relevant papers contain none of the words in the main search strings. The search strings used will be recorded and descriptions of how each database was searched will be provided.

Data bases for the searches

Google scholar (search engine) and ISI Web of Knowledge were used for the original scoping exercise and were cross referenced against some recent literature reviews on HT crops. For the results of the scoping see below.

The systematic review will search the following abstracting literature databases:

- Web of Science (ISI Web of Knowledge 1900 – 2013) (Thomson Reuters, New York, USA), contains peer reviewed scientific publications in English.
- CAB Abstracts (1984–2013) (CABI, Wallingford, UK), database that includes local and non-English publications.
- AGRICOLA (1970–2013) (National Agricultural Library, U.S. Department of Agriculture, Beltsville, USA): contains records of materials in agricultural and related sciences in the National Agricultural Library and cooperating institutes.
- AGRIS (1975–2013) (Food and Agriculture Organization of the United Nations FAO, Rome, Italy): International System for Agricultural Sciences and Technology provides an international bibliographic database of national, and international centers.

Full text search engines:

- GOOGLE SCHOLAR: online search engine of scientific and other scholarly works.
- JSTOR: Digital library of academic journals, reports, books, and other primary sources.

NTIS (ProQuest), Cochrane Library, SCOPUS (for scientific reports), Medline & EMBASE may also be considered.

The following webpages containing information on effects of GMOs will also be searched:

- Gmo-safety.eu: webpage established by the German Federal Ministry of Education and Research describing research projects and with links to reports etc.

- EFSA.europa.eu: webpage of the European Food Safety Authority containing information related to GMO applications in EU.
- EU GMO Register (http://gmoinfo.jrc.ec.europa.eu/). Lists all deliberate releases of GM plants for any experimental purposes.
- International Survey of Herbicide resistant weeds [24]. Available at http://www.weedscience.org.

Review papers

Previously published review papers on GMHT crops will be listed and screened for any references that were not picked up in literature searches.

Direct contacts

Scientists or organisations that have conducted field studies of GMHT crops will be contacted directly to obtain any relevant data. All contacts and responses will be recorded.

Non-English searches

Full-text databases, such as Google Scholar and Scientific Electronic Library Online (http://www.scielo.org) are also suitable for non-English searches (e.g. Spanish, Portugese).

Field studies of GMHT crops in South America and Spain published in English often contain references to Spanish and Portuguese language reports of field studies. These will also be examined for relevance to the systematic review. Any papers that appear to be relevant will be sent to: Professor Ramon Albajes, Universitat de Lleida, Department of Plant Production and Forest Sciences

Ramon has conducted studies of GMHT crops and is in a good position to assess the quality of studies in Spanish and Portuguese against the inclusion criteria.

Field studies of GMHT crops have mostly occurred in N and S America, Australasia and Europe and so very few reports in Oriental languages are anticipated. However if they are revealed by the search and appear relevant steps will be taken to obtain translations.

Data base of publications

All publications discovered in the searches will be collated by Endnote. Subsets will be listed according main topic. Publications meeting the selection criteria (see below) will be filed separately and subjected to more detailed analysis.

Scoping the literature

Preliminary searches made between February and October 2012 using the search terms listed above in Google scholar have revealed 318 articles on GMHT crops of which 51 scientific papers and reports contain comparative data,

including 21 papers studying effects on biodiversity, 15 papers on ecological impacts, and 23 papers on weed resistance development. The papers and reports identified were compared with those listed in recent review papers of GMHT crops e.g. [2,15,18]. The published reviews did not contain papers on GMHT crops that were not also identified in the scoping exercise. It is anticipated that a more comprehensive and systematic search using the full range of search terms and the data bases listed above will reveal more publications particularly as there are several recent comparative studies of GMHT crops commercialised in N and S America.

The quality of the studies has not been assessed but the volume of existing studies indicates that there is sufficient published information to justify a systematic review of the literature. The review will also reveal where there are gaps in the type of data collected, which in turn will identify where conclusions cannot be reached or conclusions are based on data subsets which might relate more to particular forms of bias.

Study inclusion criteria

The inclusion criteria are as follows:

Population: non-crop plant populations within fields/plots including field margins and cropped areas.
Intervention: the cultivation management of a GMHT crop which includes the use of the herbicide to which it is tolerant and tillage.
Comparator: a conventional crop (i.e. not GM HT) of a similar variety grown using conventional cultivation and weed management techniques in the same or immediately adjacent plots/fields.
Outcome: changes/differences in the non-crop plant and/or seed bank populations occurring in the GMHT and conventionally managed crops.

Study design

- The studies must be comparisons between conventional weed management and GMHT management systems preferably in studies with split field or paired field systems or plots in fully randomised and replicated systems with sufficient replication or statistical power to show differences at <50% level. The compared fields should be receiving other management inputs which are similar and in similar rotational cropping systems.
- The reviewed articles contain original data that has not been published elsewhere. Reviews, summaries, abstracts, comments papers, etc. will be listed separately and examined for their sources of information, data and bibliography (see Search strategy).

- Data includes sample sizes, plot/field means and variation (SE, SD), for each treatment.
- the study provides original data in the form of tables, figures or directly from the authors if requested.
- Data are presented for each sampling time and not pooled through the season or year.
- Weed data from studies of non-GM HT systems (e.g. Clearfield) will also be compared with weed data from GMHT crops and conventional crops.

The Comparisons should be of measurement end points (outcomes) such as the following:

Measurement end points: Populations and abundance measured in the field
a. Weed Totals,
b. Broad leaved weeds species and numbers
c. Graminaceous weeds species and numbers
d. Identified plant species totals
e. Weed seed bank, species populations and total numbers
f. feed Seed rain, species numbers and total numbers.

Methods
- Plants assessed as plant numbers/m2 counted after herbicide applications and prior to harvest (earlier season data could also be looked at but the later timing is correlated with weed seed production and return).
- Seed assessed as seed rain data collected prior to or at crop harvest, and/or seed numbers/m3 soil or germinated seedlings/m3 soil collected from soil auger sampling, or germinated seedlings/m2 soil in field after harvest of HT crop but before any weed management of the next crop.
- If continuous HT cropping systems are studied then annual and accumulated data of these parameters will also be assessed.

The GMHT systems may consist of:
- Glufosinate (glu) or glyphosate (gly) only used at Manufacturers Recommended rate
- glu or gly only used at other rates and application timings,
- glu or gly at various rates and timings used in programmes with other herbicides or weed management practices (eg hoeing)
- glu or gly used in conjunction with minimum or no tillage systems.

Each treatment difference will be taken into account when analysing comparative effects.

The conventional (comparator) systems should be any conventional management that reflects routine normal or local farm practise. Conventional weed management will also include non-spraying methods such as mechanical hoeing, tillage etc., and the management practices used in each study will be documented.

Where the effects of the comparable weed management systems could be influenced by changes in other management factors such as crop rotation, irrigation, fertilizer treatments etc., data will be analyse separately.

Screening references: applying inclusion criteria
The references revealed by the searches described above will be entered into to Endnote X5 (Thomson Reuters). Separate Endnote files will be established for each search engine for a record of each search. The files will then be combined into one database which will eliminate duplicates automatically, though this will be checked manually. Some search engines do not allow transfer of articles to Endnote so that the search results will be copied or transfered to another file and checked manually. The inclusion criteria will be applied to titles and abstracts to remove unwanted references. The result will be an Endnote X5 database containing the potentially useful articles. These data will also be transferred to CADIMA (http://www.cadima.info/index.php?r=area/centralAccessPoint) for storage, evaluation and future availability.

The selected articles will have full texts extracted and examined further according to the inclusion criteria. Any articles excluded at this stage of the study for the systematic review will be documented in an Excel table and reasons for exclusion provided.

Screening references: quality assurance process
In this screening process, a random subset of 10% of the articles identified by the search procedure will be independently assessed by a second team member, applying the same inclusion criteria. This will be done at the abstract phase and full text phase as the abstracts often do not describe the type of comparative study conducted and the nature of the data collected. The results of the independent assessments will be analyzed and documented using Kappa statistics (http://www.vassarstats.net/kappa.html). Articles excluded in one assessment, but included by the other reviewer will be recorded and the reasons for the differences will be discussed by the review team. If the kappa-value falls below 0.6 indicating significant differences, a review will be undertaken so that inclusion criteria can be revised and tested for improved reviewer agreement and to optimise the screening process.

Study quality assessment

In order to assess the reliability of the presented results, each study will be assessed for the following possible sources of bias:

- Selection bias. This can occur in field study design whereby the design of the study eg the plot size, herbicide treatment timings, sampling timing and method may favour one treatment.
- Performance bias. Systematic differences between groups e.g. exposure to factors other than the intended intervention, such as the change in other management inputs (e.g. fertilizer, pest control, irrigation) associated with one treatment but not the other.
- Detection bias (the way outcomes are measured differs between treatment groups, application of statistical analysis).
- Attrition bias may occur in differences in sampling times following herbicide treatment, frequency, sizes between study groups, selective choice of parameters, and in missing data.
- Reporting bias (e.g. preferential reporting of positive or desired outcomes).

Studies will be examined for a clear hypothesis and designs which are appropriate to achieve outcomes which test the hypothesis by testing both for similarities (equivalences) and differences.

The following sources of bias and influences on the studies are identified:

1. Research experimental studies conducted on farm will tend to have farmers using normal managed weed control on conventional crops which give acceptable levels of weed control (to the farmer) while optimising inputs and costs. By contrast the management of the GMHT crops will tend to follow a rather prescriptive or formulaic protocol for the herbicide applications as requested by the researcher and/or biotech company. Thus there will be performance bias in these studies and they will contrast with studies conducted on farms where GMHT systems are well established. In this latter case farmers will be familiar with the requirements of GMHT crops and will thus optimise their inputs. Thus the effects on weed populations may be different between experimental studies and field studies of commercial GMHT crops.

2. In areas where there are concerns about weed resistance development (eg where there is intensive and repeated use of GMHT crops), farmers will tend to go for maximum weed control to prevent survival of resistant weeds and so this performance bias may result in different GMHT treatments from those where resistance is not perceived as a problem.

3. Sequential cultivation of GMHT crops is likely to have different consequences than rotational cultivation and so this should be evaluated separately in the analysis.

4. Experimental and Reporting bias is likely as some studies will be sponsored or conducted by interested organisations (eg: agrochem Cos, biotech Cos) or by organisations antagonistic to GMOs or modern agricultural practices, and therefore there may be some selection of recorded parameters, data, results or reports to show a particular effect or show a "desired" effect or result.

For example v.low or zero weeds would be considered a desirable result by an Agchem company but an adverse effect (reduction in botanical diversity) by an environmental organisation.

The study quality assessments will be independently assessed, as in the references screening process, by a second team member, applying the same quality criteria. The results of the independent assessments will be analyzed and documented using Kappa statistics (http://www.vassarstats.net/kappa.html). Studies assessed differently by reviewers will be recorded and the reasons for the differences will be discussed by the review team. If the kappa-value falls below 0.6 indicating significant differences, a review will be undertaken so that quality criteria can be revised and tested for improved reviewer agreement and to optimise the quality assessment process.

The levels of bias in studies will be characterized and studies which contain bias which has clearly influenced results will be assessed separately and recorded in the systematic review. This will be documented to ensure transparency and reconstructability.

Data extraction strategy

Original data will be described in each article, the method of extraction (including contacts with authors) will also be provided. Any estimated or calculated values will be described or explained. Where presented data are insufficient or inadequate, authors will be directly contacted and requested to provide the appropriate data.

Data will be stored in CADIMA that will be built for this purpose. Tables in Excel format can be easily extracted from the database.

Random samples of 20% of the extracted data will be checked by the second member of the review team.

The variables extracted and recorded are shown in Table 3 in Appendix and will be similar to those proposed by Meissle M, Naranjo SE, Riedel J, Romeis J: Does the growing of Bt maize change populations or ecological functions of non-target animals compared to the growing of conventional non-GM maize? 2014: Submitted to Environmental Evidence. This is in order to have conformity and similarity of data bases for later extraction and interrogation purposes.

Data analysis

The extracted datasets will be collated and analyzed and a narrative summary of the extracted outcomes will be produced. Tables and figures illustrating and summarizing the evidence will be produced to show the products and outcomes of the literature analysis. The data will be grouped and presented in different ways in order to allow a range of analyses and demonstrate a range of outcomes. For example:

- Number of years of study.
- Study year.
- Location of study.
- Crop type.
- Spatial scale of experiment.
- Experimental design.
- Commercial field studies.
- In crop or field margin.
- Taxonomic groups of weeds (e.g. graminae, broadleaved).
- Types of weeds (e.g. annual, perennial).
- Parameters measured.
- HT system studied (glufosinate or glyphosate).
- Studies comparing GM HT crops with minimal tillage vs conventionally cultivated and treated crops.
- Weed herbicide resistance.

Suitable quantitative datasets will be analyzed statistically as described below.

For each measurement endpoint the effect size and direction in relation to comparator (ie + or − ve) will be assessed.

Statistical meta-analyses will be conducted for data sets of populations of weed taxonomic groups occurring frequently in studies with similar designs and measured parameters. The scoping exercise indicates that there may be sufficient similar data sets for conducting meta-analyses, but this will not be confirmed until a qualitative evaluation of these data sets has been conducted. The range of weed types and parameters analysed will depend on the quality and amount of appropriate data to allow quantitative assessments.

For meta-analyses, the following inclusion criteria will be applied:

- Clearly defined and similar parameters (e.g. abundance of weed taxa) were measured in the field under similar circumstance of time in the season

- The taxonomic group of the weed measured was clearly described.
- Sampling strategy (e.g. size, frequency, timing) is given for each treatment.
- Raw data as well as plot, treatment and site means are presented together with a measure of variation (SE, SD).
- Convention crops and their normal routine herbicide treatments are compared with GMHT crops and their specific herbicide treatments (e.g. gly and glu).
- Mean measures of effect and variance are reported in the article or can be acquired from the author.
- Data are recorded on an annual/seasonal basis (pooled and cumulative data sets from more than one year may be separately analysed).

Measures of treatment effect

The weed response variable to herbicide treatments is abundance. If enough data are available, response variables of individual weed species and weeds of certain taxonomic groups will be analysed. The mean effect size is calculated as the difference between the effects on weed abundance of herbicide treatments applied to GMHT crops and the herbicide treatments applied to conventional crops. This is reported is Hedges' d weighted mean response, and is calculated as the mean response divided by a pooled standard deviation and multiplied by a correction term to allow for small sample size bias.

GMHT and conventional crop fields will be compared for

- Different crop species
- different weed species and groups
- herbicide resistance development
- geographical regions
- different herbicide treatments
- differences in commercial crop studies and experimental studies

Where there are comparisons of the effects of GMHT systems with those of HT systems for crops developed through non-GM techniques (e.g. imidazolinone tolerance developed through mutation breeding), separate analyses will be performed comparing the effects on weed populations of the GMHT and non-GMHT herbicide programmes.

Assessment of statistical power of included studies

For each measurement parameter in a study the statistical power of the treatment effect will be assessed either as an ex-post calculation of the power of the study for the effect, or as a prospective power of the study to detect pre-established significant effects magnitudes (e.g.

the power of the study to detect a significant 10%, 20% and 50% impact effect on the outcome).

Data which is linked or consequential or sequential will be assessed separately and relationships studied. Interactions with other external factors such as management, season and environment will be considered.

Unit of analysis issues

A data issue that is likely to occur is the multiple use of the same dataset for different comparisons (for example the Farm Scale studies conducted in UK, [21,25]. Data on higher taxonomic levels or groupings may include data derived from other studies on individual species and so might result in multiple use of the same data. In addition data on a species or effect may be used from a study designed to study another different species or effect. The design may therefore not be appropriate or optimal. Secondary datasets will be identified and related to primary (original) data in the database during the data extraction process. The number of datasets reused in analyses will be recorded and discussed. Analyses might be conducted including and excluding multiple use datasets to determine their impacts as described below.

Dealing with missing data, and non-interpretable data

Where the nature of the data is not apparent or the article has introduced methods for allowing for missing data or there are missing or unusual/unlikely values, then attempts will be made to contact the authors or others who may be able to comprehend and interpret the data sets. Adjustments for missing data will only be made when the reasons for the absence are clear.

The reviewers will note data that is missing or has been adjusted or substituted.

Synthesis

Quantitative synthesis The outcomes of the systematic review and meta analyses, in terms of the sizes of effects on weed abundance in comparisons of GMHT systems with control crop systems, including 95% confidence intervals, will be presented in tabular and graphic form. Total numbers of observations and effect sizes that are significantly different greater or less than 0 will be shown. Negative effect sizes are associated with lower abundances of weeds in GMHT crops compared with non-GMHT controls.

The nature and quality of the results will be discussed and conclusions drawn where appropriate.

Assessment of heterogeneity

The heterogeneity of the data across studies will be examined using appropriate statistical methods. The effects

of studying different weed taxa, different regions, crop managements (minimum tillage, herbicide application methods, rotation), commercial and experimental scales, and experimental designs will be examined for their influence on variation.

Sub sets of studies containing similar parameters such as tillage, herbicide timing will be separately assessed to see how these influence outcomes. Where it is identified that certain parameters cause high heterogeneity, separate analyses of data subsets will be conducted.

Sensitivity analyses will be conducted to determine whether the outcome of the systematic review and meta-analyses might be influenced disproportionally by:

Studies that provide a high proportion of the datasets used in the meta-analyses.
Studies with high levels of precision, statistical power and/or replication which will tend to have higher weightings.
Studies with low levels of statistical power, poor replication or high variability which diverge from weighted means to a greater extent than studies with high precision, power or low variability.
The funding sources for the overall results.
studies where final weed numbers were measured in comparison to studies where weed abundance was measured at different times.
Data on particular common, rare or unevenly distributed weed taxa.

Assessment of publication bias

Effect sizes from different studies will be compared for articles produced by authors or journals with different funding types. Consistent differences may indicate a publication bias or a bias depending on funding source of a study (see 3.3) or a Journal.

Appendix to "What are the effects of the cultivation of GM herbicide tolerant crops on botanical diversity? A systematic review protocol"

By Jeremy Sweet and Kaloyan Kostov

This appendix lists the people and organizations (Tables 1 and 2) involved in the consultations when developing the conceptual model, the prioritization of the topic and this protocol for a systematic review of the available information on the effects of GM herbicide tolerant crops on botanical diversity in farmland.

This appendix lists the variables extracted from scientific reports and papers for the systematic review on GMHT crops (Table 3) and which are used as the primary data base for analyzing the effects GMHT crops and their management on botanical diversity. The variable name in the database, the definition of the variable,

Table 1 List of participants to the stakeholder consultation workshop on good review practice in GMO impact assessment

GRACE team members

1   Gloria Adduci, IFZ- Inter-University Research Centre for Technology, Work and Culture, Austria

2   Wendy Craig, International Centre for Genetic Engineering and Biotechnology (ICGEB), Italy

3   Jaqueline Garcia-Yi, Centre of Life and Food Sciences Weihenstephan, Technical University of Munich (TUM), Germany

4   Maria Garrone, Centre for European Policy Studies (CEPS), Belgium

5   Achim Gathmann, Federal Office of Consumer Protection and Food Safety (BVL), Germany

6   Paul Kenning Krogh, Department of Bioscience, Aarhus University, Denmark

7   Steffen Kecke, Julius Kühn Institut (JKI), Germany

8   Gijs Kleter, RIKILT-Institute of Food Safety, Wageningen University and Research Centre, The Netherlands

9   Christian Kohl, Julius Kühn Institut (JKI), Germany

10 Kaloyan Kostov, AgroBioInstitute (ABI), Bulgaria

11 Klaus Minol, Genius, Germany

12 Kai Priesnitz, Federal Office of Consumer Protection and Food Safety (BVL), Germany

13 Patrick Rüdelsheim, PERSEUS, Belgium

14 Joachim Schiemann, Julius Kühn Institut (JKI), Germany

15 Greet Smets, PERSEUS, Belgium

16 Armin Spök, IFZ- Inter-University Research Centre for Technology, Work and Culture, Austria

17 Jeremy Sweet, Sweet Environmental Consultants, UK

18 Stefan Unger, Julius Kühn Institut (JKI), Germany

19 Justus Wesseler, Centre of Life and Food Sciences Weihenstephan, Technical University of Munich (TUM), Germany

20 Ralf Wilhelm, Julius Kühn Institut (JKI), Germany

Competent authorities

21 Rita, Andorkó, Hungarian Ministry of Rural Development, Hungary

22 Zsuzsanna Bardócz, Hungarian Ministry of Rural Development, Hungary

23 Martin Bencko, Ministry of Agriculture and Rural Development of the Slovak Republic, Slovakia

24 Doris Bühler, Federal Office for Agriculture (BLW), Switzerland

25 Victoria Colombo Rodríguez, Ministerio de Agricultura, Alimentación y Medio Ambiente, Spain

26 Omar del Río Fernández, Ministerio de Agricultura, Alimentación y Medio Ambiente, Spain

27 Yann Devos, European Food Safety Authority (EFSA), Italy

28 Lyubina Donkova, Ministry of Agriculture and Food, Bulgaria

29 Catherine Gerard, French Embassy in Germany, Germany

30 Niall Gerlitz, European Commission, DG SANCO, Belgium

31 Petra Heinze, Federal Office of Consumer Protection and Food Safety (BVL), Germany

32 Rangel Krastanov, Ministry of Agriculture and Food, Bulgaria

Table 1 List of participants to the stakeholder consultation workshop on good review practice in GMO impact assessment (Continued)

33 Louise Lundstrøm Nielsen, Danish Ministry of the Environment, Environmental Protection Agency, Denmark

34 Odeta Pivoriene, Ministry of Environment of Lithuania, Lithuania

35 Annette Pöting, Federal Institute for Risk Assessment (BfR), Germany

36 Eline Rademakers, Federal Public Service (FPS) Health, Food Chain Safety and Environment, Belgium

37 Leif Erik Rehder, U.S. Embassy in Berlin, Germany

38 Jane Richardson, European Food Safety Authority (EFSA), Italy

39 Neringa Sarkauskiene, Ministry of Environment of Lithuania, Lithuania

40 Andrea Scheepers, Federal Office of Consumer Protection and Food Safety (BVL), Germany

41 Paul Spencer, U.S. Embassy in Berlin, Germany

42 Beatrix Tappeser, Federal Agency for Nature Conservation (BfN), Germany

43 Valentina Zoretic-Rubes, Ministry of Health, Croatia

Civil Society Organisations

44 Rosa Binimelis, GenOk – Centre for Biosafety, Norway

45 Walter Haefeker, European Professional Beekeepers Association, Germany

46 Lise Nordgård, GenOk – Centre for Biosafety, Norway

47 Christoph Then, Testbiotech - Institute for Independent Impact Assessment in Biotechnology, Germany

48 Kristina Wagner, Eurogroup for Animals, Belgium

49 Dirk Zimmermann, Greenpeace e. V., Germany

Industry

50 Manuel Gómez-Barbero, EuropaBio, Belgium

51 Andrew Tommey, Pioneer Overseas Corporation, Belgium

Academia

52 Didier Breyer, Institut Scientifique de Santé Publique (WIV-ISP), Belgium

53 Sylvia Burssens, Ghent University, Belgium

54 Barbara De Santis, National Institute of Health (ISS), Italy

55 Geoff Frampton, School of Medicine University of Southampton, UK

56 Hrvoje Fulgosi, Ruđer Bošković Institute, Croatia

57 Boet Glandorf, National Institute for Public Health and the Environment, The Netherlands

58 Richard Goodman, University of Nebraska—Lincoln, USA

59 Linde Inghelbrecht, Department of Agricultural Economics, University of Ghent, Belgium

60 Harry Kuiper, Wageningen University and Research Centre, The Netherlands

61 Annalisa Paternò, Istituto Zooprofilattico Sperimentale delle Regioni Lazio e Toscana, Italy

62 Monica Racovita, International Centre for Genetic Engineering and Biotechnology (ICGEB), Italy

Table 2 List of organisations submitting written comments

German Federal Agency for Nature Conservation (BfN).	
Boet Glandorf	Dutch National Institute for Public Health and the Environment (RIVM).
Richard E. Goodman	University of Nebraska, Lincoln.
EuropaBio.	
European Professional Beekeepers Association.	
Linde Inghelbrecht	Ghent University.
Testbiotech.	

Table 3 List of variables extracted for the systematic review on GMHT crops

Variable Name	Definition	Type	Closed terms
article_id	Unique identification number assigned to each publication	Integer	No
Author	Author(s) of the listed publication.	Text	No
Crop	Common and species name	Text	Yes
publication_year	Year of publication of study	Integer	No
citation	Citation, e.g. journal name, volume and page numbers	Text	No
title	Title of the publication	Text	No
author_affiliation	Type(s) of institutions/organisations that the author(s) are affiliated with	Text	Yes
author_institute	Institution of corresponding author (or of first author if no corresponding author was listed).	Text	No
funding	Funding source	Text	No
peer_reviewed	Indicates whether study was published in a peer reviewed journal.	string	yes
Location	Geoposition or region/country where field study was performed	string	yes
Site_info	Information on previous cropping, rotation, soil type, irrigation etc.	string	yes
expmt_num	Number of experiments within a study (e.g. different locations, years, etc.)	string	no
data_location	Figure number, table number or page number where means and variation were found.	string	no
was_data_scanned	Indicates whether figures were scanned to obtain data values.	YesNo	yes
GM trait	Herbicide tolerance gene engineered into the transgenic crop.	string	yes
Conv. herb	Conventional herbicide comparator treatment	string	yes
event	Transgenic event of the crop tested.	string	no
transgenic_hybrid_or_var	Transgenic hybrid or variety name.	string	no
nontransgenic _var	Non-transgenic comparator variety name.	string	no
Weed/plant_class	taxonomic class	string	yes
Weed/plant_sub class	Monocotyledon or dicotyledon	string	yes
plant_order	taxonomic order	string	yes
plant_family	taxonomic family	string	yes
plant_genus	taxonomic genus	string	yes
plant_species	taxonomic species	string	yes
plant_finest grouping	Finest level of taxonomic resolution reported for the plant(s).	string	yes
Annual/perennial		string	yes
Strata	Emergence/establishment E= early, M= midseason, L=late	string	yes
replicate_data_issues	Codes flag for non-independence to be considered for analyses: TGLE=taxonomic group lumped elsewhere; EMUE= experimental means used elsewhere; CMUE=control means used elsewhere;	string	yes
Plant _stage	Stage.	string	yes
field_location	Location of field(s) to the level of specificity provided by the author	string	no

Table 3 List of variables extracted for the systematic review on GMHT crops *(Continued)*

number_of_fields	Number of fields as described by the author.	integer	no
cultivation	Cultivation practices used within the fields (notes on tillage, herbicides, etc.)	string	no
plot_size	Size of replicate plots (in hectares)	real	no
plot_size_explanation	Explanations for any calculations done to obtain plot size	string	no
is_plot_size_avg	Indicates whether the listed plot size is an average or an estimate	string	yes
was_study_randomized	Indicates whether the authors indicated that they randomly assigned replicates to treatments.	string	yes
planting_date	Date on which field plots were planted.	string	no
first_sample	Date on which first sample was taken	string	no
last_sample	Date on which last sample was taken	string	no
study_duration	More detailed description on study duration	string	no
herbicide_name	Brand name of herbicides used.	string	yes
herbicide_active_ingr	Active ingredient for herbicides used.	string	yes
_spray_rate	Amount of active ingredient per spray.	string	no
mechanism_of_herbicide_app	Indicates if herbicide was applied as spray, granule, LV,ULV	string	yes
num_of_herbicide_app	Number of herbicide applications.	real	no
is_num_of_herbicide_app_avg	Indicates if the number of herbicide applications is an average	string	yes
sampling_method_abbrev	Abbreviated description of sampling method	string	yes
sampling_method_detailed	Detailed description of sampling method.	string	no
sampling_frequency	Frequency of repeated samples per replicate field or plot.	string	yes
number_of_sample_days	The number of times that each replicate field or plot was repeatedly sampled over the duration of the experiment.	real	no
num_subsamples	Number of subsamples per true replicate.	real	no
response_variable_abbrev	Major category of response variable.	string	yes
response_variable_detailed	Detailed description of response variable.	string	no
true_control_sample_size	True sample size for control treatment.	real	no
true_expmtl_sample_size	True sample size for experimental treatment.	real	no
authors_control_sample_size	Sample size for the control treatment as stated by the author.	real	no
authors_expmtl_sample_size	Sample size for the experimental treatment as stated by the author.	real	no
seasonal_or_peak	Indicates whether values represent seasonal means across multiple sample days or means from peak days	string	yes
did-we-calc	Indicates whether we calculated the seasonal mean or peak days.	YesNo	yes
calc_method_seas_mean	Explains how we calculated the seasonal mean or peak days	string	yes
comparison_type	Indicates whether the comparison is with conventional treated HT or conventional treated control, minimum tillage vs conventional tillage	string	yes
control_mean	Mean for the control treatment.	real	no
expmtl_mean	Mean for the experimental treatment.	real	no
control_std_err	Standard error for the control treatment.	real	no
expmtl_std_err	Standard error for the experimental treatment.	real	no
control_std_dev	Standard deviation for the control treatment.	real	no
expmtl_std_dev	Standard deviation for the experimental treatment.	real	no
mean_unit	Unit of measurement for the response variable. E.g. weed/seed numbers, biomass, ground cover.	string	yes
statistical_test_used	Statistical test used by author.	string	yes
is_effect_significant	Indicates whether a significant effect was detected by the author.	string	yes
warning1	Space for remarks for this record	string	no

Given is the variable name in the database, the definition of the variable, the type, and whether the variable content is restricted to closed (predefined) terms.

the type, and whether the variable content is restricted to closed (predefined) terms are presented.

Competing interests

The authors have no financial or other conflicting interests in GM crops or herbicides and no interest in the success or failure of any particular GM crops or herbicides.

Authors' contributions

JS: review protocol writing, study quality assessment, search, data extraction, narrative synthesis, and project management. KK: review protocol writing, study quality assessment, data-analysis and narrative synthesis. KK is also involved in data extraction. Both authors read and approve the final manuscript.

Acknowledgements

This review is funded within the GRACE consortium, a EU-FP7 programme project, Grant Agreement KBBE-2011-6-311957. We acknowledge the support of Ramon Albajes with interpretation of Spanish and Portugese articles. We thank Christian Kohl of JKI for his support and guidance in developing this protocol. We are grateful to Michael Meissle, Steven E. Naranjo, Judith Riedel and Jörg Romeis for allowing us to use their manuscript and data base in formulating this protocol and to all collaborators within GRACE who provided valuable comments on drafts of this review protocol.

Author details

[1]Sweet Environmental Consultants, 6 Green Street, Willingham, CB24 5JA Cambridge, UK. [2]Agrobioinstitute, Dragan Tzankov 8, 1164 Sofia, Bulgaria.

References

1. Boatman ND, Parry HR, Bishop JD, Cuthbertson AGS: **Impacts of Agricultural Change on Farmland Biodiversity in the UK.** In *Biodiversity Under Threat*, Issues in Environmental Science and Technology, Volume 25. Edited by Hester RE, Harrison RM. The Royal Society of Chemistry; 2007. Chapter 1, 1-32.
2. Sweet J, Bartsch D: **Synthesis and overview studies to evaluate existing research and knowledge on biological issues on GM plants of relevance to Swiss environments.** *SNSF Nationales Forschungsprogramm "Nutzen und Risiken der Freisetzung gentechnisch veränderter Pflanzen" (NFP 59)* 2012: vdf Hochschulverlag, 194 Seiten ISBN 978-3-7281-3498-1.
3. Sweet JB, Lutman PJW: **A commentary on the BRIGHT programme on herbicide tolerant crops and the implications of the BRIGHT and farm scale evaluation programmes for the development of herbicide tolerant crops in Europe.** *Outlooks on Pest Management* 2006, **2006:**249-254.
4. Brookes G: **The farm-level impact of herbicide-tolerant soybeans in Romania.** *Agbioforum* 2005, **8**(4):235-241. Available on the worldwide web at http://www.agbioforum.org.
5. Carpenter J: **GM crops and patterns of pesticide use.** *Science* 2001, **2001**(292):637-638.
6. EFSA: European Food Safety Authority: **Application of systematic review methodology to food and feed safety assessments to support decision making.** *EFSA Journal* 2010, **8**(6):1637. Available online: http://www.efsa.europa.eu.
7. Geisy JP, Dobson S, Solomon KR: **Ecotoxicological risk assessment for Roundup herbicide.** *Rev Environ Contam Toxicol* 2000, **167:**35-120.
8. Kleter GA, Harris C, Stephenson G, Unsworth J: **Comparison of herbicide regimes and the associated potential environmental effects of glyphosate-resistant crops versus what they replace in Europe.** *Pest Manag Sci* 2008, **64:**479-488.
9. Kovach J, Petzoldt C, Degni J, Tette J: *A Method to Measure the Environmental Impact of Pesticides*, New York's Food and Life Sciences Bulletin, Volume 139. Geneva, NY: NYS Agricul. Exp. Sta. Cornell University; 1992:8. Annually updated http://www.nysipm.cornell.edu/publications/EIQ.html.
10. Pleasants JM, Oberhauser KS: **Milkweed loss in agricultural fields because of herbicide use: effects on the monarch butterfly population.** *Insect Conservation and Diversity* 2012, **2012:** doi: 10.1111/j.1752-4598.2012.00196.x.14.
11. Sankula S, Blumenthal E: *Impacts on US Agriculture of Biotechnology-Derived Crops Planted in 2005- an Update of Eleven Case Studies*. Washington: NCFAP; 2006. http://www.ncfap.org.
12. Squire GR, Hawes C, Begg GS, Young MW: **Cumulative impact of GM herbicide-tolerant cropping on arable plants assessed through species-based and functional taxonomies.** *Environ Sci Pollut Res* 2009, **16**(1):85-94. doi:10.1007/s11356-008-0072-6.
13. Sweet J, Bartsch D: **Guidance on risk assessment of herbicide tolerant GM plants by the European Food Safety Authority.** *Journal für Verbraucherschutz und Lebensmittelsicherheit (Journal of Consumer Protection and Food Safety)* 2011, **6**(supplement 1):65-72. DOI:10.1007/s00003-011-0686-3.
14. Hurley TM, Mitchell PD, Frisvold GB: **Effects of weed resistance concerns and resistance management practices on the value of Roundup Ready crops.** *Agbioforum* 2009, **12**(3 & 4):291-302.
15. Benbrook C: **Impacts of genetically engineered crops on pesticide use in the U.S. – the first sixteen years.** *Environ Sci Eur* 2012, **24:**1-24.
16. Hartzler RG: **Reduction in common milkweed (Asclepias syriaca) occurrence in Iowa cropland from 1999 to 2009.** *Crop Prot* 2009, **29:**1542-1544 (2010).
17. Council NR: *Impact of Genetically Engineered Crops on Farm Sustainability in the US*. Washington DC: National Acadamies Press; 2010.
18. Beckie HJ, Harker KN, Hall LM, Warwick SI, Légère A, Sikkema PH, Clayton GW, Thomas AG, Leeson JY, Séguin-Swartz G, Simard M-J: **A decade of herbicide-resistant crops in Canada.** *Can J Plant Sci* 2006, **86:**1243-1264.
19. Bonny S: **Genetically modified glyphosate-tolerant soybean in the USA: adoption factors, impacts and prospects. A review.** *Agron Sustain Dev* 2008, **28:**21-32.
20. Carpenter J: **Impact of GM crops on biodiversity.** *GM Crops, Landes Bioscience* 2011, **2**(1):7-23.
21. Johnson WG, Davis VM, Kruger GR, Weller SC: **Influence of glyphosate-resistant cropping systems on weed species shifts and glyphosate-resistant weed populations.** *Europ J Agronomy* 2009, **31:**162-172.
22. Graef F: **Agro-environmental effects due to altered cultivation practices with genetically modified herbicide-tolerant oilseed rape and implications for monitoring: a review.** *Sustainable Agriculture* 2009, **2:**229-242. Part 2, 229-242, DOI: 10.1007/978-90-481-2666-8_15.
23. Monquero PA: **Plantas transgênicas resistentes aos herbicidas: situação e perspectivas.** *Bragantia, Campinas* 2005, **64**(4):517-531.
24. Heap, I: **The international survey of herbicide resistant weeds.** 2014: Online. Internet. Saturday, January 11, 2014. Available http://www.weedscience.org.

4

Systematic review of effects on biodiversity from oil palm production

Sini Savilaakso[1*], Claude Garcia[2,5], John Garcia-Ulloa[2], Jaboury Ghazoul[2], Martha Groom[3,4], Manuel R Guariguata[1], Yves Laumonier[1,5], Robert Nasi[1], Gillian Petrokofsky[6], Jake Snaddon[6] and Michal Zrust[7]

Abstract

Background: During the past decade there has been a growing interest in bioenergy, driven by concerns about global climate change, growing energy demand, and depleting fossil fuel reserves. The predicted rise in biofuel demand makes it important to understand the potential consequences of expanding biofuel cultivation. A systematic review was conducted on the biodiversity impacts of three first-generation biofuel crops (oil palm, soybean, and jatropha) in the tropics. The study focused on the impacts on species richness, abundance (total number of individuals or occurrences), community composition, and ecosystem functions related to species richness and community composition.

Methods: Literature was searched using an a *priori* protocol. Owing to a lack of available studies of biodiversity impacts from soybean and jatropha that met the inclusion criteria set out in the systematic review protocol, all analyses focused on oil palm. The impacts of oil palm cultivation on species richness, abundance, and community similarity were summarized quantitatively; other results were summarized narratively.

Results: The searches returned 9143 articles after duplicate removal of which 25 met the published inclusion criteria and were therefore accepted for the final review. Twenty of them had been conducted in Malaysia and two thirds were on arthropods.

Overall, oil palm plantations had reduced species richness compared with primary and secondary forests, and the composition of species assemblages changed significantly after forest conversion to oil palm plantation. Abundance showed species-specific responses and hence, the overall abundance was not significantly different between plantations and forest areas. Only one study reported how different production systems (smallholdings vs. industrial estates) affect biodiversity. No studies that examined the effects on ecosystem functions of reduced species richness or changes in community composition met the inclusion criteria. Neither were there studies that reported how areas managed under different standards (e.g. different certification systems) affect biodiversity and ecosystem function.

Conclusions: Our review suggests that oil palm plantations have reduced species richness compared with primary and secondary forests, and the composition of species assemblage changes significantly after forest conversion to oil palm plantation. Effects of different production systems on biodiversity and ecosystem function are clear knowledge gaps that should be addressed in future research.

Trial registration: CEE10-013

Keywords: Land use change, Mitigation, Oil palm, Species diversity, Tropical forest

* Correspondence: s.savilaakso@cgiar.org
[1]Center for International Forestry Research, P.O. Box 0113 BOCBD, Bogor 16000, Indonesia
Full list of author information is available at the end of the article

Background

Over the last decade there has been a growing interest in bioenergy, especially biofuels, that has been driven by concerns about global climate change, increasing energy demand, reducing dependence on fossil fuel [1]. Energy derived from plant material, such as sugarcane and oil palm, offers, at least in theory, a promising way to answer energy demand without increasing greenhouse gas (GHG) emissions. In addition, biofuel production can create additional income for the rural poor and advance economic development [2].

Nevertheless, biofuel based opportunities do not come without concerns. Direct or indirect land use change resulting from expansion of biofuel cultivation can cause deforestation and destroy natural habitats [3,4], which in turn may lead to the loss of biodiversity [5,6]. Reduced biodiversity may have further negative impacts on ecosystem functions [7].

To respond to the concerns about potential negative social and environmental impacts, several voluntary standards have emerged since the beginning of the millennium. The most prominent have emerged from the Roundtable on Sustainable Palm Oil (RSPO) [8], which was formally established in 2004, the Roundtable on Responsible Soy Association (RRSA) in 2006 [9], and the Roundtable on Sustainable Biofuels (RSB) [10] in 2007. There have also been legislative efforts (e.g. Directive 2009/28/EC of the European Parliament and of the Council) to ensure that the production of imported is considered sustainable. However, there have been concerns that the standards are not effective enough to reduce the threat biofuel production poses to tropical forest ecosystems [11].

Currently palm oil and soybean are produced mainly for food, and thus cultivation for biofuel production has contributed little to the land-use change patterns for these crops [1,6]. Nevertheless, biofuel production has been predicted to grow [12] and it is important to know what the potential consequences of expanding biofuel cultivation are for biodiversity and biodiversity-related ecosystem functions, and to understand how well the standards in their current form might help to mitigate those impacts.

Objective of the review

The purpose of this review was to assess objectively the current state of knowledge of the impact of three first-generation biofuel crops (oil palm, soybean, and jatropha) on biodiversity in the tropics. The focus was on the direct impacts of forest conversion for crop plantations (resulting in forest fragmentation and deforestation) on species richness, abundance (i.e. overall number of individuals or occurrences) and community composition, and on ecosystem functions related to biodiversity (such as pollination, seed dispersal, biocontrol, nutrient cycling, soil fertility, decomposition). In addition to impacts, different standards

related to oil palm, jatropha, and soybean were assessed for their potential to mitigate the impacts. The specific study questions were:

- Does cultivation of oil palm, soybean, and jatropha in the tropics lead to the loss of biodiversity and ecosystem functions due to deforestation and fragmentation?
- Is there a difference in the impacts on biodiversity between industrial plantations and smallholder plantations per volume of fuel produced?
- Do different standards related to oil palm, jatropha and soybean mitigate the negative impacts?

Methods

Search strategy

Design of review

An *a priori* protocol was established, peer reviewed and posted on the website of the Collaboration for Environmental Evidence (CEE) after acceptance by CEE [13]. The protocol was followed with one change: the secondary study question on standards was revised after publication of the protocol and is presented in this review in the form used.

Search sources

The original literature search was conducted between May and November 2011 and updated between October and November 2012 to retrieve articles published after November 2011. The search included academic literature databases, internet search engines, as well as websites of specialist organizations. In addition, bibliographies of articles included in the review and previously published reviews were checked for references. The following is the full list of sources searched:

Literature databases
- Biofuels abstracts database by CAB
- Directory of Open Access Journals
- Web of Science

Internet search engines
- Google: www.google.com
- Google Scholar: www.scholar.google.com
- Scirus: www.scirus.com

Websites of specialist organizations
- European Biofuels Technology Platform: www.biofuelstp.eu,
- Center for International Forestry Research: www.cifor.org
- Food and Agriculture Organization of the United Nations: www.fao.org
- Forest Trends: www.forest-trends.org
- Global Bioenergy Partnership: www.globalbioenergy.org

- The International Fund for Agricultural Development: www.ifad.org
- International Finance Corporation: www.ifc.org
- International Food Policy Research Institute: www.ifpri.org
- International Institute for Environment and Development: www.iied.org
- International Union for Conservation of Nature: www.iucn.org
- WWF: www.panda.org
- Rainforest Alliance: www.rainforest-alliance.org
- Rights and Resources Initiative: www.rightsandresources.org
- Roundtable on Sustainable Palm Oil: www.rspo.org
- Tropenbos International: www.tropenbos.org
- United Nations Framework Convention on Climate Change: www.unfccc.int
- World Resources Institute: www.wri.org

The internet search engines typically returned several thousand results. Therefore, the searches were restricted to the first fifty hits and links to potentially relevant material were followed only once from the original hit. At the websites of specialist organizations, the search was limited to the publications section of the website if there was one. At the website of the European Biofuels Technology Platform the search was restricted to sustainability articles.

Search terms and languages

Search strings were created using three categories (exposure, location, and outcome) with Boolean operators AND between categories and OR within categories (Table 1). No specific search terms were used for the study population, i.e. faunal and floral species, as they are inherent in the outcome category. A wildcard character, i.e. the asterisk, was used in the location category to include alternative word endings. When the search string could not be used in its complete form, combinations of the search terms were used so that one term from each three categories was included, e.g. oil palm AND tropic* AND species richness. Owing to the limitations of the search engine, two search strings were used for the Directory of Open Access Journals: *(Oil palm OR jatropha OR soybean) AND tropic** and *(Oil palm OR jatropha OR soybean) AND tropical.* Similarly, only terms *Oil palm OR jatropha OR soybean* were used at the website of Forest Trends –organization owing to the limitation on number of words imposed by the search engine. The search terms were also translated into French, Spanish, German, Swedish, and Finnish (Additional file 1) and searches conducted using the same logic.

Table 1 Search terms in different categories

Exposure	Location	Outcome
Oil palm		Species diversity
Soybean	Tropic*	Species richness
Jatropha		Species abundance
		Species similarity
		Species composition
		Community composition
		Deforestation
		Land use change
		Fragmentation
		Habitat loss
		Connectivity
		Functional diversity
		Ecosystem
		Displacement

*Denotes a wildcard character that was used to include alternative word endings.

Study inclusion criteria

In collaboration with stakeholders, a set of inclusion criteria was developed. Studies that had data about relevant subject, exposure and outcome, together with a valid comparator were included if they fulfilled the quality criteria discussed in the section on study quality assessment.

Studies related to the primary study question were included according to the following criteria:

- Geographical location: Study area within the tropics (23.438°S to 23.438°N).
- Relevant subject(s): Faunal and floral species.
- Type of exposure: Conversion of the land to cultivate oil palm, soybean, and jatropha for any purpose.
- Type of comparator: Other land use or land cover (primary forest, logged-over forest, secondary forest (*i.e.* regrowth forest), scrubland, grassland, cropland). Both before-after and site comparison studies were accepted.
- Types of outcome: Change in species richness, abundance (the overall number of individuals or occurrences), community composition, and ecosystem functions (pollination, seed dispersal, biocontrol, soil processes).
- Types of study: Qualitative and quantitative primary studies as well as descriptive studies and reports.

For the secondary study question "Is there a difference in the impact on biodiversity between industrial plantations and smallholder plantations per volume of fuel produced?", location, subjects and outcome were the same, but the types of exposure and comparator were different:

- Type of exposure: Conversion of the land to industrial plantations for the cultivation of biofuel crops
- Type of comparator: Smallholder plantations

For the secondary study question "Do different standards related to oil palm, jatropha, and soybean mitigate the negative impacts?" the following criteria were used:

- Relevant subject(s): Faunal and floral species.
- Types of exposure: Standard in place should mitigate the impact of crop cultivation on biodiversity.
- Types of comparator: Standards were compared against each other to clarify how they mitigate the impact on biodiversity.
- Types of outcome: Any reported change within and nearby production area.
- Types of study: Standards related to oil palm, jatropha, and soybean, *i.e.* international legislation, industry standards, ISO management standards, NGO standards

Articles were assessed for relevance first by title, as well as keywords if these were available, then by abstract, and finally, by full text. If the inclusion of an article was in doubt in either of the first two stages, the article was included and the suitability determined at a later stage.

To assess the consistency in the use of inclusion criteria a kappa test was performed. Two reviewers applied the inclusion criteria to a random set of 108 articles at the abstract filter stage. The kappa statistic was calculated to measure the level of agreement between the reviewers. A score of 0.704 was achieved, which indicates substantial strength of agreement [14].

Potential effect modifiers and study quality assessment

Studies do not happen in a vacuum and hence, a number of variables that have the potential to affect study outcomes were recorded when available. The focus was on variables that can influence reliability and generalization of the findings. The following variables were recorded:

- Temporal and spatial scale. The temporal and spatial aspects of sampling were recorded, as well as whether sampling effort was evaluated.
- Comparator features: before/after or site comparison.
- Methodology used to collect data.
- Environmental features of the site: soil type, original vegetation, and the type of surrounding landscape
- Variables related ecological interactions: competition and predation.

- Variables related to plantation management: use of herbicides, insecticides, and fertilizers.
- Plantation type (industrial vs. smallholder), age, size, and certification status.

To avoid misleading conclusions by including studies with inappropriate design, the studies were evaluated according to the hierarchy of quality of evidence (Table 2). Studies that fell into the category VI were excluded from analysis.

Data extraction and synthesis

Originally we planned to categorize the data for the analyses using the following five categories: Mammals, birds, amphibians and reptiles, invertebrates, and plants. However, as there were relatively few studies overall, the data were not categorized in this way for the analysis.

There were enough data on species richness (i.e. number of species) and abundance (i.e. overall number of individuals or occurrences) to perform meta-analysis. The purpose of meta-analysis is to quantitatively summarize the results of individual studies using specific statistical methods [16]. The concept at the heart of a meta-analysis is the effect size, which is a statistical measure that portrays the magnitude of which given effect is present in a sample. It makes it possible to determine whether the overall effect is greater than expected by chance [17]. There are several effect size estimates that measure the standardized mean difference between two samples and are thus suitable for species richness and abundance data. Hedges' d was chosen because it corrects for a small sample size [18] (for the equations used in this section see the Additional file 2). The heterogeneity of the effect sizes was estimated using the Q-statistic. The I^2-statistic was used to describe the proportion of the observed variance that reflects real differences in effect sizes [19].

Table 2 Hierarchy of quality of evidence based on the information provided in the documents

Category	Quality of evidence presented
I.	Randomized controlled trials of adequate spatial and temporal scale for the study species.
II.	Controlled trials without randomization with adequate spatial and temporal scale for the study species.
III.	Comparisons of differences between sites with and without controls with adequate spatial and temporal scale for the study species.
IV.	Evidence obtained from multiple time series or from dramatic results in uncontrolled experiments.
V.	Opinions of respected authorities based on qualitative field evidence, descriptive studies or reports of expert committees.
VI.	Evidence inadequate owing to problems of methodology e.g. sample size, spatial or temporal scale.

Modified after [15].

To perform a quantitative meta-analysis on species richness and abundance the estimates of mean species richness and abundance, the corresponding estimates of standard deviations, and sample sizes were tabulated. If the estimate of standard deviation was not provided it was calculated from the estimate of standard error and sample size. In some cases the estimates of mean and standard deviation or standard error were measured from the published figures. The measurements were made by one person, so any measurement error is expected to be consistent. In cases where the estimates of mean and standard deviation were not provided but a t-statistic was, this was used to calculate Hedges' d by transforming the t-statistic first to Hedges' g and the g then to Hedges' d [18].

The effect sizes were analyzed using a random effects model. This was chosen because the subject groups and data collection methods varied between the studies and hence, there may be real differences among effect sizes of studies on different subjects [19,20]. Different taxa and taxa that were collected using different methods within the same study were treated as independent samples. Also, data that had significant differences between sampling occasions [21,22] were included as independent samples. Studies by different authors from the same location, regardless of the taxa studied, were treated as separate cases. Although originally we wanted to include explanatory variables into the model, this was not feasible owing to the small number of studies that met the inclusion criteria and hence, only the average effect sizes were estimated, along with 95% bias-corrected confidence intervals. The bias-corrected confidence intervals were chosen because of the relatively small sample sizes. The analyses were performed using MetaWin 2.1 release 5.10 [23].

One of the well-known problems associated with meta-analysis is that studies with higher effects are more likely to be published; relying only on results published in academic journals can potentially lead to misleading conclusions about the effect [19]. To address this problem, an extensive search was performed to uncover "grey" (variously defined, but here we mean conference papers, book chapters, reports that are no part of established Series, etc.) and unpublished literature. Another reported source of publication bias is that non-significant results may not be published at all. We did not test for publication bias for two reasons. First, a variety of responses are expected in ecological studies dealing with different taxa and we therefore did not expect suppression by Editors of studies of smaller effects or non-significant results. Secondly, existing statistical tests require reasonable numbers of cases and dispersion in sample sizes, two conditions which the meta-analyses we performed do not fully meet.

A variety of different methods used for examining changes in species composition makes it difficult quantitatively to assess the effects of habitat modification on species composition. Hence, to have a standardized measure to assess changes in species composition, a simple averaging method following Nichols *et al.* [24] was used to calculate the mean change in the number of shared species between forest and oil palm habitats, standardized by the total number of species recorded in forest. In addition to the mean response, 95% confidence intervals were calculated. The value was considered significant when the confidence interval did not include one. Primary and secondary forest data were combined in the analysis. When both primary and secondary forests were sampled, only primary forest data were used. The analysis was performed using SPSS version 17.0 [25].

Results
Review statistics
The searches returned 9143 articles after duplicate removal (Figure 1). Of these articles, approximately 13 per cent had a relevant title and keywords and were therefore examined further. At the abstract-assessment stage 9.8 per cent of articles satisfied the inclusion criteria and were read in full. Of those, 25 articles (21 per cent of those read in full) reported single studies with an appropriate comparator (Additional file 3). All of the selected studies belonged to category III (Table 2), which meant that none were excluded on the grounds of weak methodology.

Description of studies
Source
All 25 articles included in the review were published in peer-reviewed journals. Only three articles were published before 2000, and the majority of the articles were published after 2005 (Figure 2). The figure for 2012 is not fully representative of the whole year because the search was conducted on articles published by the bibliographic databases up to November 2012.

Context of the studies
Study location Most of the studies were conducted in Asia: 20 of them in Malaysia. Of the studies conducted in Malaysia, 10 were from one State Sabah. There were only single studies from other tropical regions, Africa (Ghana), Oceania (Papua New Guinea), and Latin America (Dominican Republic).

Study comparator Only studies of oil palm were retrieved using our search strategy. Typically, oil palm plantations were compared with forest, either primary ($n = 20$) or secondary forest ($n = 14$). All except one study were site comparisons. None of the studies were experimental. Only one of the studies examined outcomes before and after forest conversion.

Captured by the search:
9143

Relevant title and keywords:
1201

Relevant abstract:
118

Relevant articles:
27

Articles that were relevant and could be retrieved:
25

Figure 1 The number of articles at different assessment stages.

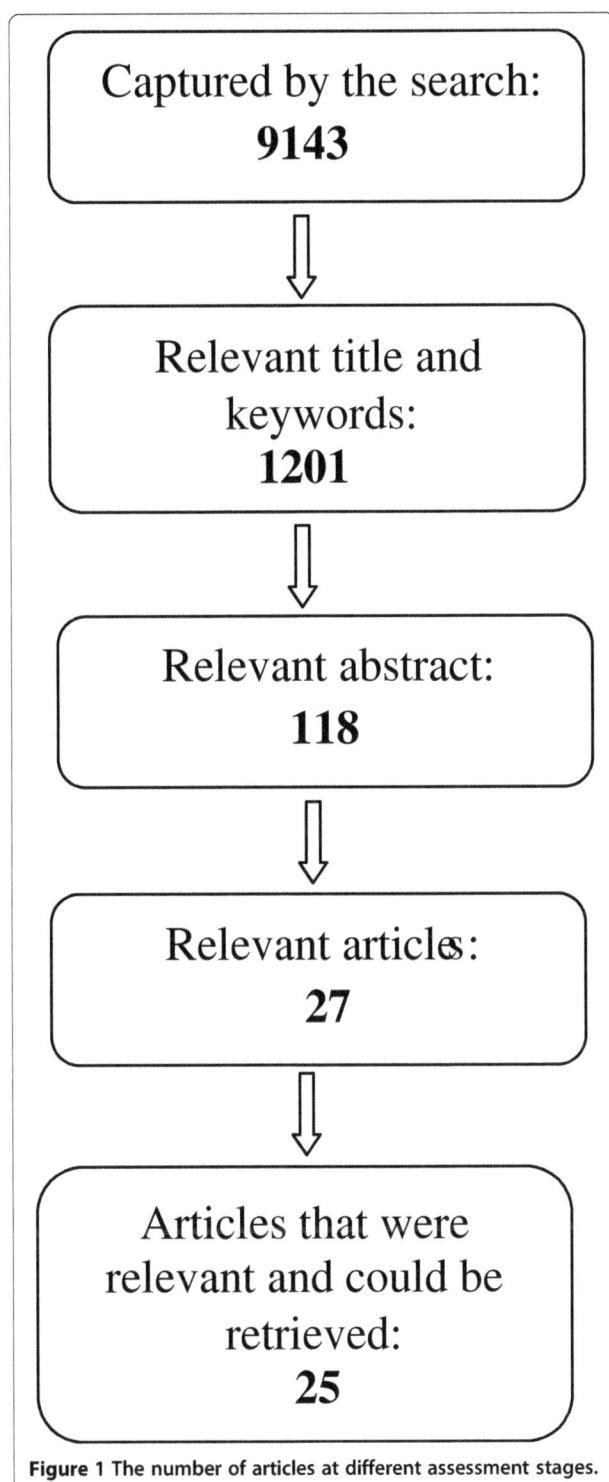

Study outcomes The 25 studies reported a total of 58 outcomes. All studies had examined faunal species richness/diversity ($n = 25$); almost all had examined abundance ($n = 21$), but only 12 had looked at species composition. Almost two thirds of them studied invertebrates (Figure 3).

The age and size of the plantations The age of the plantations was reported in 15 studies; two additional studies mentioned that the plantation was 'mature'. The age of the plantations varied from one year to more than 25 years. Nine studies collected data from plantations aged less than ten years, eight studies collected data from plantations aged ten years or more, including the study by Azhar *et al.* [26] that collected data from oil palm plantations of varying ages. Only ten studies mentioned plantation area, which ranged from 36 to 16000 hectares, with the majority of studies having studied plantations of several thousand hectares (Figure 4).

Study designs and methodology
All studies included in the review used quantitative methods. All except one study were site comparisons between oil palm plantation and primary or secondary forest or both. In the one before-and-after study Chang *et al.* [27] studied changes in abundance of mosquitoes induced by land use change during the development of an oil palm plantation.

All site comparison studies selected sites that could be paired and, except for Koh and Wilcove [28], collected data from the sites during same time period. Koh and Wilcove [28] used butterfly data collected from primary and logged forest in two earlier studies [29,30] and compared with the data they collected from an oil palm plantation. The exact method for site selection or pairing was described in only four studies [26,31-33]. It was impossible to assess the robustness of the selection in the other studies. Similarly, the selection of sub-sites within the studied habitats was unclear in most of the studies as even the studies that selected sub-sites randomly did not explain the exact method for randomization.

Half of the studies reported distance between the sites and only ten studies discussed leakage effects from or to adjacent areas. One of these [32] was specifically focused on spillover of butterflies and ants from forest to oil palm plantations and found that although vagrant forest butterflies were found in the plantations, recapture data did not reveal dispersal of butterflies across the forest-plantation ecotone. No spillover of ant species was reported. In addition, it was reported that leakage from adjacent areas was unlikely owing to behavioral characteristics [34] to dispersal capabilities [27,35] or ecological conditions [36]. In three studies on birds it was reported that nearby primary forest areas either 'probably' [37] or 'certainly' [26,38] contributed to the species richness in oil palm landscapes. Similarly, Gillespie *et al.* [39] suggested that it is possible that the occurrence of arboreal amphibian species (tree frogs), specifically *Rhacophorus appendiculatus, Rhacophorus dulitensis* and *Rhacophorus pardalis*, in the plantation resulted from local dispersal from nearby forest habitats. Juliani *et al.* [40] suspected that the lack of

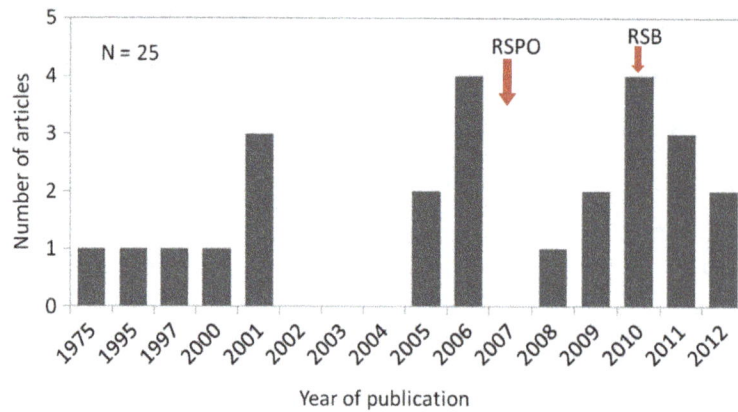

Figure 2 Number of articles published in different years. The articles shown are limited to those included in this systematic review. For articles published before 2000 only those years in which an article was published are shown. Arrows indicate the years when standards from Roundtable on Sustainable Palm Oil (RSPO) and Roundtable on Sustainable Biofuels (RSB) were first published.

shelter or roosting sites in areas adjacent to the oil palm plantation studied could have contributed to the high abundance of bats in the plantation.

The species studied in the faunal studies varied considerably, and therefore the data collection methods also differed (Table 3). Sampling effort was statistically evaluated in almost two thirds of the studies (58%) and in addition one more study [41] reported that it was 'low'. The most frequently reported method of evaluating sampling effort was by use of species accumulation curves; comparisons between observed and predicted species richness were used in three studies [22,26,36]. Generally, the studies that had statistically evaluated the sampling effort deemed it to be satisfactory to show the differences (or lack of differences) between the sites, and 11 of the 14 studies specifically discussed that point.

Nine of the studies explicitly reported efforts aimed at minimizing or controlling for the effect of extrinsic

variables. For example, sampling at the same time of the day, or only in fine weather conditions, collecting samples away from the edges of the habitat, and sampling birds at a limited spatial scale to ensure visibility.

Temporal and spatial scale of the studies

Temporal and spatial scales are important in several contexts. Although the spatial scale of data collection can influence the results of faunal studies [51], this was rarely discussed in the studies. Only two studies [32,35] discussed the results in the context of spatial scale, specifically in relation to the dispersal abilities of the species in question.

None of the studies collected long-term data and hence, the studies are based on a rather limited time scale. In addition, only two studies assessed the effects of seasonality. Fukuda *et al.* [48] conducted censuses on bats four times within 17 months and did not detect any significant differences between the seasons. Lucey and Hill [32] compared

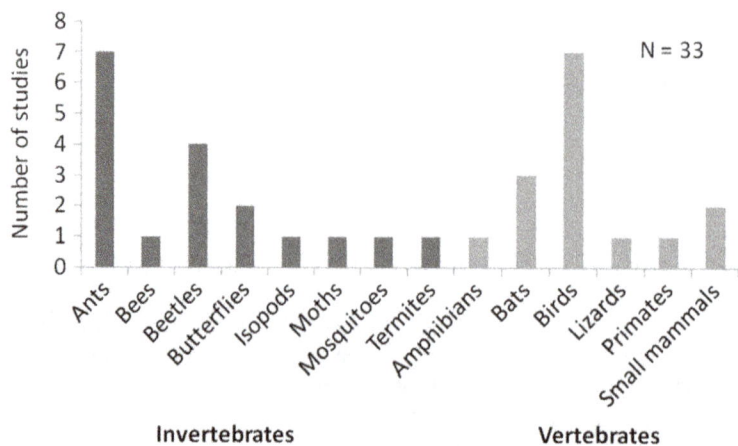

Figure 3 Taxonomic groups studied in the 25 studies on biodiversity included in the review. Some of the studies studied several taxonomic groups.

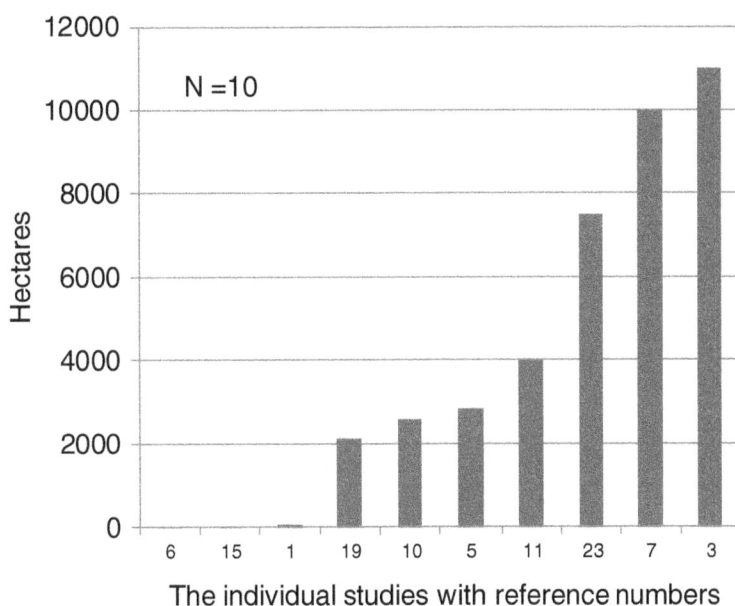

Figure 4 The size of plantations studied. The bars represent individual studies and the labels refer to the study numbers in Additional file 3.

similarity of species assemblages between first and second sampling occasion and concluded that for butterfly species temporal turnover contributed substantially to overall diversity. For ant species the similarity of species assemblages was higher for both forest and oil palm habitats and thus, temporal turnover had less impact on the diversity of ants than butterflies.

Quantitative synthesis
Species richness
We found 11 studies that provided suitable data for conducting meta-analysis to compare species richness in oil palm plantations and primary forest, and 8 whose data could be used for comparison between oil palm plantations and secondary forests. Owing to the limited amount of suitable data we focused on overall effects. Although examining only overall effects can mask differences in responses between taxa, it was done out of necessity to retain power in the analyses. As primary and secondary forests can be biologically very different environments, the analyses were done separately.

There was relatively uniform negative response as shown in the forest plots of differences in species richness between oil palm plantation and either primary or secondary forest (Figures 5 and 6). The estimated mean effect size was significantly different from zero (primary forest: E++ = -1.41, 95% bias-corrected CI -2.06 to -0.90; secondary forest: E++ = -3.02, 95% bias-corrected CI -4.42 to -1.84) indicating that oil palm plantations have fewer species than either primary or secondary forest. As the effect sizes got larger, the confidence intervals were also wider.

There was heterogeneity in the effects when the species richness of plantation was compared to that of primary forest ($Q = 29.76$, $p = 0.02$), but not when the comparison was between plantation and secondary forest ($Q = 16.19$, $p = 0.24$). The I^2 index indicated that 43% of the variance considering the effects regarding plantations and primary forests reflects real differences in the effect sizes. Correlations between effect and sample sizes were not significant (Spearman's rank correlation, $p > 0.05$) for either primary or secondary forest implying that larger effects in one direction were not reported more often than other effects, but at low samples sizes the power of the correlation is rather low [19].

Abundance
There was more dispersion in the direction of effect sizes of abundances (i.e. the overall number of individuals or occurrences) than of species richness and the mean effect size was not significantly different from zero for the comparison of an oil palm plantation to either primary forest (E++ = -0.92, 95% bias-corrected CI -2.03 to 0.01) (Figure 7) or secondary forest (E++ = -0.21, 95% bias-corrected CI -1.58 to 0.75) (Figure 8). However, it is important to note that the results for the secondary forests were based on only four independent studies, and that owing to the limitations in data available, we aggregated all taxa in these analyses. As with species richness, larger effect sizes had larger confidence intervals.

There was heterogeneity in the effect sizes when the abundance of plantations was compared with primary forest ($Q = 31.88$, $p = 0.02$) as well as with a secondary forest ($Q = 19.35$, $p = 0.01$). The I^2 index indicated that

Table 3 Summary of methods used in the studies included in the review

Study	Taxonomic group	Collected data	Sampling method	Methodology
Invertebrates				
Brühl & Eltz [36]	Ground-dwelling ants	Species richness	Tuna baits	Baits along 105 transects of various lengths (10–100 m)
Chey [31]	Moths	Species richness, abundance and composition	Light traps	One light-trap at each site for 3 consecutive nights.
Chang et al. [27]	Mosquitoes	Species richness and abundance	Human baits	All-night human landing collections on 5 consecutive nights each year.
Chung et al. [21]	Subterranean beetles	Species richness, abundance and composition	Winkler sampling	Ten 1 m² samples of leaf litter and soil at each site.
	Understorey beetles	Species richness, abundance and composition	Flight-interception-trapping	3 traps per site. Two weeks of sampling. Only samples from alternate days used.
	Arboreal beetles	Species richness, abundance and composition	Mist-blowing	10 trees at least 10 m apart
Davis & Philips [22]	Dung beetles	Species richness and abundance	Pitfall traps	4 sites per habitat, 3 traps per site at least 10 m apart, two 24-hour periods
Fayle et al. [42]	Canopy ants	Species richness, abundance and composition	Fogging	20 transects per habitat
	Ants in the ferns	Species richness, abundance and composition	Entire ferns collected, litter and core fragments processed.	20 transects per habitat
	Leaf litter ants	Species richness, abundance and composition	Litter samples	20 transects per habitat
Hashim et al. [41]	Ants	Species richness	Hand-collecting and pitfall traps	3 randomly-distributed 0.25 m² subplots within each of three 10 m × 10 m plots and 5 pitfall traps per habitat.
Hassall et al. [35]	Terrestrial isopods	Species richness and abundance	Quadrats	Plots sampled on a stratified random basis.
Koh & Wilcove [28]	Butterflies	Species richness	Banana-baited traps	98 trapping sites with total of 48 hours of trapping
Liow et al. [43]	Bees	Species richness, abundance and composition	Honey-baited traps in transects	Non-randomly selected 1–3 transects per site. On average 12.85 hours surveyed per transect
Lucey & Hill [32]	Ground-dwelling ants	Species richness, abundance, and composition	Pitfall traps	2000 m transects, five traps per trap station, six trap stations in forest and in oil palm plantations, 100 m between trap stations. Sampled twice for 12 consecutive days.
	Butterflies	Species richness, abundance, and composition	Fruit-bated traps	Two 2000 m transects, 10 trap stations in forest and in oil palm plantation, 100 m between trap stations. Sampled twice for 12 consecutive days at both occasions.
Room [44]	Ground foraging ants	Species richness, abundance and composition	Quadrats	30 samples per habitat.
Vaessen et al. [33]	Termites	Species richness, abundance and composition	Transects	One transect established randomly at each site.

Table 3 Summary of methods used in the studies included in the review (*Continued*)

Vertebrates

Study	Taxon	Response variable	Method	Details
Aratrakorn et al. [45]	Birds	Species richness and relative abundance	Timed Species Counts	30 oil palm plantations selected from aerial photographs. The number of sites based on preliminary counts. Two counts of 20 min divided into five 4-minute blocks.
Azhar et al. [26]	Birds	Species richness, abundance and composition	Transect counts	470 various-length transects: 418 in plantation estates, 52 in smallholdings and 20 in peat swamp forest.
Bernard et al. [34]	Non-volant small mammals	Species richness, abundance and composition	Live cage traps with baits	50 traps per trapping site arranged into 5 200 m long trap lines.
Danielsen & Heegaard [46]	Birds, primates, tree-shrews, and squirrels	Species richness, abundance, and composition	Variable-distance line-transect	2000 m straight line; surveyed for 40 hours in forest areas and for 20 hours in oil palm.
	Bats	Species richness, abundance, and composition	Mist nets	15-20 nets (totaling 150-250 m).
Edwards et al. [47]	Birds	Species richness and abundance	Timed point-counts along transects	5 sites per habitat, 12 sampling points at 250 m intervals at each site.
Fukuda et al. [48]	Bats	Species richness and abundance	Mist nets and harp traps	2-4 mist nets per night, 3-6 census points per habitat.
Gillespie et al. [39]	Amphibians	Species richness and composition	Transects	400 m transects; 6 in wet forest, 5 in dry forest, and 3 in oil palm plantation. Each sampled 3-4 times.
Glor et al. [49]	Lizards	Species richness and abundance	Glue traps	Non-randomly selected 10 x 10 m trapping grids with 20 traps each, 3 plots in oil palm, 4 in *mogote*.
Juliani [40]	Bats	Species richness and abundance	Mist nets	10 mist nets randomly placed in each habitat type.
Peh et al. [38,50]	Birds	Species richness and abundance	Point counts	240 point counts arbitrary chosen. At least 200 m from each other. 127 sites in the oil palm.
Sheldon et al. [37]	Birds	Species richness, abundance and composition	Point counts	20 three-minute point counts at 50 m intervals along the transects.

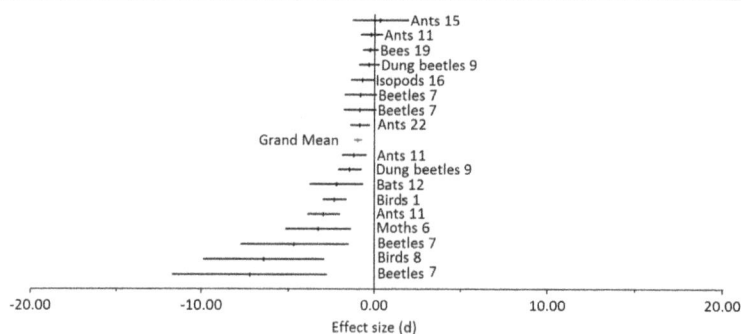

Figure 5 Forest plot of effect sizes for species richness (mean standardized difference between primary forest and oil palm plantation). The grand mean is the summary effect of all the individual effect sizes. The error bars represent 95% confidence intervals. The numbers after the taxa refer to the study number in Additional file 3.

47% of the variance considering the effects regarding plantations and primary forests reflects real differences in the effect sizes. The figure was 59% when faunal abundance of secondary forests and plantations are compared. Correlations between effect and sample sizes were not significant for either primary or secondary forest (Spearman's rank correlation, $p > 0.05$).

Species composition
The similarity of species composition was statistically assessed in 12 of the original studies while a further 11 studies provided some information about species composition (Tables 4 and 5). Species composition differed between forest and oil palm plantation areas in all except one of the 23 studies. In most of the studies that had statistically assessed the difference, the similarity between plantation and forest areas was either low or zero. However, the statistical methods used differed between the studies and results are therefore not directly comparable.

To have comparable results, a mean of shared species between oil palm plantation and forest was assessed. There were 10 studies on invertebrates and 9 studies on vertebrates that provided suitable data for the comparison. On average only 29% of the invertebrate species and 22% of

the vertebrate species were shared between oil palm plantation and forest after the values were standardized (Table 6, Figure 9). This represents significant change in community composition for both invertebrates and vertebrates.

Narrative synthesis
Biodiversity in industrial versus smallholder plantations
Only one study [26] addressed differences in species richness and community composition between smallholder and industrial plantations. The results showed that, on average, smallholdings with mixed-age stands supported higher bird species richness than industrial plantation estates that had uniform age structure (range from <6 years old to >25 years old). The average dissimilarity of bird assemblages between the plantation estates and smallholdings was 47.6%. However, since yields were not taken into account in the analyses, it is not known whether the impact is similar when compared for equivalent amounts of fuel produced under different management systems.

Explanatory factors for differences in species richness and community composition
Only four studies had statistically analyzed the causes of differences in either species richness or community

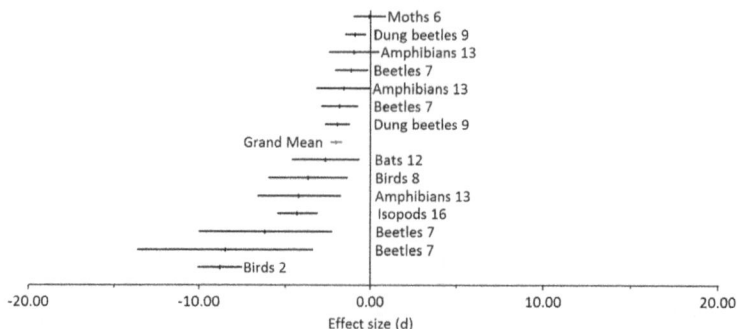

Figure 6 Forest plot of effect sizes for species richness (mean standardized difference between secondary forest and oil palm plantation). The grand mean is the summary effect of all the individual effect sizes. The error bars represent 95% confidence intervals. The numbers after the taxa refer to the study number in Additional file 3.

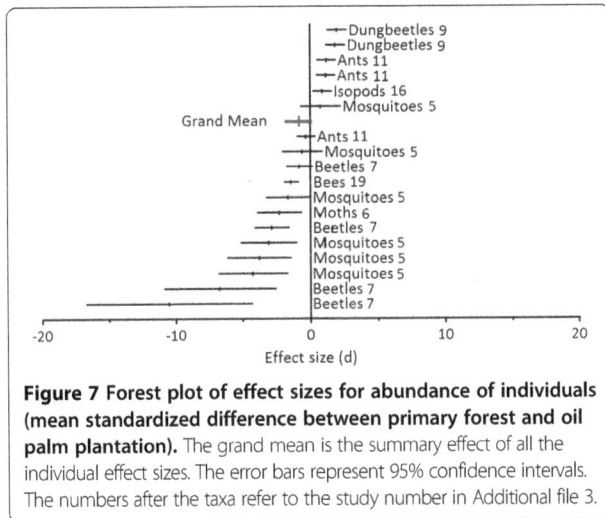

Figure 7 Forest plot of effect sizes for abundance of individuals (mean standardized difference between primary forest and oil palm plantation). The grand mean is the summary effect of all the individual effect sizes. The error bars represent 95% confidence intervals. The numbers after the taxa refer to the study number in Additional file 3.

composition. For birds, the statistical analyses showed that increased ground vegetation and undergrowth height, as well as decreased canopy cover, were all correlated with higher species richness [26]. In addition, increased proximity to a forest patch, cumulative area of natural forest patches, and decreased isolation distance positively influenced bird species richness [26]. The role of food resources was speculated about in the discussion but not tested.

In the case of invertebrates, the hotter and drier conditions in oil palm plantations were the main cause of changes in community compositions (ants [42]; beetles [21]; bees [43]). Soil pH was a significant factor for isopods [35], whereas the amount of leaf litter, tree and sapling densities, and plant species richness were significant factors for primary forest beetle species [21].

Ecosystem function

None of the studies had specifically looked at biodiversity-related ecosystem functions, such as pest control,

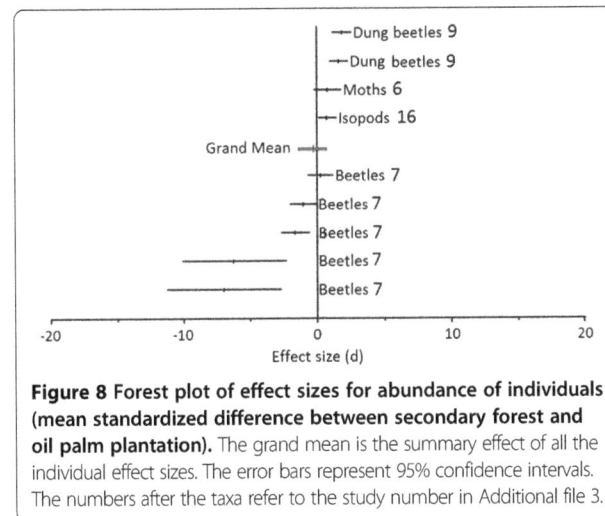

Figure 8 Forest plot of effect sizes for abundance of individuals (mean standardized difference between secondary forest and oil palm plantation). The grand mean is the summary effect of all the individual effect sizes. The error bars represent 95% confidence intervals. The numbers after the taxa refer to the study number in Additional file 3.

pollination and soil processes that might have included supporting data. However, we found some discussions about concern for the continuity of pollination processes after expansion of oil palm habitats, and the changed communities between primary forest and other areas [40,43]. In summary, these postulated that there would be negative consequences for forest regeneration if remaining forest areas cannot support large enough pollinator populations and pollinators are also absent in the surrounding oil palm matrix.

Biofuel-related standards

There were no studies that had tried to assess the impact of the standards on biodiversity. In fact, only a few of the studies reported whether the oil palm plantations studied were complying with standards. None of these had been structured to compare impacts before and after standards were applied (for a qualitative assessment of the standards see Additional file 4).

Discussion
Evidence of impact

Although the number of studies that met the inclusion criteria was small relative to the amount of literature broadly related to the review topic, the evidence on species richness and community similarity from the included studies showed clearly that oil palm plantations have reduced species richness compared with primary and secondary forests, and the composition of species assemblage changes significantly after forest conversion to oil palm plantation. Species-specific responses would be expected to vary, but based on the studies included in the review, regardless of the taxa, forest specialists do not, in general, succeed in oil palm plantations. The findings are consistent with previous reviews that have addressed similar questions [5,28,52,53].

With respect to abundance, responses appear to vary depending on species and there is no clear overall effect in one direction. When the abundance results are considered in the light of the results on species richness and similarity, it appears that certain invertebrate species, e.g. generalist species, increase in abundance after forest conversion whereas others decline. However, it is possible that the responses may differ for vertebrates, as none of the studies in the meta-analysis looked at abundance of vertebrate taxa in forest compared with plantation.

Reasons for variation in impact

The variation in effect sizes observed in the meta-analysis most likely reflects different ecological requirements of different taxa and different species within these taxa. Part of the variance in the effect sizes was due to real differences between taxa rather than general heterogeneity, but the small number of studies included in the analyses did

Table 4 Summary of information on species composition provided in the reviewed studies*

Authors	Year published	Taxonomic group	Similarity between primary forest and plantation	Similarity between secondary forest and plantation	Statistics used	Changes in communities between forest and plantation	Notes on similarity	Causes
Invertebrates								
Brühl & Eltz [36]	2010	Ground-dwelling ants	-	-	-	Yes	Communities of plantations dominated by a small number of, partly invasive, non-forest taxa. Highly impoverished in regard to forest taxa.	Absence of leaf litter. Hot and dry conditions possibly prevent colony establishment and reduce survival.
Chang et al. [27]	1997	Mosquitoes	100%	-	-	No	Lower abundances but same species composition.	na
Chey [31]	2006	Moths	0.278	-	Preston's coefficient of faunal resemblance	Yes	Noctuid and arctiid species dominated the assemblages.	Low floristic diversity. Lichens and other host plants. Open habitat (many noctuid and arctiid species favor open habitat).
Chey [31]	2006	Moths	0.228					
Chey [31]	2006	Moths	0.970					
Chung et al. [21]	2000	Subterra-nean, understorey and arboreal beetles	-	-	Detrended Correspondence Analysis and Canonical Correspondence Analysis	Yes	Species composition significantly different between sites (primary forest, secondary forest and oil palm). A few species dominated the assemblage at the plantation site.	**The amount of litter, tree and sapling densities, and plant species richness.**
Davis & Philips [22]	2005	Dung beetles	22.5%	-	Steinhaus similarity coefficient; Persentage disagreement distance measure; Cluster analysis and ordination	Yes	Similarity between both forest types and plantation.	Physiognomic differences.
Fayle et al. [42]	2010	Ants (canopy)	S: 0.191, C: 0.301	-	Sørenson's classic similarity index; Chao's incidence-based measure with a correction for unseen species	Yes	Only a small proportion of forest ant species were present in oil palm plantation. Non-native species were much more widespread.	**Temperature nearly significant factor (P = 0.073). Simplification of the canopy structure. Competitive interactions.**
Fayle et al. [42]	2010	Ants (ferns)	S: 0.056, C: 0.070	-	Sørenson's classic similarity index; Chao's incidence-based measure with a correction for unseen species	Yes	Only a small proportion of forest ant species were present in oil palm plantation. Non-native species were much more widespread.	Competitive interactions.
Fayle et al. [42]	2010	Ants (leaf-litter)	S: 0.213, C: 0.555	-	Sørenson's classic similarity index; Chao's incidence-based measure with a correction for unseen species	Yes	Only a small proportion of forest ant species were present in oil palm plantation. Non-native species were much more widespread.	**Temperature. Hotter and drier environment. Competitive interactions.**
Hashim et al. [41]	2010	Ants	-	-	-	Yes	Four species found in the plantation were absent from mangrove forest and two species found in the mangrove were absent from the plantation.	na
Hassall et al. [35]	2006	Terrestrial isopods	-	-	-	Yes		na

Table 4 Summary of information on species composition provided in the reviewed studies* (Continued)

Study	Year	Taxon	Method		%		Findings	Correlates
Liow et al. [43]	2001	Bees	Cluster analysis and canonical correspondence analysis	–	–	Yes	Families Halictidae and Anthophoridae were more commonly caught in oil palm plantation.	The occurence of families Halictidae and Anthophoridae were correlated with higher temperatures and light intensity, lower humidity levels and greater flowering intensities.
Lucey & Hill [32]	2012	Ants	Non-metric multidimensional scaling	–	–	Yes	NMDS differentiated between the habitats.	Air and soil temperature.
Lucey & Hill [32]	2012	Butterflies	Non-metric multidimensional scaling	–	–	Yes	Two distinct clusters, one for forest and one for plantation.	Air and soil temperature.
Room	1975	Ground foraging ants	Percentage similarity expressed as 100 × [(2 × number of occurences common to both)/(sum of occurences present in each)]	–	25.0%	Yes	Only a small proportion of forest ant species were present in oil palm plantation. Non-native species were much more widespread.	na
Vaessen et al. [33]	2011	Termites	–	–	–	Yes	The assemblage dominated by *Schedorhinotermes*.	Decrease in the amount of dead wood.
Vertebrates								
Aratrakorn et al. [45]	2006	Birds	–	–	–	Yes	Plantations dominated by few species. 60% of the species recorded only in the forest, 3% only in the oil palm plantation. Species recorded only in the forest had significantly smaller ranges. Species that were recorded in both forest and plantations had smaller body size than species recorded only in forest.	na
Bernard et al. [34]	2009	Non-volant small mammals	Proportional difference calculated following a formula by Thiollay (1992); a hierarchical cluster analysis	–	12.0%	Yes	Both forest types (primary and secondary) combined. Oil palm plantations may act as an effective barrier to the dispersal of small mammals.	na
Danielsen & Heegaard	1995	Birds	Proportional difference calculated following a formula by Thiollay (1992)	–	38.7%	Yes	Widespread, generalist, and common species much more abundant in plantations than in the primary forest.	Plantation age, proximity to forest, microhabitat structure, and level of human disturbance.
Danielsen & Heegaard	1995	Primates	Proportional difference	–	0.0%	Yes		na
Danielsen & Heegaard	1995	Squirrels and tree-shrews	Proportional difference	–	0.0%	Yes	No squirrels or tree-shrews observed in the plantation.	na
Danielsen & Heegaard	1995	Bats	Proportional difference	–	13.0%	Yes	Insectivorous bats appear to be more susceptible to conversion than frugivores/nectarivors.	na
Edwards et al.	2010	Birds	Analysis of Similarity	–	10.0%	Yes		na
Fukuda et al. [48]	2009	Bats	–	–	–	Yes	Certain species absent in the oil palm plantation: Two frugivorous species were	The absent frugivorous species rarely use agricultural lands for feeding.

Table 4 Summary of information on species composition provided in the reviewed studies* (Continued)

Study	Year	Group	Statistic	Method		Info	Species composition	Causes
							not recorded at all, only two insectivorous species recorded.	
Gillespie et al. [39]	2012	Amphibians	0.592 (p = 0.0002)	Analysis of Similarity between all forest transects and plantation and non-forest transects combined.	-	Yes	The assemblages reflect the strong affinities of certain species with particular habitat types. Plantation assemblages dominated by terrestrial, non-endemic, generalist species.	Absence of suitable microhabitats. The simple structure and open canopy of plantations results in greater temperature flux between day and night, increased evaporation rates and lower humidity.
Glor et al.	2001	Lizards	-	-	-	Yes		Microhabitat availability in regard to, at least, two species (grass-bush anole and Cochran's dwarf gecko). Oil palm plantation lacks the perch availability and understory microhabitat of natural forest.
Peh et al.	2005, 2006	Birds	-	Multiresponse permutation procedure	-	Yes	Forest species constituted only 26% of the total individuals observed in plantation. Nearby primary forest may act as a source habitat.	Simplification of the vertical vegetational structure.
Juliani	2010	Bats	-	-	-	Yes	Almost all species that were found in the oil palm plantation can be classified as common species in disturbed areas.	na
Sheldon et al. [37]	2010	Birds	-	-	-	Yes	Most species in oil palm plantation were open country and scrub species that are common throughout Borneo.	Simple botanical structure.

*The causes marked bold were statistically significant.

Table 5 Summary of information on species composition between logged peat forest and smallholder plantations*

Author	Year published	Taxonomic group	Similarity between logged peat forest and plantation	Similarity between logged peat forest and smallholdings	Statistics used	Changes in communities between forest and plantation	Notes on similarity	Causes
Azhar et al. [26]	2011	Birds	21.40%	19.10%	Analysis of similarity, Similarity percentage procedure	Yes	Oil palm management regimes had a similar species composition but both differed from the forest. The bird community in oil palm consisted of non-forest resident, forest-dependent, migratory, and wetland species.	**Extensive canopy cover which in turn may suppress ground layer vegetation that can provide refuge from predators and provide food sources.**

*The causes marked bold were statistically significant.

Table 6 Total species richness in forests and plantations, the number of shared species, and the proportion of species remaining

Authors	Year published	Taxonomic group	Forest species	Plantation species	Number of shared species	Proportion of species remaining
Invertebrates						
Brühl	2001	Ground-dwelling ants	31	23	14	0.45
Chang et al. [27]	1997	Mosquitoes	6	6	6	1.00
Chey [31]	2006	Moths	75	85	28	0.37
Chey [31]	2006	Moths	133	73	28	0.21
Chey [31]	2006	Moths	78	90	11	0.14
Davis & Philips [22]	2005	Dung beetles	25	20	8	0.32
Fayle et al. [42]	2010	Ants (canopy)	120	58	17	0.14
Fayle et al. [42]	2010	Ants (ferns)	36	35	2	0.06
Fayle et al. [42]	2010	Ants (leaf-litter)	216	56	29	0.13
Hashim et al. [41]	2010	Ants	5	7	3	0.60
Hassall et al. [35]	2006	Terrestrial isopods	12	4	0	0.00
Koh & Wilcove [28]	2008	Butterflies	69	15	12	0.17
Room	1975	Ground foraging ants	49	29	11	0.22
Vaessen et al. [33]	2011	Termites	11	6	2	0.18
Mean						0.29
SD						0.26
n						14
95% CI						0.14
Vertebrates						
Aratrakorn et al. [45]	2006	Birds	108	41	21	0.19
Bernard et al. [34]	2009	Non-volant small mammals	6	1	0	0.00
Danielsen & Heegaard	1995	Primates	5	1	0	0.00
Danielsen & Heegaard	1995	Bats	8	1	1	0.13
Fukuda et al. [48]	2009	Bats	19	5	4	0.21
Gillespie et al. [39]	2012	Amphibians	21	12	10	0.48
Glor et al.	2001	Lizards	11	5	4	0.36
Juliani	2010	Bats	9	7	3	0.33
Peh et al.	2005, 2006	Birds	159	40	36	0.23
Azhar et al. [26]	2011	Birds	194	55	49	0.25
Mean						0.22
SD						0.15
n						10
95% CI						0.09

not warrant further exploration, mainly because the cases could not be categorized based on a taxon.

Both temporal and spatial aspects of sampling can create variation in effect sizes, which is why the importance of scale has been emphasized in biodiversity studies [51]. As none of the studies addressed biodiversity changes at landscape level, scale-dependent variation in effect sizes could not be evaluated. Variation in impacts due to seasonality could not be evaluated because the available evidence was based on short term data collection.

The small number of studies did not allow us to conduct quantitative examination of the importance of environmental variables, or variables related to plantation management, such as clearing of ground vegetation or

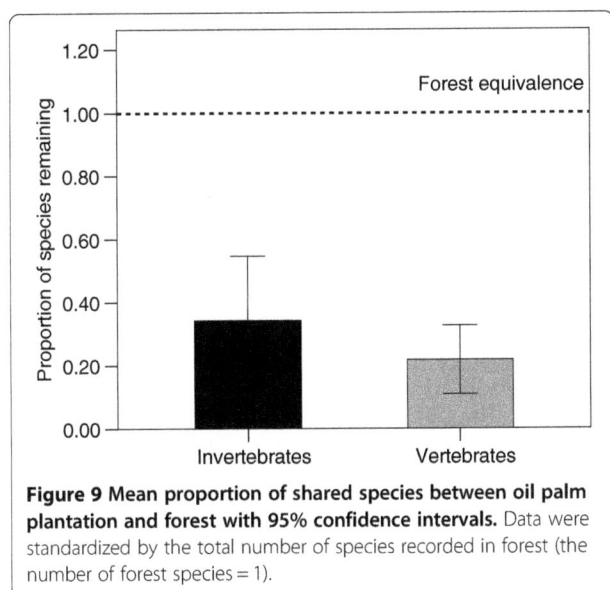

Figure 9 Mean proportion of shared species between oil palm plantation and forest with 95% confidence intervals. Data were standardized by the total number of species recorded in forest (the number of forest species = 1).

type of plantation ownership (smallholdings versus industrial estates). However, there was an indication that both types of variables had some effect [21,26,35,42,43,45] and probably contribute to variation in the effect sizes, as they are most unlikely to be constant from one area to another, or even constant temporally within the same area (for example, because management practices can differ between plantations).

There are also natural processes such as competition and predation that can influence the results and create variation. Competitive interactions were mentioned, though not analysed, in one of the studies [42] but in general the influence of competition and predation were not reported.

Review limitations

This review was based on only one crop, oil palm, with the majority of studies conducted in Malaysia and almost half of the studies in one Malaysian State. We would therefore not want to generalize our findings outside South East Asia.

When biodiversity is compared across natural and human-modified landscapes, there are many factors that can limit the generality of conclusions. Variability is an inherent component of biological systems, and human actions in the studied area as well as in the surrounding landscape can add further variability. One way to account for the variability is to include replication in the study design. Unfortunately, the majority of the studies included in the review included insufficient reporting of study conditions and details, or were poorly replicated or pseudo-replicated, which is common for biodiversity studies [54]. Although it is assumed that site comparison studies pair sites that share common attributes, this is not necessarily the case in practice. For example, only a few studies

reported on the type of surrounding landscape or on the original vegetation. A number of unreported factors could therefore have contributed to the true effect sizes.

One significant limitation of the review is the lack of landscape level comparisons. Although comparing production areas with forest provides information of the extent of losses at the management unit level, it does not provide information about whether there is a loss in biodiversity at the landscape level. A landscape level approach would be required to incorporate differences between different landscape mosaics as well as their historical backgrounds, into the analysis.

The 25 papers identified in this review compared oil palm plantations with forest. However for us to understand the differences between management systems and the link between management practices and biodiversity, we also need studies that make further comparisons between differently-managed areas. In this review such stratification was not possible because of the dearth of information. To move beyond comparing forest ecosystem with oil palm plantation, there is a need to conduct a robust impact evaluation of differently-managed areas.

The lack of information also prevented analysis of species or taxa-specific responses which is a limitation of the current review. We combined different taxa in the analyses out of necessity, but can recognized that this can mask responses that are specific to certain groups or taxa. As metrics of biodiversity, species richness and abundance suffer from a similar kind of blindness as they consider all the species and individuals to be equal. The inclusion of community similarity in the review alleviates this limitation to some extent.

Publication bias cannot be wholly discounted, even though there are grounds to assume that it is not a significant problem for this body of literature. Grey and unpublished literature was extensively searched in several languages. Correlations between sample sizes and effects were not significant. Finally, considering the nature of the subject, non-significant findings have the same value as significant ones.

Conclusions
Potential implications for biodiversity conservation, policy, and plantation management

The available evidence suggests that oil palm plantations support lower species richness than primary or secondary forest. Also, forest conversion to oil palm plantation leads to significant changes in community composition, which indicates that oil palm plantations are not suitable habitats for the majority of forest species. Unfortunately, very little information was available about the impacts of smallholder plantations or different standards, which makes it difficult to evaluate their usefulness.

Potential implications for research

The review identified several knowledge gaps about the impacts of biofuel crop cultivation on biodiversity and ecosystem function:

- Landscape level studies that would contribute better knowledge of the impacts at larger scale beyond simple habitat comparisons.
- Research on how reduced species richness or changes in community composition affect ecosystem functions. The lack of knowledge about this topic was also a conclusion of a recent review by Foster *et al.* [53].
- Research on differences in biodiversity and ecosystem function in response to different production systems, (smallholdings vs industrial estates) and different management practices (certified and non-certified plantations).
- Studies on jatropha and soybean and oil palm beyond Malaysia.

To provide a sound evidence base for land-use management decisions, future studies should pay careful attention to study designs, for example by defining the sampling population of land-uses and then using stratified randomization to select study sites, as well as ensuring that seasonality effects are taken into account, and that there are enough replicates. Methodologies should be shared across plantations, users and experiments to identify groups for future monitoring and to make use of crowdsourced identification (e.g. Ispot [55]).

Finally, there are a number of recommendations for authors and publishers that relate to the reporting of biodiversity studies. First, descriptions of methods should be more detailed, including exact explanations for site selection, and descriptions of plantation sizes, ages and management histories. Failure to include such basic information precludes subsequent analysis, and lowers the value of such studies for guiding policy. Second, details of management practices are needed, particularly whether the plantation is certified, and details about which standards are adhered to within the plantation. Finally, crop yields in plantations under different management regimes should be reported to facilitate comparisons that can support policy- and decision-making.

Additional files

> **Additional file 1:** Search terms in different languages.
>
> **Additional file 2:** Equations used in the quantitative meta-analysis.
>
> **Additional file 3:** List of studies included in the review.
>
> **Additional file 4:** Qualitative assessment of standards related to oil palm, jatropha, and soybean.

Competing interests

The authors declare that they have no competing interest.

Author's contributions

SS carried out the literature survey and assessment, extracted data, designed and carried out the statistical analysis. She drafted the systematic review and revised it following reviewers' comments. RN had the idea for the study. SS, MRG, RN, and YL contributed to the design and focus of the study. SS, MG, and MZ assessed the standards. SS, CG, JUG, JG, MG, GP, JS, MZ participated in the workshop where the first draft of the review was discussed and subsequently improved. All authors participated in the revisions of the manuscript. GP and MG edited the language. All authors read and approved the final manuscript.

Acknowledgements

We thank all the authors who responded to our queries and provided additional information. We acknowledge our superb librarian, Wiwit Siswarini, for her help in finding articles and other information resources. We are thankful for Bruno Locatelli for his guidance on statistical issues but note that he bears no responsibility for any of our decisions regarding the statistics. We thank Ankara M. Chen for her help in organizing the superb workshop in Zürich that provided the push to finalize the review. We acknowledge Wen Zhou for her support. Three anonymous reviewers provided useful suggestions to improve the protocol to conduct this systematic review. We thank Andrew Pullin and two anonymous reviewers for their comments that made the final review more focused and improved its quality. The review was supported by the Center for International Forestry Research, the Government of Finland, ETH Zürich, and CIRAD.

Author details

[1]Center for International Forestry Research, P.O. Box 0113 BOCBD, Bogor 16000, Indonesia. [2]Department of Environmental Systems Science, ETH Zurich, Universitaetstrasse 16, 8092 Zurich, Switzerland. [3]Interdisciplinary Arts & Sciences, University of Washington, Bothell, 18115 Campus Way NE, Bothell, WA 98011-8246, USA. [4]Program on the Environment, University of Washington, Seattle, WA 98195-1800, USA. [5]CIRAD, Research Unit Goods and Services of Tropical Forest Ecosystems, Avenue Agropolis, 34398 Montpellier, Cedex 5, France. [6]Biodiversity Institute, Department of Zoology, University of Oxford, South Parks Road, Oxford OX1 3PS, UK. [7]Zoological Society of London, Indonesia Programme, Jl. Gunung Gede 1 No.11A, Bogor 16151, Indonesia.

References

1. Rajagopal D, Zilberman D: **Review of environmental, economic and policy aspects of biofuels.** Washington DC, USA: World Bank Policy Research Working Paper; 2007:4341.
2. Feintrenie L, Chong WK, Levang P: **Why do Farmers Prefer Oil Palm? Lessons Learnt from Bungo District, Indonesia.** *Small-scale Forestry* 2010, 9:379–396.
3. Koh LP, Ghazoul J: **Biofuels, biodiversity, and people: understanding the conflicts and finding opportunities.** *Biol Conserv* 2008, 141:2450–2460.
4. Lewandowski I, Faaij APC: **Steps towards the development of a certification system for sustainable bio-energy trade.** *Biomass Bioenergy* 2006, 30:83–104.
5. Danielsen F, Beukema H, Burgess ND, Parish F, Bruhl CA, Donald PF, Murdiyarso D, Phalan B, Reijnders L, Struebig M, Fitzherbert EB: **Biofuel plantations on forested lands: double jeopardy for biodiversity and climate.** *Conserv Biol* 2009, 23:348–358.
6. Phalan B: **The social and environmental impacts of biofuels in Asia: an overview.** *Appl Energy* 2009, 86(Supplement 1):S21–S29.
7. Aerts R, Honnay O: **Forest restoration, biodiversity and ecosystem functioning.** *BMC Ecol* 2011, 11:29.
8. *Roundtable on Sustainable Palm Oil.* http://www.rspo.org/.
9. *Round Table on Responsible Soy Association.* http://www.responsiblesoy.org/.
10. *Roundtable on Sustainable Biomaterials.* http://rsb.org/.
11. Laurance WF, Koh LP, Butler R, Sodhi NS, Bradshaw CJA, Neidel JD, Consunji H, Mateo Vega J: **Improving the performance of the roundtable on sustainable palm oil for nature conservation.** *Conserv Biol* 2010, 24:377–381.

12. FAO: **Forests and Energy**. In *FAO Forestry Paper 154*. Rome, Italy: FAO; 2008.

13. Savilaakso S, Laumonier Y, Guariguata MR, Nasi R: **Does production of oil palm, soybean, or jatropha change biodiversity and ecosystem functions in tropical forests?** *Environmental Evidence* 2013, 2:17.

14. Edwards P, Clarke M, DiGuiseppi C, Pratap S, Roberts I, Wentz R: **Identification of randomized controlled trials in systematic reviews: accuracy and reliability of screening records**. *Stat Med* 2002, 21:1635–1640.

15. Pullin AS, Knight TM: **Support for decision making in conservation practice: an evidence-based approach**. *J Nat Conserv* 2003, 11:83–90.

16. Arnqvist G, Wooster D: **Meta-analysis: synthesizing research findings in ecology and evolution**. *Trends Ecol Evol* 1995, 10:236–240.

17. Rosethal R: **Parametric measures of effect size**. In *The handbook of research synthesis*. Edited by Cooper H, Hedges LV. New York, USA: Russell Sage Foundation; 1994:232–244.

18. Rosenberg MS, Adams DC, Gurevitch J: *MetaWin: Statistical software for meta-analysis. Version 2.0*. Sunderland, Massachusetts: Sinauer Associates; 1999.

19. Borenstein M, Hedges LV, Higgins JPT, Rothstein HR: *Introduction to meta-analysis*. John Wiley & Sons Ltd: United Kingdom; 2009.

20. Cooper H, Hedges LV: *The Handbook of Research Synthesis*. New York, USA: Russell Sage Foundation; 1994.

21. Chung AY, Eggleton P, Speight MR, Hammond PM, Chey VK: **The diversity of beetle assemblages in different habitat types in Sabah, Malaysia**. *Bull Entomol Res* 2000, 90:475–496.

22. Davis ALV, Philips TK: **Effect of deforestation on a southwest Ghana dung beetle assemblage (Coleoptera : Scarabaeidae) at the periphery of Ankasa conservation area**. *Environ Entomol* 2005, 34:1081–1088.

23. Rosenberg MS, Adams DC, Gurevitch J: *MetaWin: Statistical software for meta-analysis. Version 2.1*. Sunderland, Massachusetts: Sinauer Associates; 2007.

24. Nichols E, Larsen T, Spector S, Davis AL, Escobar F, Favila M, Vulinec K: **Global dung beetle response to tropical forest modification and fragmentation: a quantitative literature review and meta-analysis**. *Biol Conserv* 2007, 137:1–19.

25. Inc SPSS: *SPSS Statistics for Windows, Version 17.0*. Chicago: SPSS Inc. 2008.

26. Azhar B, Lindenmayer DB, Wood J, Fischer J, Manning A, McElhinny C, Zakaria M: **The conservation value of oil palm plantation estates, smallholdings and logged peat swamp forest for birds**. *For Ecol Manage* 2011, 262:2306–2315.

27. Chang MS, Hii J, Buttner P, Mansoor F: **Changes in abundance and behaviour of vector mosquitoes induced by land use during the development of an oil palm plantation in Sarawak**. *Trans R Soc Trop Med Hyg* 1997, 91:382–386.

28. Koh LP, Wilcove DS: **Is oil palm agriculture really destroying tropical biodiversity?** *Conserv Lett* 2008, 1:60–64.

29. Dumbrell AJ, Hill JK: **Impacts of selective logging on canopy and ground assemblages of tropical forest butterflies: implications for sampling**. *Biol Conserv* 2005, 125:123–131.

30. Hamer KC, Hill JK, Benedick S, Mustaffa N, Sherratt TN, Maryati M, Chey VK: **Ecology of butterflies in natural and selectively logged forests of northern Borneo: the importance of habitat heterogeneity**. *J Appl Ecol* 2003, 40:150–162.

31. Chey V: **Impacts of Forest Conversion on Biodiversity as Indicated by Moths**. *Malay Nat J* 2006, 57:383–418.

32. Lucey JM, Hill JK: **Spillover of insects from rain forest into adjacent oil palm plantations**. *Biotropica* 2012, 44:368–377.

33. Vaessen T, Verwer C, Demies M, Kaliang H, Van Der Meer PJ: **Comparison of termite assemblages along a landuse gradient on peat areas in Sarawak, Malaysia**. *J Trop For Sci* 2011, 23:196–203.

34. Bernard H, Fjeldså J, Mohamed M: **A case study on the effects of disturbance and conversion of tropical lowland rain forest on the non-volant small mammals in North Borneo: management implications**. *Mammal Study* 2009, 34:85–96.

35. Hassall M, Jones DT, Taiti S, Latipi Z, Sutton SL, Mohammed M: **Biodiversity and abundance of terrestrial isopods along a gradient of disturbance in Sabah, East Malaysia**. *Eur J Soil Biol* 2006, 42(Supplement 1):S197–S207.

36. Brühl CA, Eltz T: **Fuelling the biodiversity crisis: species loss of ground-dwelling forest ants in oil palm plantations in Sabah, Malaysia (Borneo)**. *Biodivers Conserv* 2010, 19:519–529.

37. Sheldon FH, Styring A, Hosner PA: **Bird species richness in a Bornean exotic tree plantation: a long-term perspective**. *Biol Conserv* 2010, 143:399–407.

38. Peh K, Sodhi N, de Jong J, Sekercioglu C, Yap C, Lim S: **Conservation value of degraded habitats for forest birds in southern Peninsular Malaysia**. *Divers Distrib* 2006, 12:572–581.

39. Gillespie GR, Ahmad E, Elahan B, Evans A, Ancrenaz M, Goossens B, Scroggie MP: **Conservation of amphibians in Borneo: relative value of secondary tropical forest and non-forest habitats**. *Biol Conserv* 2012, 152:136–144.

40. Juliani NS, Anuar MSS, Salmi ALN, Munira AN, Liyana KN: **Diversity pattern of bats at two contrastinng habitat types along Kerian River, Perak, Malaysia**. *Trop Life Sci Res* 2011, 22:13–22.

41. Hashim NR, Jusoh WFAW, Nasir MNSM: **Ant diversity in a Peninsular Malaysian mangrove forest and oil palm plantation**. *Asian Myrmecology* 2010, 3:5–8.

42. Fayle TM, Turner EC, Snaddon JL, Chey VK, Chung AYC, Eggleton P, Foster WA: **Oil palm expansion into rain forest greatly reduces ant biodiversity in canopy, epiphytes and leaf-litter**. *Basic Appl Ecol* 2010, 11:337–345.

43. Liow LH, Sodhi NS, Elmqvist T: **Bee diversity along a disturbance gradient in tropical lowland forests of south-east Asia**. *J Appl Ecol* 2001, 38:180–192.

44. Room PM: **Diversity and organization of the ground foraging ant faunas of forest, grassland and tree crops in Papua New Guinea**. *Australian Journal of Zoology* 1975, 23:71–89.

45. Aratrakorn S, Thunhikorn S, Donald PF: **Changes in bird communities following conversion of lowland forest to oil palm and rubber plantations in southern Thailand**. *Bird Conserv Int* 2006, 16:71–82.

46. Danielsen F, Heegaard M: **Impact of logging and plantation development on species diversity: a case study from Sumatra**. In *Management of tropical forests: towards an integrated perspective*. Edited by Sandbukt O. Oslo: Centre for Development and the Environment. University of Oslo; 1995.

47. Edwards DP, Hodgson JA, Hamer KC, Mitchell SL, Ahmad AH, Cornell SJ, Wilcove DS: **Wildlife-friendly oil palm plantations fail to protect biodiversity effectively**. *Conserv Lett* 2010, 3:236–242.

48. Fukuda D, Tisen OB, Momose K, Sakai S: **Bat diversity in the vegetation mosaic around a lowland dipterocarp forest of Borneo**. *Raffles Bull Zool* 2009, 57:213–221.

49. Glor R, Flecker A, Benard M, Power A: **Lizard diversity and agricultural disturbance in a Caribbean forest landscape**. *Biodiversity & Conservation* 2001, 10:711–723.

50. Peh KSH, Jong J, Sodhi NS, Lim SLH, Yap CAM: **Lowland rainforest avifauna and human disturbance: persistence of primary forest birds in selectively logged forests and mixed-rural habitats of southern Peninsular Malaysia**. *Biol Conserv* 2005, 123:489–505.

51. Hamer KC, Hill JK: **Scale-dependent effects of habitat disturbance on species richness in tropical forests**. *Conserv Biol* 2000, 14:1435–1440.

52. Fitzherbert EB, Struebig MJ, Morel A, Danielsen F, Brühl CA, Donald PF, Phalan B: **How will oil palm expansion affect biodiversity?** *Trends Ecol Evol* 2008, 23:538–545.

53. Foster WA, Snaddon JL, Turner EC, Fayle TM, Cockerill TD, Ellwood MD, Broad GR, Chung AY, Eggleton P, Khen CV, Yusah KM: **Establishing the evidence base for maintaining biodiversity and ecosystem function in the oil palm landscapes of South East Asia**. *Philos Trans R Soc Lond B Biol Sci* 2011, 366:3277–3291.

54. Ramage BS, Sheil D, Salim HM, Fletcher C, Mustafa NZ, Luruthusamay JC, Harrison RD, Butod E, Dzulkiply AD, Kassim AR, Potts MD: **Pseudoreplication in tropical forests and the resulting effects on biodiversity conservation**. *Conserv Biol* 2013, 27:364–372.

55. *iSpot*. http://www.ispotnature.org/.

5

What are the socio-economic impacts of genetically modified crops worldwide? A systematic map protocol

Jaqueline Garcia-Yi[1*], Tiptunya Lapikanonth[2], Hanum Vionita[2], Hanh Vu[2], Shuang Yang[2], Yating Zhong[2], Yifei Li[2], Veronika Nagelschneider[2], Birgid Schlindwein[3] and Justus Wesseler[1,4]

Abstract

Background: Genetically modified (GM) crops have generated a great deal of controversy. Since commercially introduced to farmers in 1996, the global area cultivated with GM crops has increased 94-fold. The rapid adoption of GM technology has had substantial socio-economic impacts which a vast amount of technical and non-technical literature has addressed in the last two decades. However, contradictory results between individual studies abound. Extensive and transparent reviews concerning this contentious and complex issue could help promote evidence-based dialogue among the diverse parties involved.

Methods: This protocol specifies the methodology for identifying, evaluating, and mapping evidence related to the main review question: what are the socio-economic impacts of genetically modified crops worldwide? This question has been subdivided into the following topics: (a) farm-level impacts; (b) impacts of coexistence regulations; (c) impacts along the supply chain; (d) consumer-level impacts; (e) impacts on food security; and (f) environmental economic impacts. The search strategy includes the identification of primary studies from general scientific databases; global, regional, and national specialist databases; an on-line search engine; institutional websites; journal websites; subject experts/researchers; and serendipity. Searches will be conducted in six languages (Chinese, English, French, German, Portuguese, and Spanish). Identified studies will be screened for inclusion/exclusion criteria by a group of multi-language reviewers. Finally, pre-defined data from the studies will be extracted, mapped, and presented in a report. Potential research gaps will be identified and discussed, and the review process will be documented in an open-access database (*i.e.* CADIMA, http://www.cadima.info/).

Keywords: Systematic map, Socio-economics impacts, Genetically modified organism, Crop, Chinese, English, French, German, Portuguese, Spanish

Background

Genetically modified (GM) crops have generated a great deal of controversy. The use of biotechnology in agriculture has caused major ideological and scientific concerns that continue to be echoed in the media and academic press [1]. Since commercially introduced to farmers in 1996, the global area cultivated with GM crops has increased 94-fold, from 1.7 million hectares to 160 million hectares in 2011 [2]. The rapid adoption of this technology has had substantial socio-economic impacts [3].

Consequently, a vast amount of technical and non-technical literature addressing this topic has accumulated over the last two decades [4]. Moreover, groups of stakeholders characteristically advocate opposing opinions, which may not be based on best available evidence. Therefore, the availability of transparent and reliable reviews of studies on the socio-economic impacts of GM crops could help promote evidence-based dialogue among the diverse parties involved. Systematic maps employ structured procedures that can be particularly useful for minimizing potential biases that may arise during the process of identification, selection, and analysis of evidence involved in controversial topics. Systematic maps provide an opportunity to gather and describe

* Correspondence: jaqueline.garcia-yi@tum.de
[1]Technische Universitaet Muenchen, Chair of Agricultural and Food Economics, Alte Akademie 12, 85354 Freising, Germany
Full list of author information is available at the end of the article

evidence relevant to a broad field of policy and management relevance[a]. The breadth of the evidence captured in a systematic map helps to clearly identify potential research gaps and guide future research efforts [5]. In addition, systematic maps make relevant evidence readily accessible to researchers and stakeholders through the development of extensive databases, the content of which can be relatively easily updated as needed.

Currently, numerous literature reviews and meta-analysis studies have assessed the socio-economic impacts of GM crops (a non-comprehensive list of 20 studies is included in the Additional file 1). Nevertheless, none is a systematic map, and only one is a systematic review (see Hall et al. [6])[b]. That systematic review focused on the costs and profits of GM agriculture in comparison with conventional agriculture. One shortcoming of the document, as stated by the authors, was the exclusion of studies conducted before 2006, which disregards valuable earlier literature. The authors also clarified that [6]: "Additional time for conducting a systematic review such as this one would allow the inclusion in the search process of additional databases that were excluded because it was not possible to directly export results to Reference Manager Database. An extended review on this topic would be a potentially valuable contribution to the 'GM debate'".

Through the EU project "GMO Risk Assessment and Communication of Evidence" (GRACE, 2012–2015), comprehensive reviews of existing evidence of potential health, environmental, and socio-economic impacts of GM crops worldwide will be conducted [7]. As members of GRACE, the authors of this protocol (Technische Universitaet Muenchen, TUM) will undertake a systematic map on the socio-economic impacts of genetically modified (GM) crops. In particular, the Description of Work (DoW) for GRACE states that TUM is responsible for carrying out reviews on the following key topics: (1) farm-level economic impacts of GM crops; (2) economics of coexistence; (3) economics of segregation at the level of supply-chains; and (4) consumer acceptance of GM crops.[c] GRACE is following a participatory approach, and stakeholders are being consulted during each of the project's steps. The stakeholders include members of industry and civil society organizations, as well as competent authorities on GM crops in the EU Member States and scientific experts from academia[d]. Two new topics were added based on stakeholder requests: environmental economic impacts[e] of GM crops and the impacts of GM crops on food security (for more information about the participatory process, see GRACE [8]). Therefore, TUM will produce a systematic map covering the six topics stated above, the overall conceptual model of which is outlined in Figure 1. The extensive systematic map will address the broad review question: what are the socio-economic impacts of genetically modified crops worldwide?

The systematic map undertaken will provide an important overview of the existing literature related to the socio-economic impacts of GM crops available in six languages (Chinese, English, French, German, Spanish, and Portuguese). These languages are among the top nine used for publication of research[f] [9] and also the primary languages spoken in 23 of the 28 countries currently cultivating GM crops [10].

The description of the topics to be covered in the systematic map is provided below.

Farm-level impacts

Farmers have different socio-economic motivations for adopting GM crops. Significant socio-economic determinants include: gender associated aspects (e.g., [11]); individual and social learning (e.g., [12]); educational level (e.g., [13]); and expected benefits and uncertainty (e.g., [14-16]). For GM adopters, potential changes in yield and economic returns depend on current and previous crops and specific trait characteristics; agricultural practices; incidence of pest infestation; seed costs; and market characteristics (e.g., [17,18]). Farmers' production efficiency (farmers' ability to produce more with less than or equal inputs/resources) would also be affected (e.g., [19]), as well as the frequency of pesticide poisoning incidents and health impacts (e.g., [20]). Consumption of new bio-fortified GM crops are expected to increase farmers' nutrition status and as such, they could significantly contribute to farmers' well-being (e.g., [21]). Most of the world's poor depend mainly on farming for their subsistence. The adoption of GM crops could have different impacts on wealthier and poorer farmers (e.g., [22]), which could exacerbate/mitigate social problems. Ethical aspects may also be affected, as it has been demonstrated that ethical values can change over time (e.g., changing views on euthanasia in the U.S. and Japan [23]). A change in acceptability of GM crops may imply a change in adopters' values. Finally, cultural aspects may be impacted as well; for example, GM seeds need to be purchased, causing a disturbance in the traditional exchange of seeds among indigenous farmers (along with potential changes in identity and trust among involved farmers).

The main aspects considered within this topic are presented graphically in a conceptual model (Figure 2). This conceptual model shows that socio-economic factors influence farmer decisions regarding the adoption of GM crops. GM adoption is expected to impact aspects related to farmers' income and also intangible aspects. The potential income-related impacts include changes in the use of inputs; associated costs; output (quantity and quality); and gross income. Some farmers could experience changes in time available for conducting off-farm income-generating activities. A farm's efficiency could

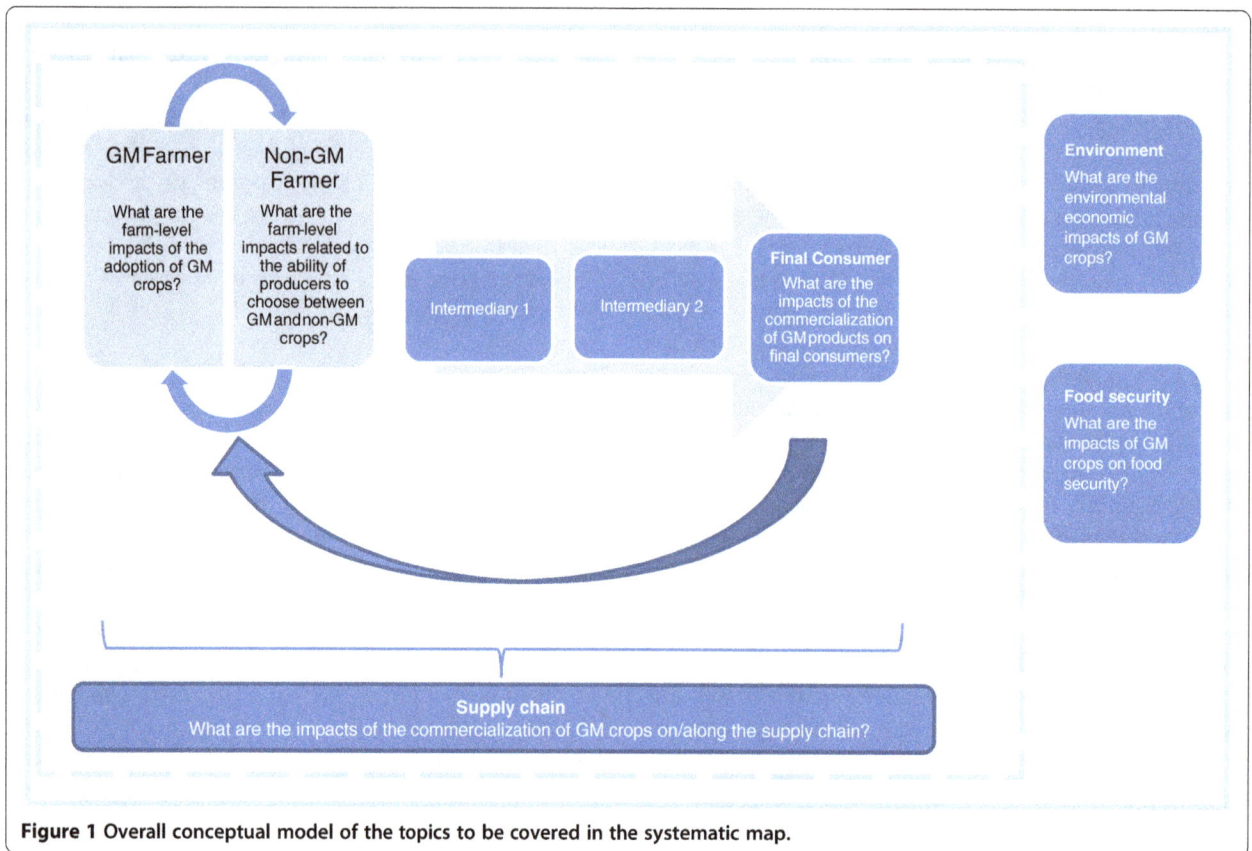

Figure 1 Overall conceptual model of the topics to be covered in the systematic map.

deteriorate or improve with use of new technologies impacting the farmer's income. Intangible aspects that may be affected after GM adoption relate to health safety issues associated with changes in pesticide use and farmers' nutritional status if they cultivate and consume bio-fortified crops. Primary social, ethical, and cultural aspects are also depicted in the conceptual model.

Coexistence related impacts

The possibility that GM farms contaminate non-GM farms via unintentional or inadvertent gene flow constitutes a challenge for the coexistence of GM farming and conventional agriculture, including organic certified agricultural systems. Several studies have analysed the effects that the introduction of ex-ante regulatory and ex-post liability aspects would have on farm-level costs and GM spatial configuration and adoption dynamics (e.g., [24-26]). In addition, potential benefits due to higher price premiums for non-GM products have also been evaluated (e.g., [27]).

The main aspects considered within this topic are presented graphically in a conceptual model (Figure 3). This conceptual model shows that GM plants and crops can be introduced under alternative coexistence systems (separation between GM and non-GM farms and dual GM/non-GM farms) and regulatory frameworks, including ex-ante (e.g., mandatory segregation, traceability,

minimum GM tolerance levels, rigid and flexible refuge areas, and voluntary GM-free zones) and ex-post liability aspects (e.g., compensation funds, insurance schemes, and marketplace liability). The different coexistence options are expected to influence in different manners GM and non-GM farm-level costs, particularly operational; transaction; opportunity; and testing and remediation costs. GM adoption dynamics could change as well, such as the rate of adoption, spatial configuration, and speed and stability of GM expansion. GM-farmers would also generate externalities and directly influence the economic benefits of non-GM farmers due to inadvertent gene flow from GM to non-GM fields which may create problems for non-GM farmers willing to sell their products in specific markets (e.g., organic certified markets). Finally, social factors, such as the level of trust between neighbors, would influence farm-level costs (e.g., lower/ higher negotiation costs) and adoption dynamics of GM crops (e.g., stronger/lower imitation or neighboring effects) in each of the ex-ante and ex-post regulatory regimes under evaluation (social aspects not pictured in the figure).

Supply chain impacts

The focus of this section is on the supply chain or organization network as unit of analysis. It aims to analyse

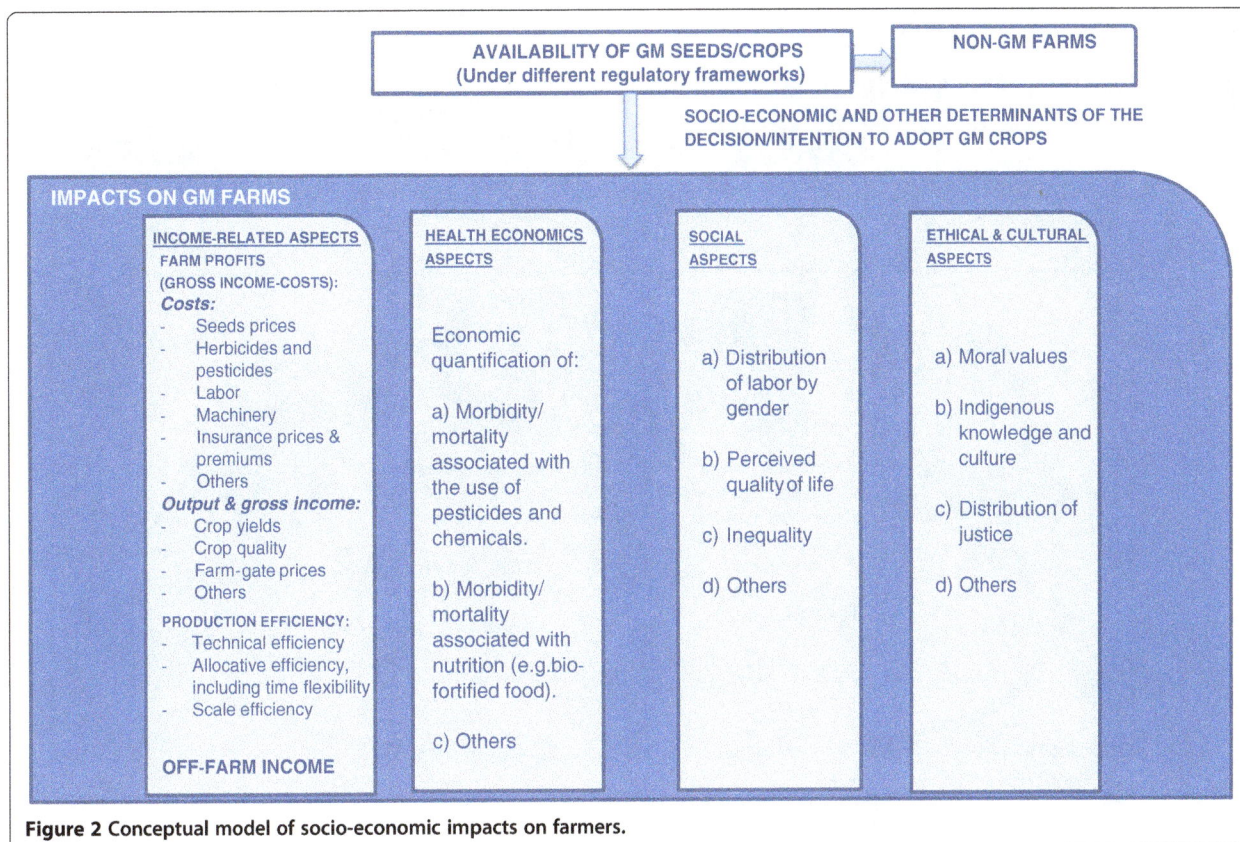

Figure 2 Conceptual model of socio-economic impacts on farmers.

the socio-economic impacts of the commercialization of GM crops on supply chain structure and performance dynamics, as well as cost and benefit distribution along different actors in the supply chain.

In general, the basic elements of the structure of the supply chain include:

(a) *Vertical relations.* These refer to the sequence of value adding activities. Actors performing different functions within the supply chain are vertically linked through buying and selling relationships. Vertical relations highlight the level of cooperation, coordination, trust, and governance (or power) along the chain.
(b) *Horizontal relations.* These reflect the relationships among actors performing the same function within the chain. Horizontal relations can be formal (e.g., cooperatives and associations) or informal.

The main factors related to supply chain performance are:

(a) *Efficiency* or the ability to deliver value at a minimum of total costs.
(b) *Effectiveness* or the ability of the chain to provide superior value.

(c) *Innovation* or the ability to respond to changes in consumer demand or the external environment.

Several studies have analysed the effect that the commercialization of GM crops would have on the supply chain structure, as well as the distribution of costs and benefits of different actors along the supply chain (e.g., [28-32]). Moreover, governance mechanisms and market power of different actors would also be affected (e.g., [33,34]). The main aspects considered under this topic are presented graphically in a conceptual model (Figure 4). This conceptual model shows that the commercialization of GM products under different enforced coexistence rules, labeling schemes, and protection of intellectual property rights would have impacts on the supply chain structure (e.g., vertical and horizontal relations) and performance (e.g., efficiency, effectiveness, and innovation ability). This in turn would affect the distribution of costs and benefits for the different actors along the supply chain, as well as their market power (ability to influence the price of a commercialized item).

Consumer-level impacts
The socio-economic determinants for consumers' acceptance of GM food and the associated price premiums for non-GM products have been evaluated under different

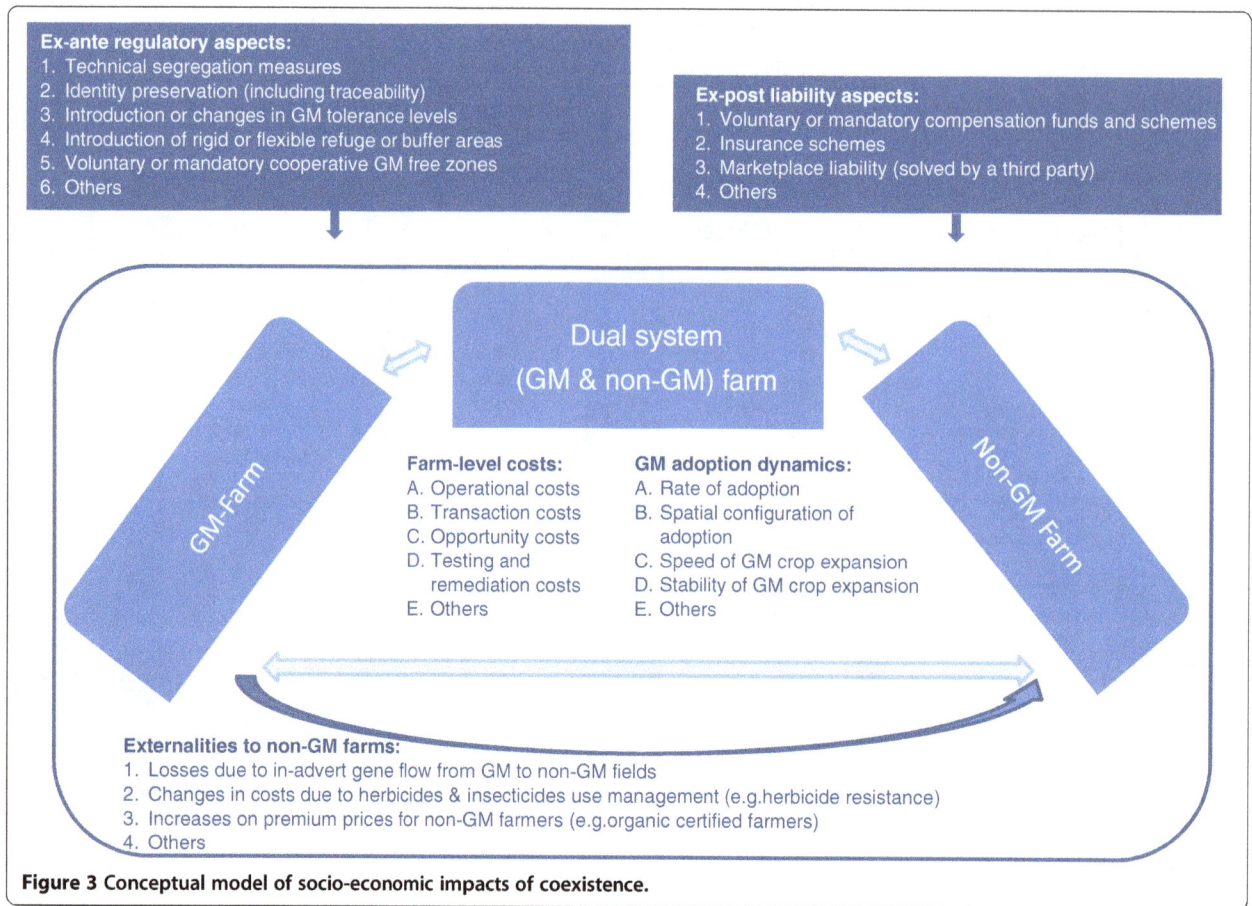

Figure 3 Conceptual model of socio-economic impacts of coexistence.

mandatory and voluntary GM-related label schemes (e.g., [35-37]). Other studies have evaluated the option values of a moratorium or ban on GM products (e.g., [38]). Those price premiums and option values have been used to calculate economic welfare effects (e.g., [39]). These and other main aspects related to the impacts of GM products on consumers are presented graphically in a conceptual model (Figure 5). The conceptual model shows that GM products can be introduced into the market under mandatory and voluntary GM-related labels, including different tolerance levels (or percentage of GM ingredients in the final products) or can be subject to moratorium or ban. The decision or intention to buy those products is based on consumers' socio-economic characteristics (e.g., age, gender, and educational level). Potential buyers can indicate their willingness to pay (WTP) for these products, and changes in social welfare can be calculated based on the differences between the WTP and actual or expected prices (price premiums). If there is a moratorium or ban on GM products, option values can be calculated based on a (hypothetical) WTP to preserve or maintain this situation. Social welfare can be estimated by the difference between the WTP and the opportunity costs of forgoing economic growth associated with the commercialization

of GM products. GM products can have an impact on consumers' health, for example in the case of bio-fortified food. Social, ethical, and cultural aspects were added as requested by stakeholders.

Environmental economic impacts of GM crops

GM crops may substitute for agricultural inputs and practices that are environmentally harmful. The study by Brookes & Barfoot [40] suggest that "since 1996 the use of pesticides (counted as active ingredients) on the GM crop area was reduced by 448 million kg (9% reduction), and the environmental impact quotient — an indicator measuring the environmental impact associated with herbicide and insecticide use on these crops — fell by 17.9%. In 2010, the total carbon dioxide emission savings associated with GM crop adoption were equal to the removal from the roads of 8.6 million cars due to reduced fuel use and additional soil carbon sequestration".

GM crops can cause environmental harm as well (although there is considerable uncertainty and no consensus among scientists) [41]. In particular, the protection of biodiversity and ecosystem services ought to be a top priority when taking into consideration the dependency on a healthy environment of all human activity, now and

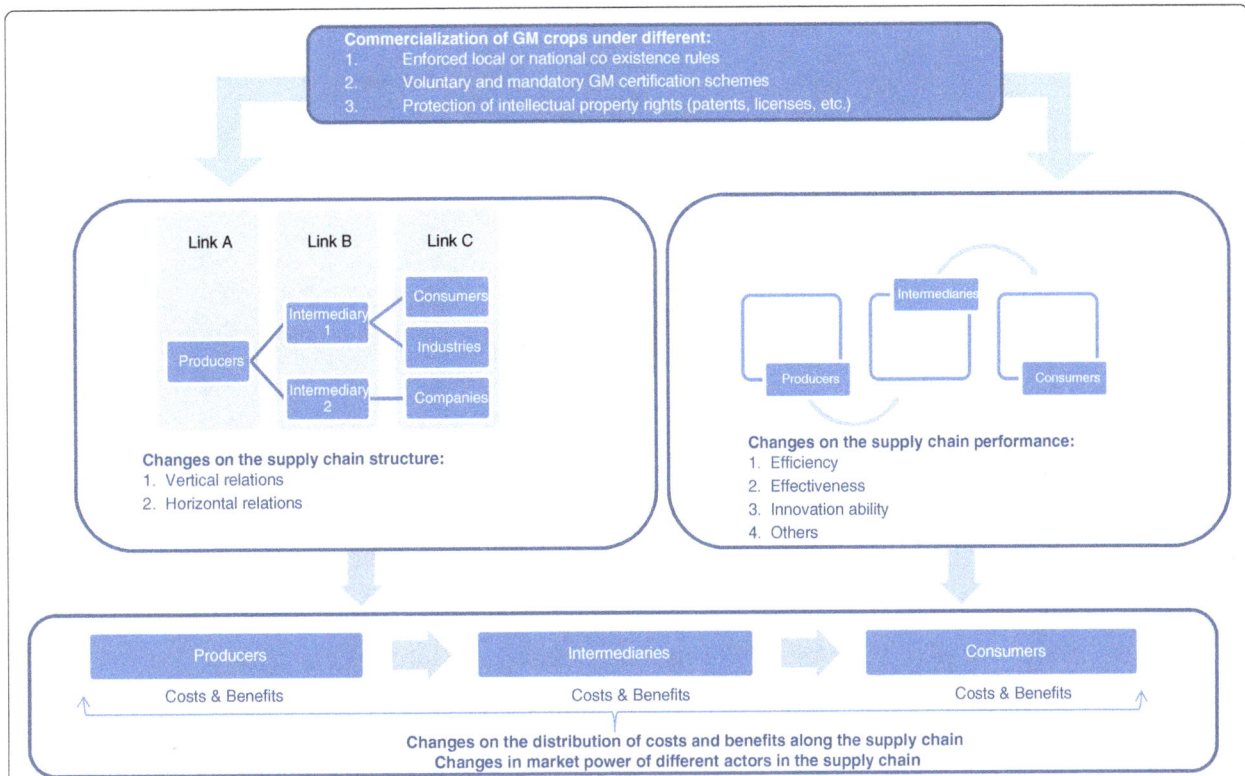

Figure 4 Conceptual model of socio-economic impacts along the supply chain.

in the future [42]. For those opposed to GM technology, GM crops are exotic species being introduced into open complex ecosystems of which we have limited understanding [43], and as such it is impossible to anticipate all impacts of GM technology on the environment.

The effects of GM crop adoption on the environment will depend not only on human behavior but on biological, ecological, and chemical interactions as well. Many disciplines are needed to evaluate these kinds of

impacts [41]. In addition, there is the possibility of irreversible ecosystem disruptions due in part to the unpredictable and novel effects of gene mixing [43].

Figure 6 shows a basic conceptual model of the potential environmental economic impacts of GM crops (based on information obtained from [40-46]). Depending on the type of genetic modification, the cultivation of GM crops can change the type or quantity of herbicide/insecticide used, improve the crops' resistance to external climate

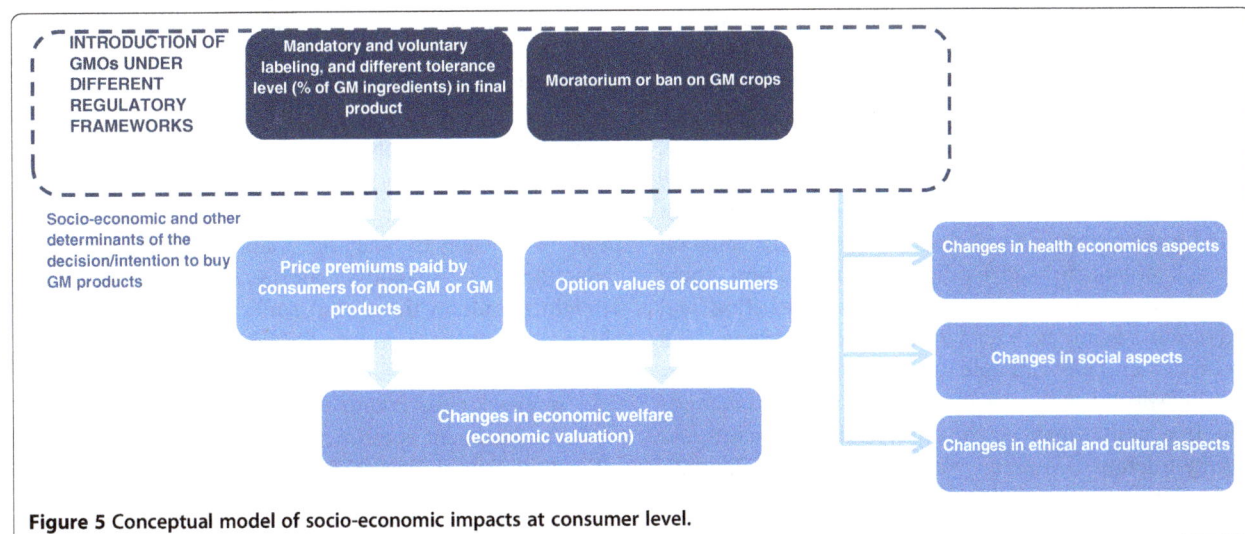

Figure 5 Conceptual model of socio-economic impacts at consumer level.

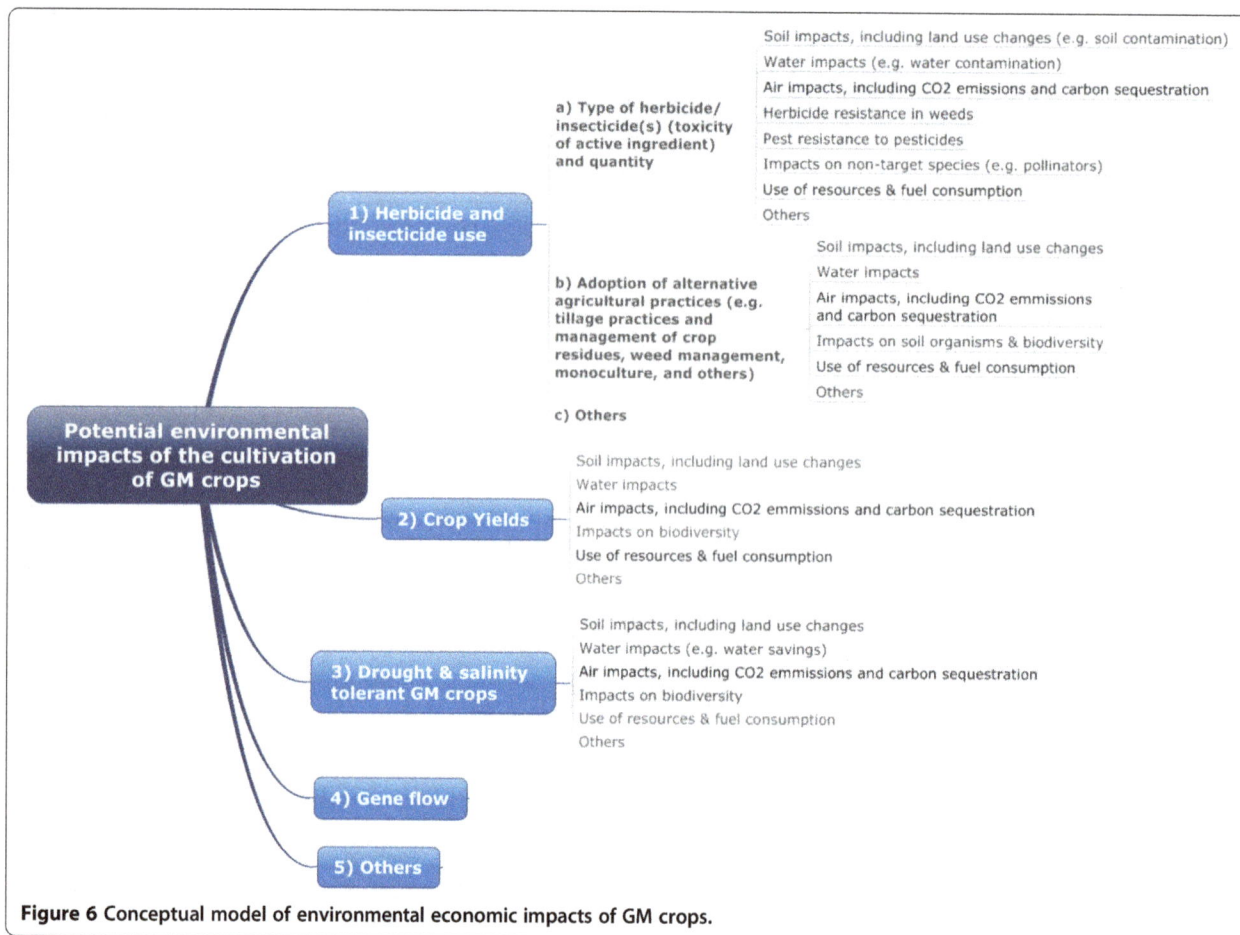

Figure 6 Conceptual model of environmental economic impacts of GM crops.

stress (e.g., drought and salinization), or cause an undesired gene flow (e.g., from GM crops to wild relatives).

Changes in the type or quantity of herbicide/insecticide could create or alter herbicide resistance in weeds or pesticide resistance in pests. Soil, water, and air contamination is reduced if the substituted herbicide/pesticide was more toxic than the new herbicide/pesticide. Further, if less herbicide/pesticide is required, resources like fuel could be saved. Changes in herbicide/insecticide use could also modify agricultural practices, such as encourage tillage, weed management, or monoculture. New alternative agricultural practices could change the use of resources and fuel consumption, which in turn would have impacts on soil, water, and air contamination and soil organisms and biodiversity. In addition, there could be improvements in crop yields using existing land and water resources, which in turn could reduce land use; water and air contamination; minimize the impacts on biodiversity; and save resources and fuel consumption. In a similar manner, the cultivation of drought- and salinity-tolerant GM crops would also impact soil, water, air, biodiversity, and modify the use of resources and fuel consumption. Finally, there could be a gene flow from

GM crops to wild relatives with unknown consequences to the environment.

It is worth mentioning that this protocol contemplates the environmental economic impacts of GM crops. Therefore, only primary studies incorporating an economic assessment of these and similar environmental impacts will be considered. The environmental impact assessment component of the included primary studies will be taken as given.

Food security at household level

The estimated number of undernourished people has continued to decrease, but the rate of progress still appears insufficient to reach international goals for hunger reduction [47]. Currently, about 842 million people (one in eight people in the world) suffer chronic hunger, unable to obtain the amount of food necessary to conduct an active life [47]. The vast majority of hungry people live in developing countries, where the prevalence of undernourishment is estimated at 14 percent [47].

Food security exists when all people, at all times, have physical and economic access to sufficient, safe, and nutritious food that meets dietary needs and food preferences for an active and healthy life [47]. There are four dimensions of food security: food availability (e.g., food production and processing); food access (e.g., having the economic resources to buy the right food); food utilization (e.g., education to individuals to make proper use of healthy food); and food system stability (e.g., adequate access to food at all times). For food security objectives to be realised, all four dimensions must be fulfilled simultaneously [47,48].

Therefore, food security is a multidimensional concept, and data on all dimensions are rarely available and frequently unreliable [49]. Moreover, the international community lacks a consensus on core household food security indicators needed in order to properly measure and monitor food security worldwide. The indicators also vary on level of analysis, ranging from the regional or national level to the household or individual level, depending on data availability and the design of the instruments used to collect the data (e.g., surveys) [49].

In relation to GM crops, reports from expert governmental and nongovernmental bodies increasingly include GM crops as part of a wider approach to food security [50]. GM crops could help to mitigate expected food shortages related to population growth and the effects of climate change in specific regions worldwide. For example, GM crops could impact *food availability* by providing seeds which are resistant to adverse climate conditions; have an effect on *food access* by increasing farmers' incomes; and, under the same *food utilization* conditions, bio-fortified crops could increase the nutritional status of households worldwide. (Figure 7 illustrates this example).

In the approach followed in this protocol, the ultimate goal of food security is to improve the nutritional status of households. It is worth mentioning that several of the multidimensional aspects of food security have been already covered by other topics in this protocol (e.g., impacts of GM crops on farm-level income). Nevertheless, there are a growing number of socio-economic studies which specifically evaluate the impacts of GM crops on (at least one component of) food security and explicitly indicate that as so.

Objective of the systematic map

The main objective of the systematic map is to identify the breadth of knowledge related to the socio-economic impacts of GM crops worldwide. Our question related to the overall objective of the systematic map is:

What research evidence exists (number of studies and the current state of research studies) on the socio-economic impacts of GM crops worldwide (in Chinese, English, French, German, Spanish, and Portuguese languages)?

Relating to the secondary objectives, the systematic map will identify the types of socio-economic impacts; populations; crops and GM traits; geographical focus; research methodologies; evidence gaps; and the particular topics that could be subjects of further analyses or subsequent systematic reviews. Therefore, the questions related to the secondary objectives of the systematic map are:

a) What types of socio-economic impacts have been addressed?
b) What types of populations have been addressed?
c) What types of crops and GM traits have been addressed?
d) What is the geographical focus of the evidence?
e) What research methods have been used to collect and analyse the evidence?
f) What evidence gaps exist that could/ should be addressed in future primary research?

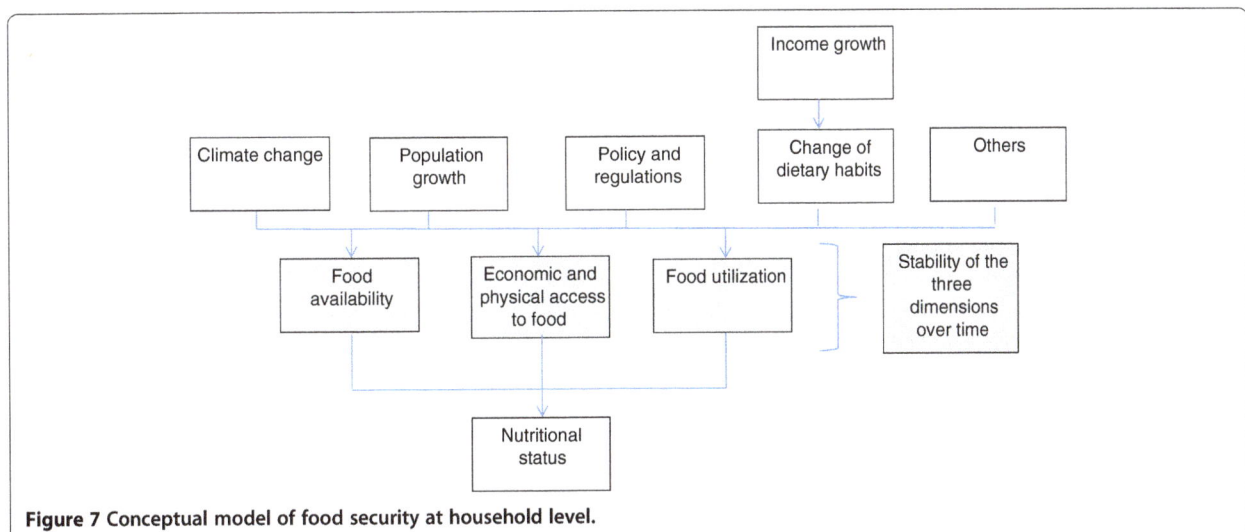

Figure 7 Conceptual model of food security at household level.

g) Which particular topics could be the subject of further analysis (e.g., meta-analyses and meta-regressions) or subsequent systematic reviews?

Methods
Search strategy
Systematic maps require an objective and reproducible search of a range of sources to identify as many relevant studies as possible (within resource and time limits). A search strategy that includes extensive search terms and a combination of multiple data sources can increase the likelihood of capturing most of the relevant references. Our search terms consider a list of intervention-, outcome- and population-related keywords in six languages. Our data sources include: general scientific databases; an on-line search engine (Google Scholar); global, regional, and national specialist databases; institutional websites (to be accessed through one of the largest institutional repository search engines: Bielefeld Academic Search Engine, BASE); journal websites; subject experts/researchers; and serendipity (e.g., finding relevant documents by accidental discovery or by chance). The reference management software to be used for exporting/importing the references is Citavi, which is freely available at TUM.

We aim to identify as many of the available relevant studies as possible (based on time and budget constraints). Sensitivity will be favored over specificity. Sensitivity implies that the emphasis of the search procedure will be in obtaining most of the relevant articles at the risk of obtaining a high number of non-relevant ones (which would need to be depurated later during the screening stage). On the other hand, specificity emphasizes the retrieval of relevant articles with the lowest number of non-relevant ones as possible (at the risk of omitting some/many of the relevant articles).

Search terms
Search terms related to the intervention The selected search terms related to the intervention (GM crops) in the English language are presented in Table 1. These search terms were derived from a preliminary list of 29 GM crop related terms compiled by experts from the GRACE project (see Additional file 2). IDEAS/REPEC, the largest freely-available bibliographic database dedicated to economics, was used to test this preliminary list. Search terms which did not retrieve relevant references (e.g., cisgenesis) or which retrieved similar references as other search terms (e.g., *glufosinate tolerant* did not retrieve additional relevant references in comparison to *herbicide tolerant*) were dropped from the list. The searches were conducted on title, abstract, and keywords. Then AGROVOC, the corporate thesaurus of the Food and Agriculture Organization (FAO), was searched

for controlled terms[g] which were also included as search terms (e.g., biosafety, biosecurity). Finally the titles from the reference lists of the reviews and meta-analysis included in the Additional file 1 were visually examined to evaluate the completeness of our selected search terms, and new search terms were added when needed (e.g., Bollgard, drought resistant).

Additional searches using the selected terms were conducted in Web of Science (All databases) in Topic (Title, abstract, keywords). Some of the terms retrieved high number of records (e.g., bt retrieves 78,027 references; and 3,310 when restricted by research domains: social sciences and arts humanities), most of them were not related to GM crops. Therefore the final intervention terms include a descriptor (e.g., bt crop*) or the type of crop (e.g., bt cotton). The types of crops were compiled from the list of approved crops by GM trait reported in the "GM approval database" by the International Service for the Acquisition of Agri-biotech Applications (ISAAA, http://www.isaaa.org/gmapprovaldatabase/default.asp). In this way, we obtained a reasonable number of references. For example, for bt as a group ("bt crop*" or "bt seed*" or "bt cotton" or "bt maize" or "bt corn" or "bt soybean" or "bt tomato" or "bt eggplant" or "bt rice" or "bt potato"), we obtained 300 references (restricted by research domains: social sciences and art humanities). Importantly, most of those references were relevant.

In the case of non-English languages, the preliminary list of GM related terms was translated to Chinese, French, German, Spanish, and Portuguese, and each term was then tested in Google Scholar, which allows for the retrieval of a comprehensive number of references in each of the non-English languages considered. Before conducting the independent test runs per keyword and language, Google Scholar was set up for retrieving results in the corresponding language, and the search history personalization (customization of results based on previous search activity) was deactivated Searches were conducted in full-text. Terms which retrieved relevant studies among the top 10 percent of the records obtained per search term (ordered by relevance) were selected as search terms, and the remaining were dropped. AGROVOC was also searched for controlled terms in those languages, but additional relevant terms were not found. In addition, reviews of the socio-economic impacts of GM crops using literature in the languages selected in this protocol were not found; however, a visual evaluation of the reference list of some relevant articles in those languages suggested that the selected terms are adequate. The list of selected search terms in non-English languages is included in the Additional file 2.

Search terms related to the outcome and population
A comprehensive list of 380 terms related to the outcome and population was compiled in the English language and

Table 1 List of search terms in English language

Intervention (GM crops)	Outcome and population
1) BT	1) Econ*
bt crop*	2) Socio*
bt seed*	3) Social*
bt cotton	4) Cost*
bt maize	5) Benefit*
bt corn	6) Regulat*
bt soybean	7) Farmer*
bt tomato	8) Consumer*
bt eggplant	9) Supply chain
bt rice	10) Coexist*
bt potato	11) Food security
2) Insect* resistan*	
insect* resistan* crop*	
insect* resistan* seed*	
insect* resistan* cotton	
insect* resistan* maize	
insect* resistan* corn	
insect* resistan* soybean	
insect* resistan* tomato	
insect* resistan* eggplant	
insect* resistan* rice	
insect* resistan* potato	
3) Biotech*	
Agricult* biotech*	
Biotech* crop*	
Biotech* food*	
Biotech* seed*	
4) Bioeng*	
Bioeng* crop*	
Bioeng* food*	
Bioeng* seed*	
5) GM	
GM* crop*	
GM* food*	
GM* seed*	
GM* free	
GM* label*	
GM* product*	
6) GMO*	
7) Transgenic*	
Transgenic* crop*	
Transgenic* food*	
Transgenic* seed*	

Table 1 List of search terms in English language
(Continued)

8) Genetic* engineer*
Genetic* engineer* agricult*
Genetic* engineer* crop*
Genetic* engineer* food*
Genetic* engineer* seed*
9) Genetic* modif*
Genetic* modif* agricult*
Genetic* modif* crop*
Genetic* modif* food*
Genetic* modif* seed*
10) HT
HT crop*
HT seed*
HT alfalfa
HT canola
HT rapeseed
HT chicory
HT cotton
HT flax*
HT maize
HT corn
HT soybean
HT sugar beet
HT rice
HT potato
HT wheat
11) Herbicide* resistan*
herbicide* resistan* crop*
herbicide* resistan* seed*
herbicide* resistan* alfalfa
herbicide* resistan* canola
herbicide* resistan* rapeseed
herbicide* resistan* chicory
herbicide* resistan* cotton
herbicide* resistan* flax*
herbicide* resistan* maize
herbicide* resistan* corn
herbicide* resistan* soybean
herbicide* resistan* sugar beet
herbicide* resistan* rice
herbicide* resistan* potato
herbicide* resistan* wheat

Table 1 List of search terms in English language
(Continued)

12) Virus resistan*

virus resistan* crop*

virus resistan* seed*

virus resistan* bean

virus resistan* papaya

virus resistan* squash

virus resistan* sweet pepper

virus resistan* plum

virus resistan* potato

13) Drought resistan*

drought resistan* crop*

drought resistan* seed*

drought resistan* corn

drought resistan* maize

drought resistan* sugarcane

14) Biofortif*

Biofortif* crop*

Biofortif* food*

Biofortif* seed*

Biofortif* cassava

Biofortif* corn

Biofortif* maize

Biofortif* rice

Biofortif* sorghum

15) Biosafe*

Biosafe* agricult*

Biosafe* crop*

Biosafe* food*

Biosafe* seed*

16) Biosecur*

Biosecur* agricult*

Biosecur* crop*

Biosecur* food*

Biosecur* seed*

17) Roundup ready*

18) Liberty link*

19) Starlink*

20) Bollgard*

21) Golden rice

*Indicates a truncation/wild card symbol (i.e. any character(s) permitted).

translated into the non-English languages considered in this protocol. Then, we selected the outcome and population terms that retrieved the largest number of relevant references, based on visual inspections from the results of

IDEAS/REPEC for English searches. The descriptors per topic considered in this protocol (farmers, consumers, supply chain, coexistence, environmental economics, and food security) were also included as additional search terms. The selected outcome and population terms for searches in the English language are included in Table 1. Note that outcome and population terms will be joined by OR while conducting the actual searches. Thus, "environmental economics" is not directly included in the list as a population term since econ*, a more general term, is already in the list of search terms. Any potential results of searches for "environmental economics" will be already captured in the results of the searches for econ*.

The original list of the 380 terms and the corresponding translations in non-English languages are included in the Additional file 3.

Database searches

Database selection Little evidence exists to guide prioritization of databases for reviewers [51]. Our criteria for database selection considered the following aspects:

a) subject area (socio-economics);
b) geographic coverage of the studies (e.g., databases covering developing countries); and
c) inclusion of primary studies.

We gave preference to databases that allow for directly exporting batch results to the reference software, can retrieve full text documents, or provide links to access those studies. However, relevant databases in non-English languages do not provide those facilities (e.g., a number of the Spanish-language databases), and therefore we will not strictly restrict our searches to "user-friendly" databases.

We excluded databases that do not correspond to our subject areas (e.g., biochemistry); do not focus on primary studies (e.g., newspapers); are redundant (included in other databases or platforms; e.g., BIOSIS Previews and Current Contents Connect which are included under TUM's subscription to Web of Science); require additional payment; or databases that are unavailable or inaccessible at the present time (e.g., databases undergoing major restructuring).

The list and a general description of the selected databases, platforms, and search engines for searches in the English language are provided in Table 2, while the selected databases for searches in non-English languages are included in the Additional file 4.

It is expected that the selected databases will offer good coverage of the literature available in each of the six languages included in this protocol. An overview of the content of all the selected databases, platforms, and

Table 2 Selected databases, platforms, and search engines for searches in English language

General scientific databases and platforms	Web of Science (WoS) (Thomson Reuters) (includes Web of Science Core Collection, BIOSIS Citation Index, BIOSIS Previews, Current Contents Connect, Derwent Innovations Index, Inspec, MEDLINE, and SciELO Citation Index)
	Scopus (Elsevier)[a]
On-line search engine	Google Scholar (GS)[b]
Organizations with focus on developing countries	British Library for Development Studies (includes African and Indian journals)
	ELDIS (information service related with international development issues)
International organizations	AGRIS (maintained by the Food and Agriculture Organization, FAO)
	IFPRI (International Food Policy Research Institute)
	JOLIS (World Bank, International Monetary Fund, IMF & International Finance Corporation, IFC)
	OECD iLibrary (Organization for Economic Co-operation)
Other organizations/institutional repositories	AGRICOLA (US National Agricultural Library)
	IDEAS/ REPEC (Largest bibliographic database dedicated to economics freely available. It contains bibliographic information from other open source databases such as AgEcon)
	Bielefeld Academic Search Engine (BASE)[c]
Grey literature	Open Grey (system for information on grey literature in Europe)

[a]Scopus is a multidisciplinary database, which along with WoS, is considered the most complete and widely used for scientific information identification and retrieval [52].
[b]Google scholar retrieves peer-review and non peer-review publications (grey literature). According to Gehanno *et al.* [53], one of the advantages of using Google scholar is that it identifies more types of literature compared to a general scientific database. The results of a study conducted by the same authors suggest that the current coverage of Google scholar allows retrieving all the high quality studies identified by other general scientific databases such as WoS, and "could be the first choice for systematic reviews or meta-analysis" [53]. On the negative side, Google scholar is "constantly-changing content, algorithms and database structure" and Google does not provide details about Google scholar's database coverage [54]. The results of the searches will be ordered by relevance and the first 1000 documents will be imported to Citavi. The reason for this is that Google scholar limits the retrieval of search results to 1000 documents for any particular search query.
[c]BASE is one of the largest institutional repository search engines [55], which allows access to 2762 content sources, such as the National Library of Australia, Institutional Repository of PhD theses from Katholieke Univ. Leuven (Belgium), EMBRAPA (Brazil), University of Saskatchewan (Canada), Peking University Institutional Repository (China), among others. The full list of sources is available at: http://www.base-search.net/about/en/about_sources_date_dn.php?menu=2.

search engines (in English and non-English) is provided in the Additional file 5.

Overall search procedure for databases Our overall search strategy for searches in the English language considers the following:

a) searches will be conducted on "title, abstract or keyword", when this option is available. If the database does not offer this facility, searches will be conducted in the common default option "all fields". Nevertheless, searches in Google Scholar, BASE, and Agricola will be conducted in "title" due to the large number of references obtainable;

b) searches will be limited to the time period from 1996 (the year GM crops were commercially introduced to farmers) to present;

c) searches will be filtered by type of document (article, chapter, book, thesis, manuscript, and conference paper) and socio-economic subjects or disciplines (if the database provides these facilities);

d) searches will be conducted using only the intervention terms in socio-economic related databases (IDEAS/REPEC, British Library for Development, IFPRI, JOLIS, and OECD iLibrary) and in databases providing socio-economic filters

(Web of Science, Scopus, ELDIS, Agricola, BASE, and Open Grey). There is no need to combine the intervention terms with the outcome and population terms in these databases. Without doing so, they provide a manageable number of results primarily related to the socio-economic issues of GM crops;

e) intervention terms will be combined with the outcome and population terms in searches conducted in databases not related to socio-economic issues or without socio-economic filters (Google Scholar and AGRIS) in order to avoid obtaining a large number of irrelevant results.

The procedure for searches in non-English languages is described in the Additional file 6.

Estimating the comprehensiveness of the search Specific search strategies must be constructed for each database indicated above. Some databases allow truncation, stemming, and searches with strings, while others only partially or do not allow doing so. When a database does not allow truncation or stemming, different words endings (or suffixes) need to be used for conducting the searches (for example, with Google Scholar). During our scoping exercise, we identified the total number of records obtained from searches in each of the databases included in our protocol. The results for the searches in English are reported in Table 3. The details of the searches per database in English, and the results of the searches for non-English languages are included in the Additional file 7.

The results suggest that it is possible to follow the proposed search strategy. The total number of references (without eliminating duplicates) from all the databases in the English language is 38,781. Based on visual inspections, we roughly estimate that about one third will be duplicates. (Many searches will be conducted per keyword and then joined, which will generate many duplicates in the final list of references). In the end, we expect to manually screen about 26,000 references from searches in databases in the English language.

Validation of search strategies Following Hausner et al., we will validate our search strategy by checking if the relevant references from the reviews and meta-analyses included in the Additional file 1 are among our included studies. In the event that the references were not retrieved, we will refine the search strategies until we are able to retrieve those references.

Searches in journals

No database is capable of exhaustively monitoring all existing journals. We retrieved a list of journals, which contained articles related to the socio-economic impacts of GM crops based on the results of a search conducted in IDEAS/REPEC using the search terms "genetically modified" and "transgenic" in 2012. It is expected that some of these journals are already indexed in one or more of the databases considered in this protocol, especially the high ranked journals. (For example, Scopus indexes the journals published by Elsevier and other selected journals based on their quality. See the full list of 33,635 journals and conferences at http://www.elsevier.com/online-tools/scopus/content-overview). We will check if the journals in our list mentioned above are already fully indexed in our selected databases, and those which are not indexed or partially indexed (e.g., no 'cover to cover' or some years missing) will be searched manually. The complete list of (93) journals (in English) is included in the Additional file 8.

Subject expert consultation

When eligible and appropriate, including completed yet unpublished studies in a systematic map helps to minimize bias [56]. Also requesting and obtaining lists of publications from experts could allow us to verify if we have retrieved all the relevant information and could help us to fill potential information gaps in our data collection. Finally, it is also important to identify ongoing studies/research related to the socio-economic impacts of GM crops worldwide. Therefore, we expect to conduct an online survey requesting experts for their published and unpublished studies; and for information about previous and current projects, which are related to socio-economic impacts of GM crops. The draft version of the questionnaire is included in the Additional file 9. The list of experts will be primarily compiled using lists of authors given in the included studies and information gained from worldwide economic organizations about authors in this field (see http://edirc.repec.org/alphabet.html).

Study inclusion criteria

Our inclusion criteria specify the types of populations, interventions, comparators, outcomes, and study designs, to be addressed in the systematic map. The identified studies will be screened against these criteria in order to be included in the systematic map. Our screening process will be conducted stepwise (see Figure 8). First, the studies will be screened against the inclusion criteria by title (and abstract when available). The studies which do not fulfill the inclusion criteria will be excluded. In case of doubt, the study will be retained for further evaluation. Second, we will review the abstract and full text of the articles, and in a similar way, the studies that do not fulfill the general inclusion criteria will be excluded. In case of doubt, the study will be retained for further analysis.

The general inclusion criteria are the following:

- Relevant Population: Global human civilizations and their economies

Table 3 Number of references identified per database in English language

Database	Number of references (with duplicates)
Web of Science	4,110
Scopus	3,434
Google Scholar	8,763
British Library for Development Studies	850
ELDIS	1,551
AGRIS	3,720
IFPRI	2,125
JOLIS	440
OECD iLibrary	339
AGRICOLA	261
IDEAS/REPEC	4,932
BASE	8,050
Open Grey	206
TOTAL	38,781

Details of the searches can be found in the Additional file 7.

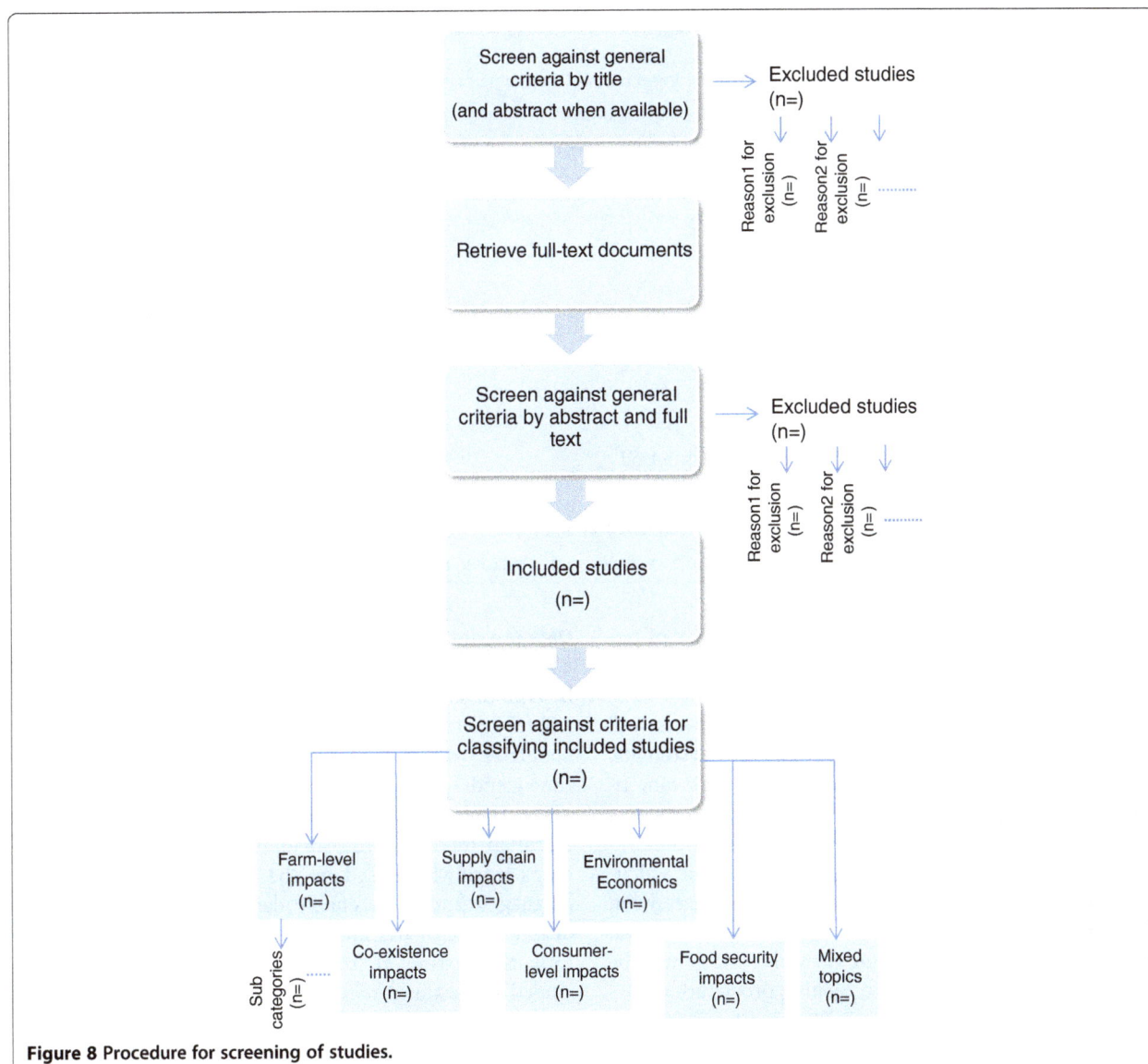

Figure 8 Procedure for screening of studies.

– Relevant Intervention: availability/adoption/ commercialization of any type of crop (e.g., maize, soybean, cotton, canola) with any type of genetic modification (e.g., herbicide resistance, insect resistance)

– Relevant Comparator: situation before the availability/ adoption/commercialization or without the intervention for a comparable group of population[h]

– Relevant Outcome: economic quantification or social analysis of the effects of the intervention

– Relevant Study design: primary study (survey/ interview, observational/ethnographic, model or experiment)

The exclusion criteria consider studies which are not related to GM crops (e.g., animals and microorganisms); do not include a comparator (which allow for an impact assessment); are in a language not considered in this protocol; are published before 1996; are not primary studies (e.g., newspapers, editorials, opinions, literature reviews); or are not accessible.

The included evidence will be coded using criteria for classifying the included studies. The elements of these criteria are related to the topics considered in this protocol and are summarized in Table 4 below.

It is important to mention that we will make our best effort to retrieve the full text of all potential relevant studies after the first screening, given time and budget constraints. The maximum amount of time we are considering for finishing collecting all the data for the systematic map is six months, which includes the time we will dedicate to contacting the authors of missing references.[i] In case we are not able find the full texts of potential relevant references, these references will be

Table 4 Criteria for classifying included studies

Group	Characteristics
Farm-level	The study mainly focuses on the impacts of GM crops at farm-level without considering co-existence issues (see below).
Co-existence	The study mainly focuses on the impacts of co-existence regulations
Supply chain	The study mainly focuses on the impacts of GM crops on/ along the supply chain
Consumer level	The study mainly focuses on the impacts of GM crops on consumers.
Environmental economics	The study mainly focuses on the impacts on environmental economics aspects, including economic quantification of:
	a) agro-biodiversity conservation
	b) land use changes
	c) climate change mitigation and production of renewable energy
	d) others
Food security	The study mainly focuses on the impacts on food security, including:
	a) physical availability of food
	b) economic and physical access to food
	c) food utilization
	d) stability of the other three dimensions over time
Mixed topics	The study indistinctly focuses on two or more topics indicated above

excluded from the analysis but included in a list of potential relevant studies (with full text not available) in the final systematic map.

Our team currently composed of 10 reviewers will screen the studies identified during the systematic searches in different languages to exclude irrelevant titles. The reviewers will also perform a random screening of 10 percent of the studies from one of the other reviewers, and a Cohen's Kappa coefficient will be calculated to measure the degree of inter-reviewer agreement. If the Kappa value is less than 0.5, the reviewers will examine their differences, and possible errors will be corrected to ensure a reliable screening procedure.

Data extraction strategy

Studies that pass the inclusion criteria will be imported into a database. Each study will be coded based on the following information:

- General information about the study (authors, year of publication, affiliation, donor)
- Type of publication (e.g., peer-review article, non peer-review manuscript/article, book, book chapters)
- Location of the study (e.g., region, state, country, locality)
- Description of the population (e.g., average age, gender, education)
- Type of crop and GM traits (e.g., Bt cotton)
- Type of evaluation method (e.g., propensity score matching, differences in differences)
- Other relevant qualitative information, especially when the study design is only qualitative (e.g., descriptions from ethnographic studies).

Data presentation

The data presentation will include descriptive statistics by type of socio-economic impact; population; geographical focus of the evidence (e.g., developed and developing countries); research methods; and changes on time (or time trends) when possible.

The final outcomes of this protocol will be a systematic map report on the socio-economic impacts of GM crops worldwide based on the evidence available in six languages, and a searchable database (including the list of references of the included studies, along with the information extracted from those studies; the list of excluded studies and reasons for exclusion; and the list of potentially relevant studies with full text not available). Also the review process will be documented and included in an open-acess database named "Central Access Database for Impact Assessment of Crop Genetic Improvement Technologies" (CADIMA, http://www.cadima.info/) which is currently under development by members of the GRACE project.

Endnotes

[a]Nevertheless, as indicated by a reviewer, systematic maps can also be used to map narrow questions, especially where the studies pertaining to a particular topic are unlikely to meet the criteria for quantitative synthesis (or meta-analysis).

[b]Systematic maps and systematic reviews follow the same structured methodologies. However, systematic reviews, in contrast to systematic maps, include an evidence synthesis and are set out to critically appraise the evidence. Both systematic maps and systematic reviews are considered stand-alone pieces of review of the

evidence. Nevertheless, systematic maps can also be undertaken as the first step before conducting systematic reviews, which would be only undertaken if there is sufficient quantity and quality of evidence on specific sub-topics (for example, the CASE project [57] conducted first a broad systematic map and then subsequent systematic reviews on particular sub-topics identified from the systematic map). TUM may conduct subsequent systematic reviews, which will depend on the amount of evidence found in particular sub-topics and time availability after finishing the systematic map. Given the case, TUM will elaborate additional independent protocols before conducting the systematic reviews. The total duration of the GRACE project is from 2012 to 2015.

[c]Another institution, the Center for European Policy Studies (CEPS), is in charge of reviewing the evidence at the "macro-level" (socio-economic impacts at the sectoral and macro level, trade impacts of GM crops, and politics of GM crops).

[d]The stakeholder consultation process is being facilitated by the working package/group "Stakeholder and user involvement" of the GRACE project (see GRACE [7]).

[e]In the context of the GRACE project, environmental economics is defined as the economic effects or consequences of current or potential environmental impacts.

[f]The other three top languages used for publication of research are Dutch, Italian, and Russian [9].

[g]Controlled terms are standardized subject terms used by a database to categorize articles based on the content. In contrast, free terms are natural language terms (i.e., terms included in the title of a document).

[h]The comparators are associated with "before-after" and "with-without" evaluations, which allows controlling for selectivity bias.

[i]We expect to finish the whole systematic map in a maximum of fifteen months.

Additional files

> **Additional file 1:** Reviews and meta-analysis studies on socio-economic impacts of genetically modified crops.
>
> **Additional file 2:** List of GM crop related terms.
>
> **Additional file 3:** List of socio-economic terms.
>
> **Additional file 4:** Selected databases for searches in non-English languages.
>
> **Additional file 5:** Overview of the content of selected databases, platforms and search engines.
>
> **Additional file 6:** Procedure for searches in non-English languages.
>
> **Additional file 7:** Results for searches per database in English and Non-English languages.
>
> **Additional file 8:** List of journals.
>
> **Additional file 9:** Draft survey.

Competing interests
The authors declare that they have no competing interests.

Authors' contributions
JGY drafted the protocol and selected the search terms in English and Spanish; and the databases in English, Spanish, and Portuguese. TL, HV, HVu, SY, and YZ conducted searches in the databases in English. YL, & VN translated the search terms from English to Chinese, and German, respectively; selected the corresponding language-specific databases; and conducted searches in those languages. BS provided support with the Citavi reference software and comments related to the search strategies in English and German. JW is the person responsible for the GRACE project at TUM and indicated the scope of the work (topics to be covered and languages). All authors read and approved the final manuscript.

Acknowledgements
Jiao Xu, Camilo Lopez, and Oliver Etzel provided support with the searches in different databases and languages. Roi Duran provided inputs related to the translation of the search terms in Portuguese. H.Al-Asadi translated the search terms to French, selected the language-specific databases and conducted searches in this language (no longer member of the GRACE team). The authors are grateful to the six reviewers from CEE who provided useful comments and suggestions to this protocol.

Sources of support
The research is funded by the FP7 project GRACE.

Author details
[1]Technische Universitaet Muenchen, Chair of Agricultural and Food Economics, Alte Akademie 12, 85354 Freising, Germany. [2]Program on Sustainable Resource Management, Hans-Carl-von- Carlowitz-Platz 2, 85354 Freising, Germany. [3]Library of the Technical University of Munich, Branch Weihenstephan, Maximus-von-Imhof-Forum 1-3, 85354 Freising, Germany. [4]Chair of Agricultural Economics and Rural Policy, Wageningen University, Hollandseweg 1, 6706KN Wageningen, The Netherlands.

References

1. Smale M, Zambrano P, Gruere G, Falck-Zepeda J, Matuschke I, Horna D, Nagarajan L, Yerramareddy I, Jones H: *Measuring the Economic Impacts of Transgenic Crops in Developing Agriculture during the First Decade. Approaches, Findings and Future Directions.* Washington: USA: Food Policy Review 10. IFPRI; 2009.
2. Chen H, Lin Y: **Promise and issues of genetically modified crops.** *Curr Opin Plant Biol* 2013, **16:**1–6.
3. National Research Council: *The Impact of Genetically Engineered Crops on farm Sustainability in the United States,* Committee on the Impact of Biotechnology on Farm-Level Economics and Sustainability. Board on Agriculture and Natural Resources. USA: The National Academic Press; 2010.
4. Smale M, Zambrano P, Falck-Zepeda J, Gruere G: *Parables: Applied Economic Literature about the Impact of Genetically Engineered Crop Varieties in Developing Countries.* Washington: USA: EPT Discussion Paper 158. IFPRI; 2006.
5. Bragge P, Clavisi O, Turner T, Tavender E, Collie A, Gruen R: **The global evidence mapping initiative: scoping research in broad topic areas.** *Med Res Methodol* 2011, **11:**1–12.
6. Hall C, Knight B, Ringrose S, Knox O: **What have been the farm-level economic impacts of the global cultivation of GM crops? Systematic Review.** *Environ Evid* 2013, CEE Review 11–002. http://www.environmental evidence.org/wp-content/uploads/2014/07/CEE11-002.pdf.
7. GRACE: *GMO risk assessment and communication of evidence - GRACE. (Description of Work - Annex I, Part B).* FP7 Collaborative Project; 2012.
8. GRACE: *GRACE Stakeholder Consultation on Good Review Practice in GMO Impact Assessment. Part 2: Stakeholder priorities for review questions- Review questions on socioeconomic impacts;* 2013.
9. van Weijen D: **The language of (future) scientific communication.** [http://www.researchtrends.com/issue-31-november-2012/the-language-of-future-scientific-communication/]
10. ISAAA: *Global status of commercialized biotech/GM crops;* 2012 [http://www.isaaa.org/]
11. Zambrano P, Maldonado J, Mendoza S, Ruiz L, Fonseca L, Cardona I: *Women Cotton Farmers. Their Perceptions and Experience with Transgenic Varieties. A*

Case Study for Colombia. Washington, USA: International Food Policy Research Institute. Discussion Paper 01118; 2011.

12. Yoo D: *Individual and Social Learning in Bio-Technology Adoption: The Case of GM Corn in the U.S.* Seattle, USA: Paper presented at Agricultural and Applied Economics Association Annual Meeting; 2012.

13. Uematsu H, Mishra A: *Net Effect of Education on Technology Adoption by U.S. Farmers.* Orlando, USA: Paper presented at the Southern Agricultural Economics Association Annual Meeting; 2010.

14. Jaramillo P, Useche P, Barhan B, Foltz J: *The State Contingent Approach to Farmers' Valuation and Adoption of New Biotech Crops: Nitrogen-Fertilizer Saving and Drought Tolerance Traits.* Denver, Colorado: Paper presented at the Agricultural and Applied Economics Association Annual Meeting; 2010.

15. Gaurav S, Mishra S: *To Bt or not to Bt? Risk and Uncertainty Considerations in Technology Assessment.* Mumbai, India: Working paper from Indira Gandhi Institute of Development; 2012.

16. Birol E, Villalba E, Smale M: **Farmer preferences for millpa diversity and genetically modified maize in Mexico: a latent class approach.** *Environ Dev* 2009, **14**:521–540.

17. Krishna V, Qaim V: **Bt cotton and sustainability of pesticide reductions in India.** *Agric Syst* 2012, **107**:47–55.

18. Mutuc M, Rejesus R, Pan S, Yorobe J: **Impact assessment of Bt corn adoption in the Philippines.** *J Agric Appl Econ* 2012, **44**:117–135.

19. Huang J, Hu R, Rozelle S, Qiao F, Pray C: **Transgenic varieties and productivity of smallholder cotton farmers in China.** *Aust J Agric Resour Econ* 2002, **46**:367–387.

20. Bennett R, Morse S, Ismael Y: **The economic impact of genetically modified cotton on South African smallholders: yield, profit and health effects.** *J Dev Stud* 2006, **42**:662–677.

21. Anderson K, Jackson L, Nielsen C: *Genetically Modified Rice Adoption: Implications for Welfare and Poverty Alleviation.* World Bank Policy Research. Working Paper 3380; 2004.

22. Lipton M: **Plant breeding and poverty: can transgenic seeds replicate the 'Green Revolution' as a source of gains for the poor?** *J Dev Stud* 2007, **43**:31–62.

23. Davis A, Mitoh T: **Dying in the USA and Japan: selected legal and ethical issues.** *Int Nurs Rev* 1999, **46**(5):135–139.

24. Groeneveld R, Wesseler J, Berentsen P: **Dominos in the dairy: an analysis of transgenic maize in Dutch dairy farming.** *Ecol Econ* 2013, **86**:107–116.

25. Consmüller N, Beckmann V, Petrick M: *Identifying driving factors for the establishment of cooperative GMO-free zones in Germany.* Foz do Iguacu, Brazil: Paper presented at Conference of the International Association of Agricultural Economists; 2012.

26. Gray E, Ancev T, Drynan R: **Coexistence of GM and non-GM crops with endogenously determined separation.** *Ecol Econ* 2011, **70**:2486.

27. Falck-Zepeda J: **Coexistence, genetically modified biotechnologies and biosafety: implications for developing countries.** *Am J Agric Econ* 2006, **88**:1200–1208.

28. Wilson W, Dahl B: **Costs and risks of testing and segregating genetically modified wheat.** *Rev Agric Econ* 2005, **27**:212–228.

29. Coleno F, Angevin F, Lecroart B: **A model to evaluate the consequences of GM and non-GM segregation scenarios on GM crop placement in the landscape and crosspollination risk management.** *Agric Syst* 2009, **101**:49–56.

30. Moss C, Schmitz T, Schmitz A: **The brave new world: imperfect information, segregation costs, and genetically modified organisms.** *Agrarwirtschaft* 2004, **53**:303–308.

31. Gryson N, Eeckhout T, Neijens T: *Cost and Benefits for the Segregation of GM and Non-GM Compound Feed.* Gent, Belgium: Paper presented at XII EAAE Congress; 2008.

32. Kalaitzandonakes N, Matsbarger R, Barnes J: **Global identity preservation costs in agricultural supply chains.** *Can J Agric Econ* 2001, **49**:605–615.

33. Murphy J, Yanacopulos H: **Understanding governance and networks: EU-US interactions and the regulation of genetically modified organisms.** *Geoforum* 2005, **36**:593–606.

34. Smyth S, Phillips P: **Competitors co-operating; establishing a supply chain to manage genetically modified canola.** *Int Food Agribus Man* 2001, **4**:51–66.

35. Aerni P, Scholderer J, Ermen D: *How Would Swiss Consumers Decide if they had Freedom of Choice? Evidence from a Field Study with Organic, Conventional and GM Corn Bread, 36.* Washington: USA: Food Policy Review 10. IFPRI; 2011:830–838.

36. Costa-Font M, Tranter R, Gil J, Jones P, Gylling M: *Do Defaults Matter? Willingness to Pay to Avoid GM Food vis-à-vis Organic and Conventional Food in Denmark, Great Britain and Spain.* Edinburgh, Scotland: Paper presented at the 84th Conference of Agricultural Economics; 2010.

37. Kikulke E, Birol E, Wesseler J, Falck-Zepeda J: **A latent class approach to investigating demand for genetically modified banana in Uganda.** *Agric Econ* 2011, **42**:547–560.

38. Donaghy P, Rolfe J, Bennett J: *Quasi-Option Values for Enhanced Information Regarding Genetically Modified Foods.* Melbourne, Australia: Paper presented at the 48th Australian Agricultural and Resource Economics Society; 2004.

39. Lusk J, House L, Valli C, Jaeger S, Moore M, Morrow B, Traill B: **Consumer welfare effects of introducing and labeling genetically modified foods.** *Econ Lett* 2005, **88**:382–388.

40. Brookes G, Barfoot P: *GM Crops: Global Socio-Economic and Environmental Impacts 1996–2011.* UK: PG Economics Ltd; 2013.

41. Ando A, Khanna M: **Environmental costs and benefits of genetically modified crops. Implications for regulatory strategies.** *Am Behav Sci* 2000, **44**:435–463.

42. Raven P: **Does the use of transgenic plants diminish or promote biodiversity?** *New Biotechnol* 2010, **27**(5):528–533.

43. Batie S: **The environmental impacts of genetically modified plants: challenges to decision making.** *Am J Agric Econ* 2003, **85**:1107–1111.

44. Knox O, Hall C, McVittie A, Walker R, Knight B: **A systematic review of the environmental impacts of GM crop cultivation as reported from 2006 to 2011.** *Food Nutr Sci* 2013, **4**:28–44.

45. Mannion A, Morse S: **Biotechnology in agriculture: agronomic and environmental considerations and reflections based on 15 years of GM crops.** *Prog Phys Geogr* 2012, **36**:747–763.

46. Lemaux P: **Genetically engineered plants and foods: a scientist's analysis of the issues (Part II).** *Annu Rev Plant Biol* 2009, **60**:511–559.

47. FAO: *The State of Food Insecurity in the World. The Multiple Dimensions of Food Security.* Rome: FAO; 2013.

48. Ruane J, Sonnino A: **Agricultural biotechnologies in developing countries and their possible contribution to food security.** *J Biotechnol* 2011, **156**:356–363.

49. Carletto C, Zezza A, Banerjee R: **Towards better measurement of household food security: harmonizing indicators and the role of household surveys.** *Global Food Security* 2013, **2**:30–40.

50. Dibden J, Gibbs D, Cocklin C: **Framing GM crops as a food security solution.** *J Rural Stud* 2013, **29**:59–70.

51. Beyer F, Wright K: **Can we prioritise which databases to search? A case study using a systematic review of frozen shoulder management.** *Health Inf Libr J* 2012, **30**:49–58.

52. Delgado E, Repiso R: **The impact of scientific journals of communication: comparing Google Scholar metrics, Web of Science and Scopus.** *Comunicar* 2013, **41**:45–52.

53. Gehanno J, Rollin L, Darmoni S: **Is the coverage of google scholar enough to be used alone for systematic reviews.** *BMC Med Informat Decis Making* 2013, **13**:1–5.

54. Giustini D, Boulus M: **Google Scholar is not enough to be used alone for systematic reviews.** *Online J Public Health Inform* 2013, **5**:1–9.

55. BASE: *About BASE;* [http://www.base-search.net/about/en/index.php]

56. Hammerstrom K, Wade A, Jorgensen A: *Searching for Studies: A Guide to Information Retrieval for Campbell Systematic Reviews.* UK: The Campbell Collaboration; 2010.

57. Tripney J, Newman M, Bird K, Thomas J, Kalra N, Bangpan M, and Vigurs C: *Understanding the drivers, impact and value of engagement in culture and sport: Technical report for the systematic review and database.* UK: The Case Programme; 2010.

6

What are the environmental impacts of property rights regimes in forests, fisheries and rangelands? a systematic review protocol

Maria Ojanen[1*], Daniel C Miller[2], Wen Zhou[1], Baruani Mshale[2], Esther Mwangi[1] and Gillian Petrokofsky[3]

Abstract

Background: Property rights to natural resources comprise a major policy instrument for those seeking to advance sustainable resource use and conservation. Despite decades of policy experimentation and empirical research, however, systematic understanding of the influence of different property rights regimes on resource and environmental outcomes remains elusive. A large, diverse, and rapidly growing body of literature investigates the links between property regimes and environmental outcomes, but has not synthesized theoretical and policy insights within specific resource systems and especially across resource systems. Here we provide a protocol for conducting a systematic review that will gather empirical evidence over the past two decades on this topic. We will ask the following questions: a) What are the environmental impacts of different property regimes in forests, fisheries, and rangelands? b) Which property regimes are associated with positive, negative or neutral environmental outcomes? c) How do those environmental outcomes compare within and across resource systems and regions?

Methods: We will assess current knowledge of the environmental impacts of property rights regimes in three resource systems in developing countries: forests, fisheries and rangelands. These resource systems represent differing levels of resource mobility and variability and capture much of the range of ecosystem types found across the globe. The review will use a bundle of rights approach to assess the impacts of three main property regimes—state, private, and community—as well as mixed property regimes that involve some combination of these three. Assessment of the impacts of property rights regimes across a range of different resource systems and ecosystem types will enable exploration of commonalities and differences across these systems. Our analysis will emphasize major insights while highlighting important gaps in current research.

Keywords: Property rights, Tenure, Bundle of rights, Forests, Fisheries, Rangelands, Environmental impacts

Background

Debate over the effects of different property regimes on natural resource systems has long been controversial, incited by Hardin's [1] thesis that common pool resources will inevitably suffer from overexploitation and degradation. Moreover, the dominant paradigm long held that government or private property was required for conservation and sustainable resource use. In response, a large body of scholarship has demonstrated that widening the breadth of property rights held by local-level actors in common property regimes can lead to more efficient and effective outcomes for resource sustainability [2-4]. Devolution of property rights to community and local level actors has since been used as an instrument for achieving goals as disparate as poverty alleviation [5], gender equity [6], resource conservation [7], and climate change mitigation [8]. Of course, states have also retained or claimed new property rights or allocated them to private sector actors in the name of these same goals [9].

A large, diverse, and rapidly growing body of literature has investigated the links between these property regime transitions and their environmental outcomes. A significant portion of the literature assesses recent decentralization policies, broadly described as tenure reforms, that transfer

* Correspondence: m.ojanen@cgiar.org
[1]Center for International Forestry Research (CIFOR), Jalan CIFOR, Situ Gede, Sindang Barang, Bogor (Barat) 16115, Indonesia
Full list of author information is available at the end of the article

decision-making rights and authority from central to local governments or formally recognized existing de facto rights at the local level. Another branch of literature assesses the outcomes of initiatives for community-based natural resource management and community-based conservation. Thus far, the literature has yielded mixed findings on resource conditions and sustainability such as biodiversity loss, forest cover change [10], fisheries decline [11], and rangeland degradation [12]. Despite the expanding literature, little has been done to account for the variation in environmental impacts which limits advances in policy making and management intervention. Moreover existing syntheses and reviews on property regimes focus mostly on community based management [13-15], although natural resources are governed through state, private and common property regimes across diverse ecological and political systems. Broadening the scope to examine outcomes in state and private property regimes can give us valuable theoretical and policy insights on similarities and systematic differences within and across resource systems. The increasing emphasis on landscape approaches and thinking beyond individual resource systems makes informing policy and practice at multiple scales of governance even more crucial [16].

This systematic review will synthesize extant empirical evidence of the impacts of different property rights regimes on environmental outcomes in three resource systems at local to regional scales in developing countries: forests, fisheries and rangelands. Although this review will limit itself to the assessment of environmental outcomes, it will also consider context and mediating factors and will aim to determine more systematically which contextual elements matter most decisively. Accounting for the context is especially important since property regime transitions are not always unidirectional nor fully realized, leaving ample room for discrepancies between existing de facto and newly inscribed de jure regimes and conflicts between recognized and unrecognized actors [17].

Objectives of the review

The review seeks to answer the following primary question:

- What are the environmental impacts of different property regimes in forests, fisheries, and rangelands in developing countries?

It also poses two secondary questions:

- Which property regimes are associated with positive, negative or neutral environmental outcomes?
- How do those environmental outcomes compare within and across resource systems and world regions?

Assessment of the impacts of property rights regimes across a range of different resource systems and ecosystem types enables exploration of commonalities and differences across these systems. As the review is interested in looking both short term and long term results of property rights interventions, both terms impact as well as outcome will feature in the review. Appendix explains in detail the definitions of other key terms used in this review.

This review adopts a PICO (Population-Intervention-Comparator-Outcomes) framework to structure analysis of these research questions [18], summarized in Table 1.

Population

The population refers to the three resource systems: forests, fisheries and rangelands (see Appendix for operational definitions of each of these systems). We have chosen these systems due to the importance of the ecosystem services they provide as well as their broad geographical coverage, which includes much of the range of ecosystem types found across the globe. These three resource systems also represent differing levels of resource mobility and variability, thus introducing important variation in the biophysical nature of the resources they provide. We will exclude other natural systems as well as heavily human-modified systems such as irrigation systems and cities.

Intervention

The intervention refers to the introduction of a particular property rights regime, whether state, private, and community or some combination of these (mixed regimes). The intervention could also be the establishment of a protected area for the explicit objective of resource conservation, although these cases will be treated separately. The review uses a bundle of rights approach, introduced by Schlager and Ostrom [19], to examine how the distribution of access, withdrawal, management, exclusion, and alienation rights in state, private, community, and mixed property regimes affects resource outcomes. In addition, this review considers the right to income from resource use as part of the bundle of rights that comprise a property regime, defined as a system of rules governing access to and control over resources [20], and specifying permissible and forbidden actions in relation to use and management, responsibilities and obligations [2,21]. Also in the case of protected areas, the bundle of rights approach will be used as it captures well the different joint/mixed property rights arrangements present in protected areas.

Comparator

This review compares environmental outcomes based on analysis of studies using the following three methodologies: change from before to after the intervention (temporal change), case control studies (with-without comparison),

Table 1 Research framework for Population-Intervention-Comparator-Outcomes (PICO)

POPULATION Resource systems	INTERVENTION Property regimes	COMPARATOR	OUTCOME MEASURES
Forests	State	Before and after intervention (temporal dimension)	Species diversity and abundance, forest cover, forest condition, tree density, biomass, carbon sequestration, measures of land degradation and desertification, forest loss, land conversion, measures of disturbances such as number of cut stumps, number of invasive species, etc.
	OR Private	OR With and without the intervention, from a similar setting (spatial dimension)	
Fisheries	OR Community	OR Both before and after AND with and without intervention (BACI)	Abundance of fish, fish size, diversity of species, biomass, health of coral, water quality, reproductive indicators, etc.
Rangelands	OR Mixed		Species diversity and abundance, plant and bare ground cover, proportion of different species, soil indicators, carbon sequestration, biomass, soil nutrient levels, number of supported animals etc.

or a combination of both (a BACI—before-after-control-impact—design). This latter design is based on data from before and after the intervention of interest and in sites where the intervention took place and matched control sites that are similar as possible to the intervention sites except that there was no intervention. The BACI approach seeks to rule out potentially confounding effects and to increase confidence that outcomes observed were due to the intervention [22].

Reviewers anticipate that control sites will often be characterized by open access regimes, but they may be any of the different property rights regimes identified as long as they were present in both treatment and control sites prior to the treatment (change in property rights regime). Moreover, the socioeconomic and environmental baselines of control and treatment sites should be of sufficient similarity such that divergent environmental results, if any, are attributable to the intervention or named contextual factors. In the case of protected areas, the comparison will need to present temporal comparison (before-after) within the protected area or spatial comparison with another regime outside the protected area.

Outcomes

The outcome of interest in this analysis consists of qualitative and quantitative changes in environmental measures, which may vary by resource system. Table 1 includes illustrative outcome indicators likely to be found in relevant studies. Based on information on change and/or difference in these indicators in each study, reviewers will determine whether the environmental outcomes associated with different property regimes were positive, negative or neutral. Both the original outcome measures reported in the studies under review and the reviewers' assessment of environmental outcomes will be recorded.

Methods

Searches

To capture as unbiased and comprehensive a set of relevant literature as possible within the constraints of the

review budget, time allocation, and familiarity with languages, the search will be conducted in the databases shown in Table 2. Searches conducted on Google and Google Scholar will be limited to the first 200 hits retrieved. Any links will be followed only once from the original hit. Previous systematic reviews on the topic [13,15,23-25] and literature reviews assessing tenure and environmental outcomes that are identified by the search will be hand-searched to identify further relevant studies.

Electronic search strategies have been tested using the ISI Web of Knowledge, CAB Abstracts and Google Scholar. This testing process has been documented and informs this protocol (see Appendix). The following search terms will be applied to the different databases, with search term and database specific modifications. As the operation of websites and database-specific search engines varies greatly, the reviewers will modify their search and search terms for each database in order to retrieve results that balance the needs for sensitivity and specificity to the review question [18]. If use of several search terms is impossible, the reviewers will take advantage of available topic-relevant key words and publication categorizations. Reviewers will also adjust for different word permutations or suffixes through the use of wildcard symbols, where applicable. The development and implementation of the search strategy will be recorded, including the testing process, number of hits, relevance of the results and the date of search. The reviewers will also contact individual organizations (through librarians or other information specialists) for further guidance.

To reduce language bias that may be associated with limiting the review to English language publications, the search will be also conducted in French and Spanish. Search in French and Spanish will be conducted in Google, Google Scholar as well as in primary databases where applicable. In addition, reviewers will conduct search for grey literature within institutions and NGOs whose main language is French or Spanish, e.g. CIRAD (French) and Facultad Latinoamericana de Ciencias Sociales (Spanish). In order to take benefit from all the

Table 2 The list of databases and other data sources for the systematic review

Primary research databases

Agris	http://agris.fao.org/
Agricola	http://agricola.nal.usda.gov/
CAB Abstracts	http://www.cabdirect.org/
SciELO - Scientific Electronic Library Online	http://www.scielo.org/
Scopus	http://www.scopus.com/
Web of Knowledge	http://www.webofknowledge.com

General web search engines

Google	www.google.com
Google Scholar	www.scholar.google.com

Research institutes, research networks and universities

Agriculture Network Information Center (AgNIC) at Colorado State	http://lib.colostate.edu/agnic
Center for International Forestry Research	http://www.cifor.org
Centro de Informacion de Recursos Naturales	http://www.ciren.cl/web/
Centro de Investigaciones Pesqueras MINAL	http://www.ecured.cu/index.php/Centro_de_Investigaciones_Pesqueras
CGIAR System-wide Program on Collective Action and Property Rights	http://www.capri.cgiar.org/
CIRAD	http://www.cirad.fr/
Coalition of European Lobbies for Eastern African African Pastoralism (CELEP)	http://www.celep.info
Digital Library of the Commons	http://dlc.dlib.indiana.edu/dlc/
Facultad Latinoamericana de Ciencias Sociales (FLACSO)	http://www.flacso.org/
Fondo de Conservacion de Bosques Tropicales Paraguay	http://www.fondodeconservaciondebosques.org.py/
Institut de recherché pour le développement (IRD)	www.ird.fr
Institut des sciences humaines et sociales (INSHS) du centre national pour la recherche scientifique (CNRS)	http://www.cnrs.fr/inshs/recherche/librairie/176.htm
Instituto del Mar del Peru	http://www.imarpe.pe/imarpe/
Instituto Mamiraua	http://www.mamiraua.org.br/pt-br
International Food Policy Research Institute	http://ifpri.org
International Institute for Fisheries Economics and Trade (IIFET)	http://oregonstate.edu/dept/iifet/
International Livestock Research institute	http://ilri.org
Lincoln Institute of Land Policy	http://www.lincolninst.edu/

Table 2 The list of databases and other data sources for the systematic review (Continued)

Nelson Institute Land Tenure Center	http://www.nelson.wisc.edu/ltc/
PLAAS	http://www.plaas.org.za/
The Organization for Social Science Research in Eastern and Southern Africa (OSSREA)	http://publications.ossrea.net/
Thèses.fr (French master thesis and PhD)	http://www.thèses.fr
Universidad de los Andes (Colombia)	http://www.uniandes.edu.co/
Universidad Nacional Autonoma de Mexico	http://www.unam.mx/
University of Wageningen library	http://www.wageningenur.nl/
Western Indian Ocean Marine Science Association (WIOMSA)	http://www.wiomsa.org
World Agroforestry Center	http://www.worldagroforestrycentre.org/
WorldFish	http://worldfishcenter.org

International organizations and donor agencies

US Agency for International Development (USAID)	http://usaid.gov
African development bank database (AfDB)	http://www.afdb.org/en/documents/
African Journals online	http://www.ajol.info/
Amazonia	http://amazonia.org.br/
Asian Development Bank (ADB)	http://www.adb.org/publications
Banco interamericano de desarrollo (BID)	http://publications.iadb.org
Centro Peruano De Estudios Sociales (CEPES)	http://www.cepes.org.pe/portal/
DIVERSITAS	http://www.diversitas-international.org
EuropeAid european union cooperation and development agency	http://ec.europa.eu/europeaid/multimedia/index_en.htm
European Tropical Forest Research Network	http://www.etfrn.org
Food and Agriculture Organization (FAO)	http://www.fao.org
German GIZ Gesselschaft fur Internationale Zusammenarbeit	https://www.giz.de/en/mediacenter/library.html
Institute of Development Studies	http://www.eldis.org/go/topics/resource-guides/environment
Instituto Brasileiro de Geografia e Estatica	http://www.ibge.gov.br/english/
International Geosphere-Biosphere Program (IGBP)	http://www.igbp.net
International Institute for Environment and Development (IIED)	http://www.iied.org
NEAD	http://www.nead.gov.br/portal/nead/

What are the environmental impacts of property rights regimes in forests, fisheries and rangelands? a systematic...

93

Table 2 The list of databases and other data sources for the systematic review *(Continued)*

ODI overseas development institute	http://www.odi.org.uk/publications
Permanent Institution of the International Federation of Surveyors	http://www.oicrf.org/
The International Human Dimension Programme on Global Environmental Change (IHDP)	http://www.ihdp.unu.edu
The Organisation for Economic Co-operation and Development (OECD)	http://www.oecd.org/
The World Bank	http://worldbank.org
Tierra Fundacion	http://www.ftierra.org/
UK department for international development	https://www.gov.uk/government/publications
UNESCO	http://www.unesco.org/new/fr/unesco/resources/publications/
United Nations Convention to Combat Desertification (UNCCD)	http://www.unccd.int/
United Nations Development Programme (UNDP)	http://www.undp.org
United Nations Environment Programme (UNEP)	http://www.unep.org/
United Nations University	http://unu.edu/
World bank and IMF library (JOLI)	http://external.worldbankimflib.org/external.htm
World Resources Institute (WRI)	http://www.wri.org/
NGOs, Think Tanks	
aGter	http://www.agter.asso.fr/rubrique139_fr.html
Alimenterre platform and ressources	http://www.alimenterre.org/recherche/r%C3%A9gime%20foncier
AVSF agronomes et vétérinaires sans frontiers	http://www.avsf.org/fr/recherche_avancee
Community Forestry International	http://www.communityforestryinternational.org/
Conservation Gateway (TNC)	http://www.conservationgateway.org
Conservation International	http://www.science2action.org
CORDIO	http://cordioea.net/
Equator initiative	http://www.equatorinitiative.org/
GRAF action and research on tenure group in the Sahel	http://www.graf-bf.org/spip.php?rubrique4
International Land coalition	http://www.landcoalition.org/
International Union for Conservation of Nature (IUCN)	http://www.iucn.org/wisp/resources/publications
Landesa	http://www.landesa.org/
Landportal	http://landportal.info/
le Hub Rural (west and central africa platform)	http://www.hubrural.org/base_documentaire.html?lang=fr

Table 2 The list of databases and other data sources for the systematic review *(Continued)*

LMMA Network	http://www.lmmanetwork.org
Oakland Institute	http://www.oaklandinstitute.org/
Rainforest Portal	http://www.rainforestportal.org/
ReefBase	http://www.reefbase.org/pacific
Resources for the Future	http://www.rff.org
Rights and Resources Initiative	www.rightsandresources.org
Tenure observatory of Madagascar (observatoire foncier de madagascar)	http://www.observatoire-foncier.mg/
The Center for People and Forests (RECOFTC)	http://www.recoftc.org
Tropenbos International	http://www.tropenbos.org

languages mastered by the review team an additional search in Indonesian will be conducted where applicable. The databases will be searched with following English terms and with their French, Spanish and Indonesian translations.

Population terms: Forest, fish, marine, grassland, pastoralist, pasture, rangeland

Intervention terms: Collective, common, community, customary, government, public, private, small-scale, state, public, private, company, concession, participatory, collaborative, co-operative, co-manage, shared, joint

Intervention-related terms: Decentralization, tenure, reform, allocation, ownership, property right, property rights, property regime, property system, management, access, harvest, open access

Examples how search words will be combined are presented in Table 3. Searches will also be conducted using different institutional accesses (CIFOR, University of Michigan and University of Oxford) to take advantage of different subscription databases.

Study inclusion and exclusion criteria

Inclusion criteria will be applied to select the relevant articles captured by the search. Inclusion criteria will be applied to the titles and abstracts of articles. Studies will be included if they fulfill the following criteria.

Type of study

Only primary empirical literature will be included, such as case studies, case–control studies and cohort studies, including quantitative and qualitative research.

Subjects studied: We will include studies that asses any property regime associated with forests, fisheries and rangelands. Studies need to assess outcomes from before/after change in a property regime (temporal dimension),

Table 3 List of search terms in Google Scholar[(1)], WOK[(1)] and CAB[(1)]

Google Scholar (english)	(fish OR fisheries OR rangeland OR grassland OR pasture OR forest) AND (common OR community OR government OR state OR public OR private) AND (tenure OR property OR rights) Search results were limited to 1990 onwards
WOK	Topic=(forest* or fish* or marine or grassland* or pastoralis* or pasture or rangeland*) AND Topic=(decentraliz* or tenure or reform* or allocation or ownership or "property right"or "property rights" or "property regime" or "property system" or manag* or access or harvest* or open?access) AND Topic=(collective or comm?n* or small?scale or customary or state or public or privat* or compan* or concession* or participat* or collaborative or co?operative or co-manage* or shared or joint) AND Topic=(Armenia* or Bhutan* or Bolivia* or Cameroon* or "Cape Verde" or "Côte d'Ivoire" or "Ivory Coast" or Djibouti* or Egypt* or "El Salvador" or Georgia* or Ghana* or Guatemala* or Guyan* or Hondura* or Indonesia* or India* or Kiribati*or Lao* or Lesotho* or Mauritania*or Micronesia* or Mongolia*or Morocc* or Nicaragua*or Nigeria* or Pakistan* or Papua* or Paraguay* or Philippin* or Samoa* or "Sao Tome" or Senegal* or "Solomon Islands" or "Sri Lanka" or Sudan* or Swaziland*) OR
	OR Topic=(Afghan* or Bangladesh* or Benin* or Burkina* or Burundi* or Cambodia* or Central African Republic or Chad* or Comoro* or Congo* or Eritrea*or Ethiopia* or Gambia* or Guinea* or Haiti* or Kenya* or Korea* or Kyrgyz* or Liberia* or Madagascar* or Malagasy* or Malawi* or Mali* or Mozambique* or Myanmar* or Burma* or Nepal* or Niger* or Rwanda* or Sierra Leone or Somali* or Sudan* or Tajikistan* or Tanzania* or Togo* or Uganda* or Zimbabwe*)
	OR Topic=(Angola* or Algeria* or Samoa* or Argentina* or Azerbaijan* or Beliz* or Botswana* or Brazil* or Chin* or Colombia* or "Costa Rica" or "Costa Rican" or Cuba* or Dominica* or Equatorian* or Ecuador* or Fiji* or Gabon* or Grenad* or Iran* or Iraq* or Jamaica * or Jordan* or Kazak* or Leban* or Libya* or Malaysia* or Maldives or "Marshall Islands" or Mayotte or Mauritius or Mexic* or Namibia* or Palau* or Panama* or Peru * or Seychell* or "South Africa" or "St! Lucia" or "St. Vincent" or Suriname* or Thai* or Tonga* or Tunisia* or Turk* or Turkmenistan* or Tuvalu* or Venezuela*)
	Search results were limited to 1990-2013
CAB	(forest* OR fish* OR marine OR rangeland* OR grassland* OR pasture*) AND subject:((tenure OR "property rights" OR "property regime" OR "property system" OR "common property resources")) AND yr:[1990 TO 2014]
	OR AND subject:(forest* OR fish* OR marine OR rangeland* OR grassland* OR pasture*) AND subject:("tenure systems" OR "property rights" OR "property regime" OR "property system" OR "common property resources") AND yr:[1990 TO 2014]

[(1)]The symbol asterisk (*) is a truncation operator and presents zero or more characters in a search term. Forest* would thus include forests, forestry, forester, forestal etc. It was not used in Google Scholar as the search engine does not recognize truncation symbols.

assess outcomes from different regimes in a with/without setting (spatial dimension), or a combination of these (BACI). Papers reviewing environmental outcomes without a reference to a specific property regime will not be included, nor will studies of plantation forests and aquaculture. The review will exclude commentary and position papers.

Outcomes
Studies must measure and/or qualitatively assess change and/or difference in environmental outcomes as illustrated Table 1.

Regional focus
Studies will only be included if the research focuses on the developing countries of Latin America and the Caribbean, Africa, and Asia and the Pacific (see Appendix for complete list of countries). Developing countries are those defined as either low or middle income according to the World Bank. We will use this classification rather than others (such as OECD/non-OECD), as the division into low and middle income countries enables synthesis that includes economic context.

Timeframe
Studies need to have been published between 1990-present.

Based on the inclusion criteria described above, the review will use a three-step process to identify studies for inclusion.

(1) Studies clearly not relevant will be excluded on the basis of titles only
(2) Studies with potentially relevant titles will be assessed using their abstracts.
(3) Any potentially relevant study that gets through stages 1 and 2 will be collected and assessed for inclusion as full text.

A kappa statistic – the standard measurement used in previous systematic reviews [15],[13] to check for consistency - will be calculated for all reviewers involved in screening prior to work on inclusion. This has been trialed during protocol development and moderate to high levels of consistency have been achieved during these trials. Once satisfactory kappa statistics has been achieved for our final set of screeners, reviewers will determine separately which papers fulfill the inclusion criteria at each step for batches of publications.

To check for consistency of the selection at each stage, two authors will review a 10% random subsample of abstracts and full texts at these screening stages. If too many differences between inclusion and exclusion are perceived, further discussion on interpretations and possible revision of the criteria will be done iteratively until a satisfactory kappa figure is achieved, following best practice with other published systematic reviews. We will also record the reasons for exclusion at full text for each article and provide this information as an appendix in the final systematic review.

Potential effect modifiers and reasons for heterogeneity

The following potential effect modifiers related to the environmental, socio-economic, and political context of the intervention will be considered and recorded:

Environmental context

- Location
- Ecosystem type
- Spatial extent of resource area
- Elevation
- Accessibility
- Baseline resource condition
- Existence of external environmental management intervention

Socio-economic context

- Population density in study/resource area
- Change in population in study/resource area
- Local and external market demand on resource
- Economic inequality (information stated in study, GINI coefficient, etc.)
- Presence of education initiatives
- Presence of infrastructure

Political context

- Nature of political regime (democracy, authoritarian, totalitarian)
- Decentralization (whether decentralized or decentralizing; year decentralization process began; extent of decentralization: advanced; not advanced)
- Corruption (no corruption, low corruption, high corruption according to study; other measures of corruption, e.g. WGI, Transparency International)

The following additional characteristics of property regime interventions will also be noted where information is available in the study:

- Clarity of rights
- Stability of rights
- Level of enforcement
- Legitimacy of decision-making authority over rights
- Gender equality of property rights
- External support: whether the regime is supported by external actors, such as NGOs, donors, or companies
- Formal protected area: Whether property rights regime applies to a legally protected area

The variables listed above were identified based on consultation with experts in the field of property rights and natural resource governance and knowledge among the review team of the empirical and theoretical literature relevant to this review. The reviewers have winnowed the number of potentially relevant variables to a manageable set that addresses especially salient hypotheses in this area of inquiry.

Study quality assessment

Once all relevant articles have been identified, full texts will be reviewed to assess study quality according to the questions below. These questions are based on recommendations by the Cochrane Collaboration [26] as well as previous reviews [13], but have been modified to account for the realities of available research on the impacts of property rights, which is characterized by an extensive number of qualitative case studies [27]. Two researchers will code a 10% random sample of articles to test the coding protocol and intercoder reliability. Kappa values will be calculated to assess agreement and if agreement is less than 50% the researchers will adjust the coding protocol to increase clarity and agreement.

Questions and coding system used to guide the quality assessment

1. Clarity and replicability of methods: Are the research methods clearly presented so that the research could be repeated? [clear and repeatable =1, not clear and repeatable = 0]
2. Appropriateness of methods: Are the research methods appropriate for addressing the research question(s)? [appropriate = 1, not appropriate = 0]
3. Study design category: Which of the following categories is most appropriate to describe this study? [cross sectional study or time series study =0; case control study = 1; controlled before-and-after (BACI) study = 2]
4. Sample size: Is sample size explained and well justified? [yes = 1, no = 0]
5. Confounding factors: Did the study account for and seek to minimize the effects of potential confounding factors in its design and analysis? [yes = 1, no = 0]

Studies will be assessed based on the above quality categories. The quality of each study will be scored based on the above questions, with results recorded in a separate Excel spreadsheet. Explanations for each decision will be recorded in order to keep the process transparent and repeatable.

However, for those identified studies that fail the quality requirements (scoring 0 on our quality assessment scale), a sensitivity analysis will be conducted during the data analysis stage to determine the effect of their inclusion on the results of our systematic review. Should their results be markedly different from those of studies that met quality criteria, they will be discarded from the final synthesis. These studies will be available for analysis if the sensitivity analysis indicates that the review would be richer with their inclusion. We will in any case capture the numbers that have been assessed at different quality levels in graphical representations of the state of the evidence base.

In addition, reviewers will also record the type of the data analysis used in each case study according to the following typology:

- qualitative analysis
- quantitative analysis -descriptive or observational statistics
- quantitative analysis – analysis of variance, t-test, statistical correlation or other bivariate analysis
- quantitative analysis-multivariate regression or other multivariate analysis

Data extraction strategy

Data on individual property regime interventions and their environmental outcomes will be collected in a data extraction matrix using an Excel spreadsheet. This will include information on the resource systems in which the property rights regime is implemented, the de jure and de facto nature of the regime as determined by the specific property rights accorded under each regime, stated objectives of the property rights regime intervention, intervention year, study year, the environmental outcomes of regime interventions, and confounding factors that may explain the nature and variation of the outcomes of the regime, including baseline environmental characteristics, external regime characteristics that may further enable the outcomes of an intervention (such as the stability of rights held and level of enforcement of the regime), and characteristics of the socioeconomic and political context. A coding protocol has been developed and it is presented in Table 4. As the data extraction advances, other coding systems will be developed (e.g. resource systems, countries and geographic regions).

To fully address geographic differences, we will collect not only country data but also data within subregions, and further take into account the varying ecosystem types within the broad categorizations of forests, rangelands, and fisheries by documenting individual ecosystem type. Should a paper present multiple studies of different property regimes, each of these will be recorded individually within the data extraction sheet. Thus if spatial comparisons of two regimes or comparisons of outcomes from multiple regimes are made, each regime will present an individual data entry.

Table 4 Initial coding protocol for data extraction

· bundle of rights	S = State P = Private C = Community, 0 = undefined 1 = right defined
· clarity of rights	clear = 1, unclear = 0
· stability of rights	stable = 1, unstable = 0
· level of enforcement	no enforcement = 0, low enforcement = 1, high enforcement = 2
· legitimacy of decision making authority over rights	legitimate = 1, not legitimate = 0
· external support	yes = 1, no = 0
· formally protected area x	yes = 1, no = 0
· corruption	high/low/no
· environmental condition	good/fair/poor
· environmental change	positive, negative, no change

Data synthesis and presentation

Our data synthesis will be based on the information categories mentioned in "Potential effect modifiers and reasons for heterogeneity" as well as "Data extraction strategy categories." We will synthesize the results on environmental outcomes across different resource systems, ecosystem types and geographical regions. We will synthesize the environmental results considering the allocation of bundle of rights as well as the context factors. A series of matrices will be deployed to illustrate: a) how environmental outcomes may vary according to the bundle of rights allocation as well as the institutional arrangements that support (or not) the rights regimes, for example, security of rights as determined by enforcement, clarity, stability and legitimacy of authority; b) how contextual factors influence environmental outcomes; c) how environmental outcomes vary with resource type; and d) how a to c above vary in different geographical locations. The synthesis matrices will be accompanied by a narrative. We will also include a note on the performance of different methodologies in providing a nuanced understanding of the environmental effects of property regimes. A major outcome of the synthesis will be the identification of aspects that need further, in-depth inquiry as well as policy implications of current findings.

The variety of measured outcomes and possible lack of quantitative data will delimit the applicability of statistical tools. Data will be analyzed using regression and other statistical techniques (as far as possible) to complement qualitative, narrative analysis. The review will also explore whether or not to include sources for which a significant (33%-50%) portion of the data are missing as done in other relevant systematic reviews [13]. Finally, a sensitivity analysis will be conducted to synthesize the conclusions of studies that did not meet quality assessment standards, and consider differences (if any) from the results of those studies that were included in the final review. This will be done quantitatively if the data is so presented, or otherwise will be synthesized through qualitative methods using the same data extraction matrix.

It is well known that in many research areas papers are more likely to be published if they demonstrate clear, positive results (or strong negative effects), and that papers that shown little or no effect are less likely to be published than "negative". To assess the possibility of such publication bias, we are conducting searches of 'grey' literature (much of it not formally published) in addition to studies in academic journals [28] will assess whether there is evidence of publication bias. If data allow, we will assess bias using funnel plots, which show effect sizes and standard error or sample sizes [29].

Appendix
Glossary of key definitions and terms used in the review protocol

Forest: Land with tree crown cover (or equivalent stocking level) of more than 10 percent and area of more than 0.5 hectares (ha). The trees should be able to reach a minimum height of 5 meters (m) at maturity. A forest may consist either of closed forest formations where trees of various storeys and undergrowth cover a high proportion of the ground or open forest formations with a continuous vegetation cover in which tree crown cover exceeds 10 percent [30].

Wooded lands (woodland): Land either with a crown cover (or equivalent stocking level) of 5–10 percent of trees able to reach a height of 5 m at maturity, or a crown cover (or equivalent stocking level) of more than 10 percent of trees not able to reach a height of 5 m at maturity in situ (e.g. dwarf or stunted trees), or with shrub or bush cover of more than 10 percent. Wooded lands are included in this definition of forests [30].

Fishery: A geographical place, activity, or unit that is involved in raising and/or harvesting fish. As a unit, a fishery is typically defined in terms of some or all of the following: people involved, species or type of fish, area of water or seabed, method of fishing, class of boats and purpose of the activities [31].

Rangeland: Land on which the indigenous vegetation is predominantly grasses, grass-like plants, forbs, or shrubs and is managed as a natural ecosystem. If plants are introduced, they are managed similarly. Rangelands included natural grasslands, savannas, shrublands, many deserts, tundras, alpine communities, marshes and meadows" [32].

Property rights: A property right is an enforceable claim to use, control or otherwise benefit from a resource [33,34]. Property rights is often made up of a bundle of multiple rights (and responsibilities) including [19]:

- Access is the right to enter a defined physical property

- Withdrawal is the right to enter a defined physical area and obtain resource units or products of a resource system (e.g., cutting firewood or timber, harvesting mushrooms, diverting water)
- Management is the right to regulate internal use patterns and transform the resource by making improvements (e.g., planting seedlings and thinning trees)
- Exclusion is the right to determine who will have right of withdrawal and how that right may be transferred
- Alienation is the right to sell or lease withdrawal, management, and exclusion rights.

The bundle of rights also include the right to earn income from a resource even without using it directly and is derived from permitting others to use the resource [35,36].

Property rights regime: a system of rules governing access to and control over resources [20]. Rules specify permissible and forbidden actions in relation to use and management, responsibilities and obligations [2,21]. The holder of a property right can be an individual, a corporation, a group or the state/government:

a) Private property: Individual or "legal individual" holds rights.
b) Common property: group members hold rights (e.g. community)
c) Public property: state holds the rights
d) No-property or Open access: no one has rights and everyone can use the resource as they like; no effective management or regulation

List of developing countries included in the analysis

Income groups correspond to 2012 gross national income (GNI) per capita (World Bank Atlas method) [37]. We will take account of country name changes since 1990 in searching for studies from the relevant countries [38].

Low-income economies ($1,035 or less)

Afghanistan, Bangladesh, Benin, Burkina Faso, Burundi, Cambodia, Central African Republic, Chad, Comoros, Congo, Dem. Rep, Eritrea, Ethiopia, Gambia, Guinea, Guinea-Bisau, Haiti, Kenya, Kyrgyz, Liberia, Madagascar, Malawi, Mali, Mozambique, Myanmar, Nepal, Niger, Rwanda, Sierra Leone, Somalia, South Sudan, Tajikistan, Tanzania, Togo, Uganda, Zimbabwe.

Lower-middle-income economies ($1,036 to $4,085)

Armenia, Bhutan, Bolivia, Cameroon, Cape Verde, Côte d'Ivoire, Djibouti, Egypt, El Salvador, Georgia, Ghana, Guatemala, Guyana, Honduras, Indonesia, India, Kiribati, Laos, Lesotho, Mauritania, Micronesia, Mongolia, Morocco, Nicaragua, Nigeria, Pakistan, Papua, Paraguay, Philippines,

Samoa, Sao Tome, Senegal, Solomon Islands, Sri Lanka, Sudan, Swaziland, Syrian Arab Republic, Timor-Leste, Uzbekistan, Vanuatu, Vietnam, West Bank and Gaza, Yemen, Zambia.

Upper-middle-income economies ($4,086 to $12,615)

Angola, Algeria, Samoa, Argentina, Azerbaijan, Belize, Botswana, Brazil, China, Colombia, Costa Rica, Cuba, Dominica, Dominican Republic, Ecuador, Fiji, Gabon, Grenada, Iran, Iraq, Jamaica, Jordan, Kazakhstan, Lebanon, Libya, Malaysia, Maldives, Marshall Islands, Mauritius, Mexico, Namibia, Palau, Panama, Peru, Seychelles, South Africa, St. Lucia, St. Vincent and the Grenadines, Suriname, Thailand, Tonga, Tunisia, Turkey, Turkmenistan, Tuvalu, Venezuela.

Documentation of search term testing conducted in CAB, Google Scholar and WOK

Testing process with search terms for CAB, Google Scholar and WOK. The search was conducted in the on-line CAB, Google Scholar and WOK databases by the authors WZ and MO, see table for detailed search terms, results, dates and comments on results.

CAB Database: note that the preferred term for tenure is tenure systems; other dictionary terms include: common property resources, common lands, coownership, property rights; ownership on its own not useful This search was amended 12.5.2014.to include Descriptors and geographic locations based on reviewer suggestions.Symbol * notes truncation in order to retrieve various endings

Search	Changes	Search terms	Search results
1		Subject:(forest* OR fish* OR marine OR rangeland* OR grassland* OR pasture*) AND subject:(collective OR comm? n* OR customary OR state OR public OR government OR private OR participat* OR collaborative OR cooperative OR coownership) AND subject:("tenure systems" OR property OR "property rights" OR "property regime" OR "property system" OR "common property resources" OR common lands) AND yr:[1990 TO 2013]	2,133 results; appears somewhat relevant although many soil/plant biology entries (OXFORD)
2	Removed common lands	Subject:(forest* OR fish* OR marine OR rangeland* OR grassland* OR pasture*) AND subject:(collective OR comm? n* OR customary OR state OR public OR government OR private OR participat* OR collaborative OR cooperative OR coownership) AND subject:("tenure systems" OR property OR "property rights" OR "property regime" OR "property system" OR "common property resources") AND yr:[1990 TO 2013]	2,409 results; still many irrelevant biological studies (OXFORD)
3	Searched 2nd and 3rd lines in all fields rather than restricted to subject field alone	Subject:(forest* OR fish* OR marine OR rangeland* OR grassland* OR pasture*) AND (collective OR comm? n* OR customary OR state OR public OR government OR private OR participat* OR collaborative OR cooperative OR coownership) AND ("tenure systems" OR property OR "property rights" OR "property regime" OR "property system" OR "common property resources") AND yr:[1990 TO 2013]	5,047 results; too many irrelevant results (OXFORD)
4	Changed 2nd and 3rd line fields to descriptor	Subject:(forest* OR fish* OR marine OR rangeland* OR grassland* OR pasture*) AND de:(collective OR comm? n* OR customary OR state OR public OR government OR private OR participat* OR collaborative OR cooperative OR coownership) AND de:("tenure systems" OR property OR	55 results; Very focused results but may be too narrow a search (OXFORD)

(Continued)

		"property rights" OR "property regime" OR "property system" OR "common property resources") AND yr:[1990 TO 2013]	
5	Changed 2nd and 3rd line fields back to subject; removed collective and property; added traditional	Subject:(forest* OR fish* OR marine OR rangeland* OR grassland* OR pasture*) AND subject:(comm? n* OR customary OR traditional OR state OR public OR government OR private OR participat* OR collaborative OR cooperative OR coownership) AND subject:("tenure systems" OR "property rights" OR "property regime" OR "property system" OR "common property resources")	64 results; too few even without time limits! (OXFORD)
6	Added property* and right*	Subject:(forest* OR fish* OR marine OR rangeland* OR grassland* OR pasture*) AND subject:(comm? n* OR customary OR traditional OR state OR public OR government OR private OR participat* OR collaborative OR cooperative OR coownership) AND subject:("tenure systems" OR property* OR right* OR "property rights" OR "property regime" OR "property system" OR "common property resources")	3559 results; again much noise comes in; traditional often refers to medicine/plants (OXFORD)
7	Removed traditional	Subject:(forest* OR fish* OR marine OR rangeland* OR grassland* OR pasture*) AND subject:(comm? n* OR customary OR state OR public OR government OR private OR participat* OR collaborative OR cooperative OR coownership) AND subject:("tenure systems" OR property* OR right* OR "property rights" OR "property regime" OR "property system" OR "common property resources")	2,509 results; results appear mostly relevant (many articles are from US, and should be easy to discard) (OXFORD)
8	Added yr:[1990–2013]	Subject:(forest* OR fish* OR marine OR rangeland* OR grassland* OR pasture*) AND subject:(comm? n* OR customary OR state OR public OR government OR private OR participat* OR collaborative OR cooperative OR coownership) AND subject:("tenure systems" OR property* OR right* OR "property rights" OR "property regime" OR "property system" OR "common property resources")	2446 results, showing that most relevant articles were written in review time span (OXFORD)
9	Added yr:[1990–2013]	IBID	
10	Changed yr:[2008–2013]	IBID	4,530 results
11	Removed property*	Subject:((forest* OR fish* OR marine OR rangeland* OR grassland* OR pasture*)) AND subject:((comm? n* OR customary OR state OR public OR government OR private OR participat* OR collaborative OR cooperative OR coownership)) AND subject:(("tenure systems" OR "property rights" OR "property regime" OR "property system" OR "common property resources"))	1,668 results
12	Added yr:[2008 TO 2013]	IBID	509 results
13	Removed search string of regimetype	Subject:(forest* OR fish* OR marine OR rangeland* OR grassland* OR pasture*) AND subject:("tenure systems" OR "property rights" OR "property regime" OR "property system" OR "common property resources")	2351 results; looks very relevant
14	Added yr:[2008 TO 2013]	IBID	723 results

Google Scholar (English)

Search	Changes	Search terms	Search results	Search date
1		(forest OR fish OR marine OR rangeland OR grassland OR pastoralis OR pasture) AND (collective OR common OR community OR communal OR small-scale OR customary OR state OR public OR government OR private OR privatized OR company OR concession) AND (decentralization OR decentralize OR reform OR tenure OR allocation OR ownership OR property OR property right OR property regime OR property system)	8000 hits on Google Scholar, but bias towards decentralization studies	9.2.2013

(Continued)

2	Deleted small-scale, customary, decentralization, decentralize	(forest OR fish OR marine OR rangeland OR grassland OR pastoralis OR pasture) AND (collective OR common OR community OR communal OR state OR public OR government OR private OR privatized OR company OR concession) AND (reform OR tenure OR allocation OR ownership OR property OR property right OR property regime OR property system)	Reform brings a lot of noise	9.2.2013
3	Deleted community, privatized, company, concession, reform, property right, property regime, property system	(forest OR fish OR marine OR rangeland OR grassland OR pasture) AND (collective OR common OR communal OR state OR public OR government OR private) AND (tenure OR allocation OR ownership OR property)	Good first 20–30 results	9.2.2013
4	Changed order of resource systems, added fisheries	(fish OR fisheries OR marine OR rangeland OR grassland OR pasture OR forest) AND (collective OR common OR communal OR state OR public OR government OR private) AND (tenure OR allocation OR ownership OR property)	More results for fisheries	9.2.2013
5	Added property right, property regime again	(fish OR fisheries OR marine OR rangeland OR grassland OR pasture OR forest) AND (collective OR common OR communal OR state OR public OR government OR private) AND (tenure OR allocation OR ownership OR property OR property right OR property regime)	Much fewer, narrower results (around 450,000) - but great emphasis on common property, delimitation by time range further narrows results	9.2.2013
6	Enclosed phrases "property right", "property regime" is parentheses, deleted common	(fish OR fisheries OR marine OR rangeland OR grassland OR pasture OR forest) AND (collective OR communal OR state OR public OR government OR private) AND (tenure OR allocation OR ownership OR property OR (property regime) OR (property right))	Many more results with the phrase "property right" in initial search results; some do not appear relevant to environmental issues	9.3.2013
7	Readded common	(fish OR fisheries OR marine OR rangeland OR grassland OR pasture OR forest) AND (collective OR common OR communal OR state OR public OR government OR private) AND (tenure OR allocation OR ownership OR property OR (property right) OR (property regime))	Many more results with "common property", as found in search 5	9.3.2013
8	Readded community, manage, property system Removed allocation, ownership	(fish OR fisheries OR marine OR rangeland OR grassland OR pasture OR forest) AND (collective OR common OR communal OR community OR state OR public OR government OR private) AND (tenure OR manage OR property OR (property right) OR (property regime) OR (property system))	Common property still come to the top of search results, collective seems to bring up mosty collective action results	9.3.2013
9	Removed collective, replaced manage with management	(fish OR fisheries OR marine OR rangeland OR grassland OR pasture OR forest) AND (common OR communal OR community OR state OR public OR government OR private) AND (tenure OR management OR property OR (property right) OR (property regime) OR (property system))	Results are not too different	9.3.2013
10	Added rights terms: access, withdrawal, exclusion, exclude, alienate, alienation	(fish OR fisheries OR marine OR rangeland OR grassland OR pasture OR forest) AND (common OR communal OR community OR state OR public OR government OR private) AND (tenure OR property OR (property right) OR (property regime) OR (property system)) AND (management OR access OR withdrawal OR exclusion OR exclude OR alienate OR alienation)	Results are not well targeted; limitation by time (2008-present) makes results more irrelevant	9.3.2013
11	Removed rights terms, communal, property right, property regime, property system	(fish OR fisheries OR rangeland OR grassland OR pasture OR forest) AND (common OR community OR state OR public OR government OR private) AND (tenure OR property)		9.3.2013

(Continued)

12	Added rights	(fish OR fisheries OR rangeland OR grassland OR pasture OR forest) AND (common OR community OR government OR state OR public OR private) AND (tenure OR property OR rights)	Emphasis on common property, community management, but first pages of search results look good	9.3.2013
13	Added management	(fish OR fisheries OR rangeland OR grassland OR pasture OR forest) AND (common OR community OR government OR state OR public OR private) AND (tenure OR property OR rights OR management)	Management gives broad policy results	9.3.2013
14	Removed management, added ownership	(fish OR fisheries OR rangeland OR grassland OR pasture OR forest) AND (common OR community OR government OR state OR public OR private) AND (tenure OR property OR rights OR ownership)	Ownership adds no new results of pertinence	9.3.2013

WOK database	Hits	Date	Comments
Symbol (*) notes truncation in order to retrieve various endings			
stock OR resource OR population AND common* AND property	6637	16.4.2013	Obviously needs a better description of the resource
fish* OR forest* OR rangeland* OR pasture OR cattle AND comm* AND govern* AND property OR right*	1463	16.4.2013	still too much noise
fish*, forest*, rangeland*, pasture, cattle AND common OR community OR private OR shared AND access OR management	88308	22.4.2013	Management brings noise
fish*OR forest* OR rangeland* OR pasture OR cattle AND common property OR common-pool OR privat* AND access OR management AND effect* AND benefit*	3301	22.4.2013	Seem relevant
fish*OR forest* OR rangeland*, pasture OR cattle AND common OR private OR shared AND property OR rights OR quota* OR tenure OR title OR deed	28971	22.4.2013	Too huge number
fish* OR forest* OR pasture OR rangeland* OR cattle AND tenure or reform or regime or rule* or quota* or customary AND impact* OR effect* OR effectiveness OR benefit* AND common property or "common-pool" or "community-based" or state or private "community-controlled" AND access or management or use* or withdrawal or harvest or monitor	4914	24.4.2013	Seem relevant
fish* OR forest* OR pasture OR rangeland* OR cattle AND tenure or reform or property or rule* or quota* or custom* or transfer* AND impact* OR effect* OR benefit* AND common property or "common-pool" or "community-based" or state or private "community-controlled" AND access* or management or harvest* or monitor*	5064	24.4.2013	Rules leads to misleading results
fish* OR forest* OR pasture OR rangeland* OR cattle AND tenure or reform or property or governance or quota* or custom* or transfer* AND impact* OR effect* OR benefit* AND common property or "common-pool" or "community-based" or state or private "community-controlled" AND access* or management or harvest* or monitor*	4702	24.4.2013	Replaced rules with governance, transfer still brings misleading results
fish* OR forest* OR pasture OR rangeland* OR cattle AND tenure or reform or property or govern* or quota* or custom* AND impact* OR effect* OR benefit* AND common property or "common-pool" or "community-based" or state or private or "community-controlled" or "open-access" AND access* or management or harvest* or monitor*	6919	24.4.2013	Changed governance into govern*, removed transfer
fish* OR forest* OR pasture OR rangeland* OR cattle AND tenure or reform or govern* or quota* or custom* AND impact* OR effect* OR benefit* AND common property or " common-pool" or "community-based" or state or private or "community-controlled" or "open-access" AND access* or management or harvest* or monitor*	4001	24.4.2013	Removed property because it brings noise
fish* OR forest* OR pasture OR rangeland* OR cattle AND tenure or reform or govern* or custom* AND right* or ownership* AND	938	25.4.2013	Remove state, private and quota

(Continued)

Search	Results	Date	Note
common property or " common-pool" or "community-based" or "community-controlled" or "open-access"			
fish* OR forest* OR pasture OR rangeland* OR cattle AND tenure or reform or govern* or custom* AND common property or "common-pool" or "community-based" or "community-controlled" or "open-access" AND biodivers* or divers* or loss or deplet* or deforestation or conservation AND sustainab*	316	25.4.2013	Inserted environmental outcomes and sustainability, because many articles did not mention environmental outcomes
fish* OR forest* OR pasture OR rangeland* OR cattle AND tenure or reform or govern* or custom* AND common property or "common-pool" or "community-based" or "community-controlled" or "open-access" or private or state AND biodivers* or divers* or loss or deplet* or deforestation or conservation or benefit* AND sustainab*	1264	25.4.2013	Added state and private
fish* OR forest* OR pasture OR rangeland* OR cattle AND biodivers* or divers* or loss or deplet* or deforestation or degradation or conservation AND benefit* or sustainab* or improv* or effect* AND tenure or regime or management or access or right AND tenure or reform or governance or customary	3035	25.4.2013	Testing with sustainability indicators
rangeland OR pasture or pastoralist* AND biodivers* or divers* or loss or deplet* or deforest* or degradat* or conservation AND benefit* or sustainab* or improv* or effect* AND community-based or "community-controlled" or "open-access" or enclosure	270	30.4.2013	Trying with rangeland and pasture and pastoralist. Management brings too much noise, also enclosure needs a more descriptive attribute
rangeland OR pasture or pastoralist* AND biodivers* or divers* or loss or deplet* or overgraz* or degradat* or conservation AND benefit* or sustainab* or improv* or effect* AND community-based or community-controlled or open-access or tenure or property or collective	2977	30.4.2013	Removing enclosure and adding overgrazing
rangeland OR pasture or pastoralist* AND biodivers* or divers* or loss or deplet* or overgraz* or degradat* or conservation AND benefit* or sustainab* or improv* or effect* AND community-based or community-controlled or open-access or tenure or property rights or collective	614	30.4.2013	Inserting property rights
rangeland OR pasture or pastoralist* AND biodivers* or divers* or loss or deplet* or overgraz* or degradat* or conservation AND benefit* or sustainab* or improv* or effect* AND community or open-access or tenure or property rights or collective or privatization	8154	30.4.2013	inserting just community and privatization
rangeland OR pasture or pastoralist* AND biodivers* or divers* or loss or deplet* or overgraz* or degradat* or conservation AND benefit* or sustainab* or improv* or effect* AND community managed or "community based" or "community controlled" or tenure or property rights or collective or privatization	604	30.4.2013	Community-based returns a lot of relevant looking results
rangeland OR pasture or pastoralist* AND biodivers* or divers* or loss or deplet* or overgraz* or degradat* or conservation AND benefit* or sustainab* or improv* or effect* AND community managed or "community based" or "community controlled" or tenure or property rights or collective or privatization and ranch	903	30.4.2013	Adding ranch brings some good articles not so many
rangeland OR pasture or pastoralist* AND biodivers* or divers* or loss or deplet* or overgraz* or degradat* or conservation AND benefit* or sustainab* or improv* or effect* AND community based or collective or privatization or common property or tenure or property rights	2086	30.4.2013	Seems relevant, but still not discussing environmental outcomes that much
fish* OR forest* OR rangeland* OR pasture OR grasslands or livestock or cattle AND property near (common or regime) or common?pool or community same (based or manage*) or privat* or state or compan* AND tree or wood* or environment* or ecologic* OR condition or population fish* OR forest* OR rangeland* OR pasture OR grasslands or livestock or cattle AND chang* or impact* or effect* or improv* or declin* or *crease AND tenure or allocation or ownership or right or intervention	10334	17.5.2013	Cattle and livestock bring a lot of noise, maybe a separate search needs to be done on the
fish* OR forest* or marin* AND property same (common or regime) or common?pool or community same (based or manage*) or privat* or state or compan* AND forest or timber or tree or environment* or ecologic* OR condition or population or fish* or stock AND impact* or effect* or improv* or deplet* or conserv* or sustainab* AND tenure or allocation or ownership or intervention or quota	2377	17.5.2013	Nice results for forests, fisheries irrelevant

(Continued)

fish* or marine AND customary or small?scale or (property right*) or (common property) or tenure or (open access) AND resource* or population or size or stock or catch or specie* or biomass AND access or manag* or harvest*	1007	20.5.2013	Based on keywords for fisheries article
forest* AND comm?n* or community* or privat* or state or compan* or concession* or compan* or collective AND tree* or wood* or environment* or ecologic* OR condition* or biodiversity or specie* or forest* or resource AND chang* or impact* or effect* or improv* or declin* or *crease AND decentraliz* or tenure or allocation or ownership or intervention or reform or (property rights) or property regime or management	5272	20.5.2013	Very good, easy to exclude with words USA, Canada, Sweden
forest* or property rights or tenure AND comm?n* or community* or privat* or state or compan* or concession* or compan* or collective AND tree* or wood* or environment* or ecologic* OR condition* or biodiversity or specie* or forest* or resource AND chang* or impact* or effect* or improv* or declin* or *crease AND decentraliz* or tenure or allocation or ownership or intervention or reform or property rights or property regime or management	5538	21.5.2013	By Excluding Canada, Sweden, Finland, Mediterranean, Australia 316+150+133+128+ 245
grassland* or pastoralist or pasture or livestock AND comm?n* or state or privat* AND property rights or tenure or property regime AD animal* or land or environment* or vegetation or composition or species	599	21.5.2013	Stil not too many articles discussing environmental outcomes but common property regimes in general
forest* or fish* or marine or grassland* or pastoralis* or pasture or rangeland* AND decentraliz* or tenure or reform or allocation or ownership or "property right"or "property rights" or "property regime" or "property system" or manag* or access or harvest* or open?access AND collective or comm?n* or small?scale or customary or state or public or privat* or compan* or concession* or participat* or collaborative or co?operative or co-manage* or shared or joint AND Afghan* or Bangladesh* or Benin* or Burkina* or Burundi* or Cambodia* or Central African Republic or Chad* or Comoro* or Congo* or Eritrea*or Ethiopia* or Gambia* or Guinea* or Haiti* or Kenya* or Korea* or Kyrgyz* or Liberia* or Madagascar* or Malagasy* or Malawi* or Mali* or Mozambique* or Myanmar* or Burma* or Nepal* or Niger* or Rwanda* or Sierra Leone or Somali* or Sudan* or Tajikistan* or Tanzania* or Togo* or Uganda* or Zimbabwe*	2650	2.9.2013	2652 all the years/2557 (1990–2013) and 1451 2008–2013. Contains reform
forest* or fish* or marine or grassland* or pastoralist* or pasture or rangeland AND decentraliz* or tenure or allocation or ownership or reform or (property rights) or property regime or management* AND collective or comm?n* or small?scale or customary or state or public or privat* or compan* or concession* or participat* or collaborative or cooperative or co-manage* AND Angola* or Algeria* or Samoa* or Argentina* or Azerbaijan* or Beliz* or Botswana* or Brazil* or Chin* or Colombia* or "Costa Rica" or "Costa Rican" or Cuba* or Dominica* or Equatorian* or Ecuador* or Fiji* or Gabon* or Grenad* or Iran* or Iraq* or Jamaica * or Jordan* or Kazak* or Leban* or Libya* or Malaysia* or Maldives or "Marshall Islands" or Mayotte or Mauritius or Mexic* or Namibia* or Palau* or Panama* or Peru * or Seychell* or "South Africa" or "St. Lucia" or "St. Vincent" or Suriname* or Thai* or Tonga* or Tunisia* or Turk* or Turkmenistan* or Tuvalu* or Venezuela*	7169	23.8.2013	7169 (1990–2013), 4147 (2008–2013)
forest* or fish* or marine or grassland* or pastoralis* or pasture or rangeland* AND decentraliz* or tenure or reform* or allocation or ownership or "property right"or "property rights" or "property regime" or "property system" or manag* or access or harvest* or open?access AND collective or comm?n* or small?scale or customary or state or public or privat* or compan* or concession* or participat* or collaborative or cooperative or co-manage* or shared or joint AND Angola* or Algeria* or Samoa* or Argentina* or Azerbaijan* or Beliz* or Botswana* or Brazil* or Chin* or Colombia* or "Costa Rica" or Cuba* or Dominica* or Ecuador* or Fiji* or Gabon* or Grenad* or Iran* or Iraq* or Jamaica * or Jordan* or Kazak* or Leban* or Libya* or Malaysia* or Maldives or "Marshall Islands" or Mayotte or Mauritius or Mexic* or Namibia* or Palau* or Panama* or Peru* or Seychell* or "South Africa" or "St. Lucia" or "St. Vincent" or Suriname* or Thai* or Tonga* or Tunisia* or Turk* or Tuvalu* or Venezuela*	7356	2.9.2013	7356 (1990–2013), 4241 (2008–2013). This search contains benchmark studies identified

Competing interests

The authors do not declare any conflicts of interest. This research is funded by the Center for International Forestry Research (CIFOR).

Authors' contributions

EM developed the research question. MO & WZ conducted pilot research. MO, DCM, WZ, BM and EM wrote the protocol with support from GP. All the authors MO, DCM, WZ, BM, EM and GP read and approved the final manuscript.

Acknowledgements

The review team would like to thank Eduardo Araral Jr., Eduardo Brondizio, Anne Larson, Monica Lengoiboni, Michael Mascia, Maryam Niamir-Fuller, Frank Place, and William Sunderlin, for their time and comments on an earlier draft of this research protocol. The review team would also like to thank the Center for International Forestry Research (CIFOR) and the UK Department for International Development (DfID) for financing this research through its KNOW-FOR program grant.

Author details

[1]Center for International Forestry Research (CIFOR), Jalan CIFOR, Situ Gede, Sindang Barang, Bogor (Barat) 16115, Indonesia. [2]School of Natural Resources and Environment, University of Michigan, 4024 Dana Building, 440 Church Street, Ann Arbor, MI 48109, USA. [3]Department of Zoology, The Tinbergen Building, South Parks Road, Oxford OX1 3PS, UK.

References

1. Hardin G: **The tragedy of the commons.** *Science* 1968, **162:**1243–1248.
2. Ostrom E: *Governing the Commons: The Evolution of Institutions for Collective Action.* Cambridge: Cambridge University Press; 1990.
3. Larson AM, Soto F: **Decentralization of Natural Resource Governance Regimes.** *Annu Rev Environ Resour* 2008, **33:**213–239.
4. Ostrom E, Nagendra H: **Insights on linking forests, trees, and people from the air, on the ground, and in the laboratory.** *Proc Natl Acad Sci* 2006, **103:**19224–19231.
5. Besley T, Burgess R: **Land reform, poverty reduction, and growth:evidence from India.** *Q J Econ* 2000, **115:**389–430.
6. Meinzen-Dick RS, Brown LR, Feldstein HS, Quisumbing AR: **Gender, property rights, and natural resources.** *World Dev* 1997, **25:**1303–1315.
7. Berkes F: **Community conserved areas: policy issues in historic and contemporary context.** *Conserv Lett* 2009, **2:**19–24.
8. Sunderlin WD, Larson AM, Cronkleton P: **Forest Tenure Rights and REDD+: From Inertia to Policy Solutions.** In *Realising REDD+: National Strategy and Policy Options.* Bogor, Indonesia: Center for International Forestry Research (CIFOR); 2009:139–149.
9. Lemos MC, Agrawal A: **Environmental Governance.** *Annu Rev Environ Resour* 2006, **31:**297–325.
10. Dahal GR, Larson AM, Pacheco P: **Outcomes of Reforms for Livelihoods, Forest Condition and Equity.** In *Forests for People: Community Rights and Forest Tenure Reform.* London, UK: Earthscan; 2010:183–208.
11. Costello C, Gaines SD, Lynham J: **Can catch shares prevent fisheries collapse?** *Science* 2008, **321:**1678–1681.
12. Homewood KM: **Policy, environment and development in African rangelands.** *Environ Sci Pol* 2004, **7:**125–143.
13. Brooks J, Waylen KA, Borgerhoff Mulder M: **Assessing community-based conservation projects: A systematic review and multilevel analysis of attitudinal, behavioral, ecological, and economic outcomes.** *Environ Evid* 2013, **2.**
14. Evans L, Cherrett N, Pemsl D: **Assessing the impact of fisheries co-management interventions in developing countries: a meta-analysis.** *J Environ Manage* 2011, **92:**1938–1949.
15. Hellebrandt D, Sikor T, Hooper L: *Rigorous Literature Review. Is the use of Renewable Natural Resources in the Developing World more or less Sustainable, Pro-Poor and Profitable under Controlled Access compared to Open Access.* Norwich, UK: International Development UEA –DEVCo; 2012:1–98.
16. Sayer J, Sunderland T, Ghazoul J, Pfund J-L, Sheil D, Meijaard E, Venter M, Boedhihartono AK, Day M, Garcia C: **Ten principles for a landscape approach to reconciling agriculture, conservation, and other competing land uses.** *Proc Natl Acad Sci* 2013, **110:**8349–8356.
17. Larson AM, Barry D, Dahal GR, Colfer CJP: *Forests for People: Community Rights and Forest Tenure Reform.* London: Earthscan; 2010.
18. CEE: **Guidelines for Systematic Reviews in Environmental Management (version 4.2).** *Collaboration Environ Evid* 2013, 1–80.
19. Schlager E, Ostrom E: **Property-Rights Regimes and Natural Resources: A Conceptual Analysis.** *Land Econ* 1992, **68:**249–262.
20. Waldron J: *The Right to Private Property.* Oxford: Clarendon Press; 1988.
21. Bromley DW: *Environment and Economy: Property Rights and Public Policy.* Oxford: Blackwell; 1991.
22. Smith EP: **BACI design.** *Encyclopedia Environ* 2006, **1:**141–148.
23. Porter-Bolland L, Ellis EA, Guariguata MR, Ruiz-Mallen I, Negrete-Yankelevich S, Reyes-Garcia V: **Community managed forests and forest protected areas: An assessment of their conservation effectiveness across the tropics.** *For Ecol Manag* 2012, **268:**6–17.
24. Robinson BE, Holland MB, Naughton-Treves L: **Does Secure Land Tenure Save Forests? A Review of the Relationship between Land Tenure and Tropical Deforestation.** In *CCAFS Working Paper.* CGIAR Research Program on Climate Change, Agriculture and Food Security; 2011.
25. Shahabuddin G, Rao M: **Do community-conserved areas effectively conserve biological diversity? Global insights and the Indian context.** *Biol Conserv* 2010, **143:**2926–2936.
26. Higgins J, Green S: *Cochrane Handbook for Systematic Reviews of Interventions.* The Cochrane Collaboration; 2008.
27. Sikor T, Lund C: **Access and property: a question of power and authority.** *Dev Chang* 2009, **40:**1–22.
28. Leimu R, Koricheva J: **What determines the citation frequency of ecological papers?** *Trends Ecol Evol* 2005, **20:**28–32.
29. Sterne JA, Egger M: **Funnel plots for detecting bias in meta-analysis: guidelines on choice of axis.** *J Clin Epidemiol* 2001, **54:**1046–1055.
30. **Terms and Definitions-Global Forest Resources Assessment.** [http://www.fao.org/docrep/006/ad665e/ad665e06.htm]
31. **Fisheries Glossary online.** [http://www.fao.org/fi/glossary/]
32. **Glossary of Terms used in Range Management.** [https://globalrangelands.org/rangelandswest/glossary]
33. Bromley DW, Cernea MM: **The Management of Common Property Natural Resources: Some Conceptual and Operational Fallacies.** Volume 57. World Bank Discussion Papers; 1989.
34. Macpherson CB: **Property, Mainstream and Critical Positions.** Toronto, Canada: University of Toronto Press; 1978.
35. Eggertsson T: *Economic Behavior and Institutions.* Cambridge, U.K.: Cambridge University Press; 1990.
36. Honore AM: **Ownership.** Oxford: Oxford University Press; 1961.
37. **World Bank Country Classifications.** 2013.
38. **Timeline of Country and Capital Changes.** [http://en.wikipedia.org/wiki/Timeline_of_country_and_capital_changes]

7

Evidence on the environmental impacts of farm land abandonment in high altitude/mountain regions: a systematic map

Neal R Haddaway[1*], David Styles[2] and Andrew S Pullin[1]

Abstract

Background: Many ecosystems have developed in the presence of agriculture and cessation of management resulting from land abandonment can have significant ecological impacts. Around 56 percent of the utilised agricultural area of the European Union is classified as 'less-favourable areas' and much of this is mountainous. The small-scale and extensively managed farmlands that are common in mountain areas are particularly vulnerable to marginalisation and abandonment. We conducted the first systematic global mapping of evidence to inform stakeholders and policy makers of the potential impacts of farm land abandonment in mountain areas.

Methods: Evidence was collated from a range of academic literature databases and grey literature sources. Identified articles (8,489) were screened for relevance at title, abstract and full text using predefined inclusion criteria set out in a published protocol. Relevant studies (165 across 189 articles) were then mapped using predefined coding and critically appraised for internal validity (i.e. susceptibility to bias).

Results: Mapping identified a number of interesting themes in the evidence base: the majority of research was undertaken in arable and mixed farming systems; large evidence bases were found in China, Spain and Italy; studies were mostly observational with spatial/successional comparators; biodiversity, soil and vegetation were most frequently studied. Several knowledge gaps were identified: including outcomes (socioeconomics and environmental hazards), regions (key mountain ranges including the Himalaya), and specific outcome-region groups (e.g. vegetation and soil measures in the UK). Several deficiencies in methodology were identified across studies: a lack of replication; non-random sample selection; lack of methodological detail (including details of spatial scale, replication, and sample selection).

Discussion: Systematic mapping has produced a searchable database of studies relating to high altitude farmland abandonment. The map identifies a number of potential areas for fruitful future synthesis, for example research on biodiversity, soil and vegetation in the Loess Hilly Plateau in China, and soil research in Spain. Such synthesis would be rapid given the effort expended here in identifying and screening relevant articles. It also points to several areas that were under-represented in the literature, such as natural hazards (avalanche, fire and flood risk), that would potentially benefit from increased primary research.

Keywords: Agriculture, Abandonment, Mountains, Alpine, Remote, Farming, Environmental impacts

* Correspondence: neal_haddaway@hotmail.com
[1]Centre for Evidence-Based Conservation, School of the Environment and Natural Resources and Geography, Bangor University, Bangor LL57 2UW, UK
Full list of author information is available at the end of the article

Background

Farm land abandonment can be simply defined as the cessation of agricultural activities on a given surface of land, yet there is no common precise definition of agricultural farmland abandonment in the literature [1]. Farm land abandonment occurs when income or resource generation cease to be viable or sustainable and the possibilities of adapting via changes in farming practices have been expended [2]. According to a study by Ramankutty and Foley [3], global abandonment of croplands has occurred over an estimated 1.47 million km^2 between 1700 and 1992. Meanwhile, Pointereau et al. [1] estimate that 9.09 Mha of agricultural land have been abandoned across 20 European countries between 1990 and 2000. Data cited for France for the period 1992 and 2003 show that grassland represented 57% of abandoned agricultural land; cropland 30% and vineyards and hedges/groves each 6%. However, the lack of a standardised definition of abandoned agricultural land, and the difficulty of matching this to available datasets, means that accurate estimates of abandoned area are lacking.

Land abandonment has a number of well-studied drivers, including environmental (e.g. reductions in soil fertility), economic (e.g. market globalisation) and socio-political (e.g. rural depopulation) causes [4]. From a socio-economic perspective, the abandonment of agricultural land is typically regarded as detrimental, owing to the implied loss of employment and income in rural areas. From an environmental perspective, the impacts of abandonment may be viewed as either an opportunity for ecological restoration to a state prior to agricultural establishment (typically regarded as beneficial), or as the loss of an ongoing process of land management and an associated threat to biodiversity (typically regarded as detrimental). Whether land abandonment poses an ecological opportunity or threat depends upon the agricultural history and the presence of systems that depend upon regular management for their existence. In Europe, many ecosystems have developed in the presence of agriculture and the loss of continued management resulting from land abandonment can have significant negative ecological impacts [4]. Pointereau et al. [1] suggest that abandonment of intensive agriculture often results in ecological benefits for the affected parcel of land, whilst abandonment of low intensity agricultural is more likely to result in a negative ecological impact owing to the role of such agriculture in maintaining systems classified as "high nature value" (HNV).

Around 56 percent of utilised agricultural area (UAA) of the European Union (EU) is classified as 'less-favourable areas' by the Common Agricultural Policy (CAP). According to MacDonald et al. [2], much of this is mountainous, and a report in 2004 identified mountainous regions as constituting 39.9 percent of the area of the 15 Member States at the time [5]. Mountain areas, however, are difficult to define. For the purposes of examining farm land abandonment, mountainous areas are defined by their unfavourable topography, remoteness and extreme climate. Mountainous areas are typically described by elevation and/or slope, but this can vary significantly between countries. For example, Austria defines mountain areas as being above 700 m or above 500 m if slope is greater than 20 percent, whilst Spain more strictly defines them as being above 1000 m, over 20 percent slope and a 400 m elevation gain relative to surrounding land. Some definitions include low altitude areas where low mean temperatures and alpine vegetation reflect those in the high altitude Alps, such as Sweden and Finland. Other definitions use ruggedness assessed from satellite imagery e.g. [6].

The small-scale and extensively managed farmlands that are common in mountain areas are particularly vulnerable to marginalisation and abandonment [7]. A report from the Cross-Compliance Network identified mountainous areas as key areas likely to experience farmland abandonment [8]. The causes of farmland abandonment in mountainous areas are expanded upon in more detail in Pointereau et al. [1] to include; steep slope, distance from the farm to the field, low accessibility, poor soils, land used as alpine pastures, small farms, high cultivation costs and low field size.

Resilience and adaptability in farming systems in mountain regions is limited for a number of reasons, including remoteness, climate and physical constraints, and the aversion to risk-taking, traditional cultural values and limited skill sets often held by the local population [2]. Limitations to the adaptability of agriculture in mountain regions have been compounded by the historical paucity of agricultural research in these areas and a bias towards lowland regions e.g. [9].

Consensus in the literature about the impacts of farmland abandonment in mountain regions is generally lacking. A limited review of CAB Abstracts focusing on land abandonment was published in 2007 [10]. A systematic review is currently underway on the subject of land abandonment in the Mediterranean [11]. A conceptual review of several case studies of land abandonment and EU policies responding to the problem for mountain areas was published in 2000 [2]. Systematic mapping was chosen as a suitable method for collating and cataloguing evidence of the impacts of land abandonment because the subject has not been synthesised, and the state of the evidence base was unknown. A systematic map of evidence on high altitude farmland abandonment was chosen as part of a wider project investigating best practice in agriculture as a demonstration of the suitability of the method for mapping documented knowledge. Systematic mapping appropriately identifies knowledge gaps and potential systematically reviewable questions; key benefits in this subject area. This systematic map is also intended as a demonstration of the

Evidence on the environmental impacts of farm land abandonment in high altitude/mountain...

107

suitability of the method for agricultural research synthesis in informing policy-making. The work herein forms the first systematic mapping of the evidence of impacts of farm land abandonment in mountain areas across the globe.

Objective of the review

Primary question

The primary question of this systematic map is;

What evidence exists on the environmental impacts of farm land abandonment in mountain regions?

This review is in the form of a systematic map, cataloguing the existing evidence across a wide range of variables such as setting, methodology, scale, measured outcomes etc. Mapping was undertaken to full text, meaning that the full text articles of all relevant abstracts were assessed and relevant full texts coded based on the information in these full text documents. The primary output and key objective of the systematic mapping process is an interrogatable database of research into the socio-economic and environmental impacts of high altitude/mountainous regions.

The question has the following components:

Population: All mountainous agricultural lands (global scope).
Exposure: Abandonment of agricultural land management. This definition is in accordance with that of Coppola [12] and Pointereau et al. [1] and specifies the cessation of all agricultural activity.
Comparator: Before-after land abandonment (temporal comparator), or un-abandoned nearby surrogate (spatial comparator).
Outcome: All outcomes relating to environmental and socio-economic impacts, including but not restricted to; natural hazards (fire-/flood risk, land/mud slides), soil (fertility, erosion), water (chemistry, eutrophication, sediment load, hydrology), ecosystem functioning (biodiversity, abundance, invasive species presence), human health and wellbeing (including income, employment, attitude).

Methods

The final methodology for the systematic map reported herein differs in several ways from the published protocol [13]. The title has been modified because it more accurately reflects the objective of the mapping exercise. Due to the relatively high exclusion of full texts following screening at abstract level it was decided not to generate a map based on abstracts only in addition to the full text map, since its accuracy would be questionable. Minor changes to the coding occurred due to the iterative nature of the coding and full text assessment process. Finally, rather than assign critical appraisal categories to each study it was decided to use a basic scoring system for the purposes

of assessing the collective susceptibility to bias of subsections of the evidence rather than individual studies.

Search strategy

Search terms

Scoping was undertaken in order to identify suitable relevant key terms to be included in the finalised search string (see Additional file 1). These terms include aspects of the exposure (farm land abandonment) and the population (high altitude/mountain regions) and the finalised search string is displayed in Table 1. Outcome terms were not included in the string because of the number of returns based only on exposure and population terms, which was deemed to be manageable. Furthermore, the aim of the map is to document the available literature, including the variety of forms of outcomes measured in the evidence base. Outcome documentation was therefore an iterative process, and all relevant outcomes were coded.

Search terms were only established in English language. The major academic databases detailed below catalogue non-English language research by translating titles, abstracts and keywords into English. There is a risk that this may introduce bias, but the call for evidence submitted to the identified author list and via social media (see below) should act as to include some non-English evidence if it was available. The inclusion of non-English language academic databases was outside the scope of this review, but would be a worthwhile addition during any update.

Databases

The search aimed to include the following online databases which cover the breadth and depth of available literature on the topic:

1) ISI Web of Knowledge (inc. ISI Web of Science and ISI Proceedings)
2) Science Direct
3) Directory of Open Access Journals
4) Copac
5) Agricola
6) CAB Abstracts
7) CSA Illumina/Proquest
8) GreenFile

Where databases did not accept the full search strings detailed in Table 1, search strings were modified according to the database help files, sometimes based on only pairs of exposure terms to be as sensitive as possible. All database searches and outcomes are recorded in Additional file 1.

Search engines

The following internet search engines were used to identify relevant grey literature. The first 150 hits from each

Table 1 Finalised search string following scoping in Web of Knowledge

	Search string	WoK hits
Exposure terms	((grassland OR farm* OR cropland OR agriculture* OR land OR *field OR pasture) AND (destock* OR abandon*)) AND	26,351
Population terms	("high altitude" OR "higher altitude" OR "high ground" OR "higher ground" OR *alpine OR montane OR mount* OR elevat* OR highland OR hill* OR upland OR plateau OR mesa OR tableland OR slope OR aspect OR remote* OR massif OR sierra OR steep OR rugged OR apennine OR alps OR volcano* OR Carpathian* OR Pyren* OR Caucasus OR Andes OR Rockies)	7,213

*denotes wildcard function that includes all possible word endings.

engine were screened (based on sorting by relevance of results where possible). Where abstracts were lacking articles were included to the full text assessment stage.

Google Scholar http://scholar.google.co.uk/.

Scirus http://www.scirus.com/.

Dogpile http://www.dogpile.co.uk/.

All search engine searches and outcomes are recorded in Additional file 2.

Specialist sources

The following specialist organisations were searched for relevant grey literature using manual searches of their websites and automatic search facilities using key terms (such as abandon*).

Alterra http://www.wageningenur.nl/en/Expertise-Services/Research-Institutes/alterra.htm.

Centre for Ecology and Hydrology http://www.ceh.ac.uk/.

National Farmers Union http://www.nfuonline.com/home/.

Global Environment Centre http://www.gec.org.my/.

Greenpeace http://www.greenpeace.org.uk/.

Joint Nature Conservation Committee http://jncc.defra.gov.uk/.

Macaulay Land Use Research Institute http://www.macaulay.ac.uk/.

National Soil Resources Institute http://www.cranfield.ac.uk/sas/nsri/.

Natural England http://www.naturalengland.org.uk/.

Royal Society for the Protection of Birds http://www.rspb.org.uk/.

Society for Ecological Restoration http://www.ser.org/.

DEFRA http://www.defra.gov.uk/.

Environment Agency http://www.environment-agency.gov.uk/.

PBL Netherlands http://www.pbl.nl/en/.

German Federal Ministry of Agriculture http://www.bmelv.de/EN/.

Thunen Institute http://www.ti.bund.de/en/.

ETH Zurich http://www.ethz.ch/index_EN.

European Environment Agency http://www.eea.europa.eu/.

EC Ag and Rural Dev site http://ec.europa.eu/agriculture/.

IEEP http://www.ieep.eu/.

JRC Institute for Env Sustainability http://ies.jrc.ec.europa.eu/.

JRC Institute for Prospective Tech Studies http://ipts.jrc.ec.europa.eu/.

United Nations Environment Programme http://www.unep.org/.

Food and Agriculture Organisation http://www.fao.org/index_en.htm.

Convention on Biological Diversity http://www.cbd.int/convention/.

World Wildlife Fund http://www.wwf.org.uk.

Associations des Populations des Montagnes du Monde http://www.mountainpeople.org.

Mountain Partnership http://www.mountainpartnership.org.

The International Centre for Integrated Mountain Development http://www.icimod.org.

Where organisational website search facilities did not accept the full search strings detailed in Table 1, search strings were modified according to the search help files (where provided), or subsets of key terms were searched individually. All organisational website searches and outcomes are recorded in Additional file 3.

Search comprehensiveness assessment

The comprehensiveness of the above search strategies was assessed in a number of ways. Firstly, key bibliographies from relevant reviews were compared to the search results to check that all relevant articles have been identified through searches. These bibliographic testing strategies are detailed in Additional file 4. Secondly, search results were compared with a list of includable studies, identified by subject experts prior to the review. This key article testing is described in the database search strategy in Additional file 1.

Article retrieval

Full texts were obtained digitally where possible (as PDF/Microsoft Word files), using Bangor University subscriptions if necessary. Where sources were not available through Bangor University, authors were contacted directly via email for copies of their articles. Resources that were unavailable digitally were obtained physically and scanned in (in accordance with copyright law).

Author contact

Authors and research-related stakeholders were contacted with a request for the submission of relevant evidence in a number of ways. Firstly, when authors of inaccessible articles identified through the above searches were contacted requesting unobtainable articles they were also asked to submit evidence that may not be catalogued by academic databases, particularly grey literature and non-English language articles. Secondly, calls for submission of evidence were made using the social networking web sites Linked In, Research Gate and Academia. edu. Finally, once the map database had been populated, all author email addresses extracted from full texts were contacted with a further request for the submission of relevant evidence that may have been missed by the methodology detailed in the published protocol. Evidence submitted at this time, once the map was already complete, could not be screened, coded and included in the map, but a list of potentially relevant studies submitted through this call for evidence is included in Additional file 5 as a basis for future updating of the map.

Study inclusion criteria

Study selection, according to the predefined inclusion criteria detailed below, proceeded in a three stage, hierarchical process: titles, abstracts and finally full texts were assessed against the inclusion criteria. Any doubt over the presence of a relevant inclusion criterion (or if information is absent) resulted in the articles being retained for assessment at a later stage. Title- and abstract-level assessment did not assess the presence of a comparator, which is typically not explicit. Since titles and abstracts in grey literature often do not conform to scientific standards, assessment proceeded immediately to full text assessment. Consistency checks were undertaken using a subset of 100 abstracts by two reviewers (NH and AP) independently of one another. Screening decisions were then compared using a Kappa test of agreement [14]. A score of 0.613 was obtained, which indicates substantial agreement. The few cases of disagreement were discussed and understanding of the inclusion criteria was improved before screening of the remaining abstracts.

The following aspects of the systematic review question formed inclusion criteria when assessing potentially relevant literature:

Relevant population(s): Any high altitude or mountainous region, any region with restricted access due to ruggedness, any region with possible agricultural difficulties or limits on agricultural advancement or adaptability as a result of slope, altitude, topography or ruggedness [global scope]

Types of exposure/intervention: Abandonment of agricultural land or reinstating of agricultural activity in agricultural land following abandonment

Types of comparator: Before land abandonment and/or an un-abandoned control site, or a relevant successional gradient representing change following abandonment (i.e. recent abandonment)

Types of outcome: All environmental and socioeconomic outcomes, including but not restricted to; soil chemistry (including carbon and GHG flux), soil erosion, water chemistry, hydrology, natural hazards, biological diversity and abundance, presence of invasive species, human health and wellbeing

Types of study: Both observational and experimental field studies. Experimental field studies (i.e. simulated abandonment) must investigate continued abandonment over a period in excess of one year

Unobtainable and untranslatable studies are listed in Additional file 6. Excluded studies are listed along with reasons for exclusion in Additional file 7.

Map coding

Coding of studies was undertaken based on full texts using key words and expanded comments fields describing various aspects of study design and setting. Key variables of interest were identified through scoping activities and discussion with subject experts. Coding options within these key variables were then compiled in a partly iterative process, expanding the range of options as they were encountered during scoping. The finalised coding tool for the map is displayed in Additional file 8.

Individual lines within the database represent a unit of one study-article, i.e. each individual reporting of a study. Multiple studies reported within one article are entered as independent lines in the database. Separate articles that report different outcomes from one study are entered as separate lines. This is to reflect the possible differences in reporting between different articles on the same study. These linked articles are highlighted as such, however, and are treated as one study unit.

Critical appraisal of study internal validity

Coding was used to describe the internal validity (IV) of each included study. Each study was given a score for each of the five variables listed A-E below. Studies therefore have a possible score of between 0 and 10. External validity was not taken into account since this is a systematic map and does not relate to any one context. Care was taken to avoid double scoring of individual failings, such as a lack of replication or low methodological detail. A subset (10%) of studies was scored by an additional reviewer and disagreements were resolved through discussion and clarification. A breakdown of the scoring by the following variables is displayed in the systematic map database in Additional file 9.

A. Replication

Studies were assessed for their degree of true replication. True replication exists only at the level at which the intervention is administered/the exposure experienced. Pseudoreplication was inadmissible in this scoring. Sample size refers to the smallest sample (whether in comparator or intervention). Successional gradients count as 1 group.

a. Well-replicated (>10 samples per group) 2
b. Moderate level of replication (4–10 samples per group) 1
c. Poorly-replicated/not stated (1–3 samples per group) 0

B. Sample selection

Study sample selection methods were assessed. Any form of randomised selection procedure scored highly, as did studies that deliberately spread samples evenly across clear potential confounders. Implied random or blocked designs received an intermediate score. Stated purposive selection or those that are clearly purposive (i.e. where no replication) failed to score.

a. Random/blocked/exhaustive 2
b. Not stated but possibly random/blocked 1
c. Purposive/not stated 0

C. Level of methodological detail

Articles' level of methodological detail was assessed. Well documented studies scored highly. Those with some missing information received an intermediate score. Those with very limited methodology failed to score. Understandability of text and grammar/spelling were not taken into account.

a. High 2
b. Moderate 1
c. Low 0

D. Other sources of potential bias

Sources of potential bias that do not include a lack of replication or sample selection (to avoid double scoring) were assessed. Potential confounders include; clear differences in environmental conditions, substantial spatial separation between intervention and control sites, or the presence of other variables that may confound the impact of the intervention/exposure.

a. None evident 2
b. Potential confounder 1
c. Clear confounder 0

E. Study design

Studies were assessed for the form of the comparator. If both before/after and intervention/comparator are available (BACI; before-after-control-impacts) the study scored highly. Studies with one comparator received an intermediate score. Studies with successional gradient comparators that lack a baseline failed to score.

a. BACI 2
b. Modelled 1
c. Temporal (before/after) 1
d. Spatial (intervention site/control site) 1
e. Successional gradient (no baseline)/not stated 0

Systematic map database

The systematic map output is in the form of a database of studies that describes the nature and location of evidence on the review topic. The database is provided in Additional file 10: Table S1. This database is easily searchable and freely accessible. A help file to assist readers with interrogation of the map is provided in the Appendix. The map may form the basis for further primary research by identifying key knowledge gaps, and may also form the basis for further secondary research as a starting point for the synthesis of information in focused systematic reviews.

Results

Evidence identification, retrieval and screening

A schematic showing the processes involved in this systematic map and numbers of articles and studies moving between stages in shown in Figure 1.

Searches of academic literature databases, undertaken between 7th and 13th May 2013 (detailed in Additional file 1), identified 9,355 potentially relevant articles. Web search engines added an additional 498 articles (see Additional file 2). Identification of duplicates removed 1,455 references. Screening of these results at title and abstract yielded 1,473 and 650 relevant articles, respectively. Other sources of information (bibliographic checking, web search engines, organisational websites, secondary sources and author submissions; see Additional files 3 and 4) contributed an additional 91 articles; 741 potentially relevant full texts in total. A total of 215 articles could not be obtained in full text and 26 articles were not included in further stages due to complications with translation (see Additional file 6). Table 2 details the main reasons for article inaccessibility, citing the numbers of articles in each category.

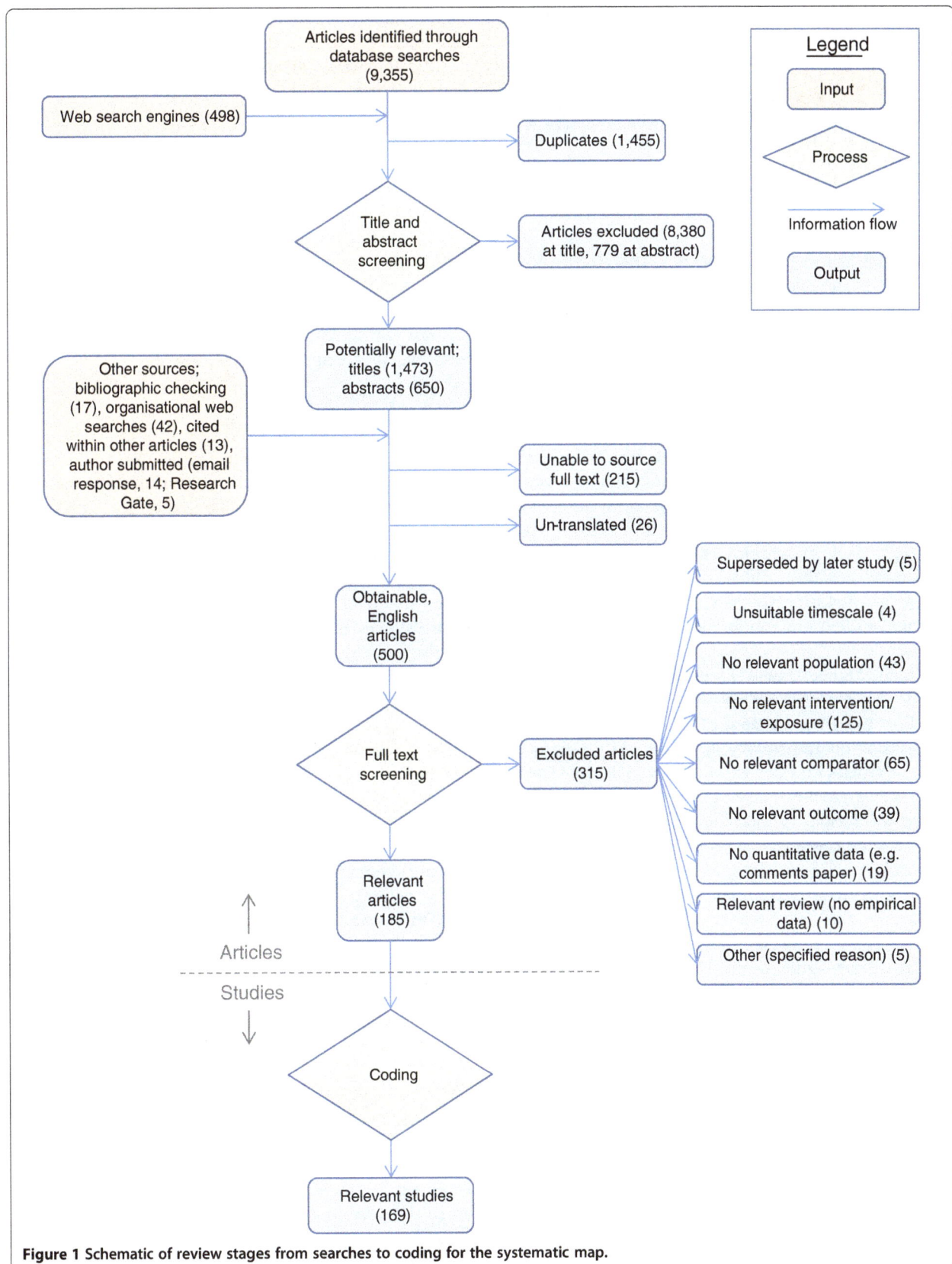

Figure 1 Schematic of review stages from searches to coding for the systematic map.

A total of 500 articles were obtained in full text (67% retrieval rate). A total of 59 contact email addresses for authors of articles that were not obtainable were contacted with a request for full texts: 21 requests were successful, 1 referred to a conference presentation, 1 was not available electronically, and 35 email addresses failed (receiving bounces/fail alerts). This retrieval rate is low for systematic reviews. We attribute this to the large proportion of research published in Chinese journals that have specific access restrictions. This low retrieval rate may, therefore, reduce the utility of the map with respect to Chinese research, but much of this research was published in Chinese, reducing its usability in this review if it were to have been obtained.

Of the 500 obtained articles, 315 were excluded (see Additional file 7). Reasons for exclusion were: article was superseded by a later article; study was undertaken over an unsuitable timescale (i.e. intervention was enacted for less than one year); lack of relevant population (i.e. non mountainous); lack of relevant intervention/exposure (including incomplete abandonment); lack of relevant comparator; lack of relevant outcome (e.g. land use change documented alone); no quantitative data (e.g. comments paper); relevant review (no empirical data); other specific reasons.

In total, 185 articles were coded for the systematic map database, corresponding to 169 individual studies.

Articles came from a range of time periods, with more publications on the subject in recent years (Figure 2). Thirteen articles included in the map were in non-English language, with the majority of these (10) being in Chinese. The earliest included publication was from 1964, with the number of articles increasing over time in an exponential manner.

Population descriptors used by articles included in the systematic map are displayed in Figure 3. The most common descriptors were 'alp*', 'hill*', 'mount*' and 'plateau'. Seven articles were identified by searching of other resources (i.e. not academic databases). These seven articles did not use a population descriptor in the title, abstract or keywords, but the studies within them

were included according to the application of the inclusion criteria in spite of this.

Critical appraisal identified a large number of studies that undertook no or little replication (Figure 4). Very few studies employed randomisation or exhaustive sampling. In general, methodological detail was limited, with 60% of studies given a '0' rating in this critical appraisal category. Studies scored highly in general for confounding factors, since methodological details were typically low and identifying potential confounders was, as a result, rarely possible. Studies typically scored moderately well for study design: few studies had strong forms of design (i.e. BACI), with most using a single comparator. A subset of 37 studies examined successional gradients following abandonment with no baseline.

Systematic map findings

Studies identified in this systematic map were undertaken across 35 countries (Figures 5 and 6). China was the most studied country (41 studies), followed by Spain (24), Italy (19), France (12) and Switzerland (11).

The altitude of study sites differed substantially between countries (Figure 7), as do the definitions of the term 'mountain' see [13]. Eighteen of the 34 countries whose studies reported altitude were from above 1,000 metres above sea level (m.a.s.l.), with an average altitude for the 127 studies from 18 countries of 1,764 m.a.s.l. The remaining studies had an average study height of 592 m.a.s.l.

There was a fairly even mixture of studies on arable abandonment and pasture abandonment, with a smaller subset of studies examining abandonment of mowing (Figure 8). A large number of studies failed to state the farming system (27 studies). Arable was the predominant farming system in China, Spain and Italy (Figure 6).

The majority of studies in the map were observational, with the remainder being predominantly experimental and some modelled studies (Figure 9a). A number of studies (13) employed BACI formats (both temporal and spatial comparators), but the majority used either spatial comparators (i.e. before and after) or successional gradients (i.e. several time periods following abandonment) (Figure 9b).

Figure 10 displays studies' time since abandonment, separated by observational, experimental and other study designs. Observational studies cover a fairly consistent range of abandonment periods, although 31–40 years appears to be underrepresented. A large number of studies failed to report abandonment period (25 studies). Experimental studies were most frequently carried out in recently abandoned farmland (1–10 years), and no experimental studies above the 31–40 year abandonment category were found.

Table 2 Unobtainable articles at full text with description of access restriction

Description of access restriction	Number of articles
No institutional subscription	134
Unpublished conference proceedings/oral presentation	11
Citation only, no indication of holdings/online access	70

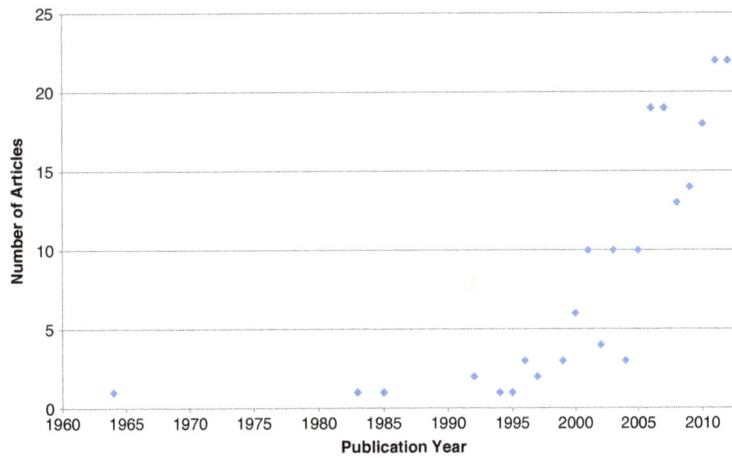

Figure 2 Publication year of articles included in the systematic map.

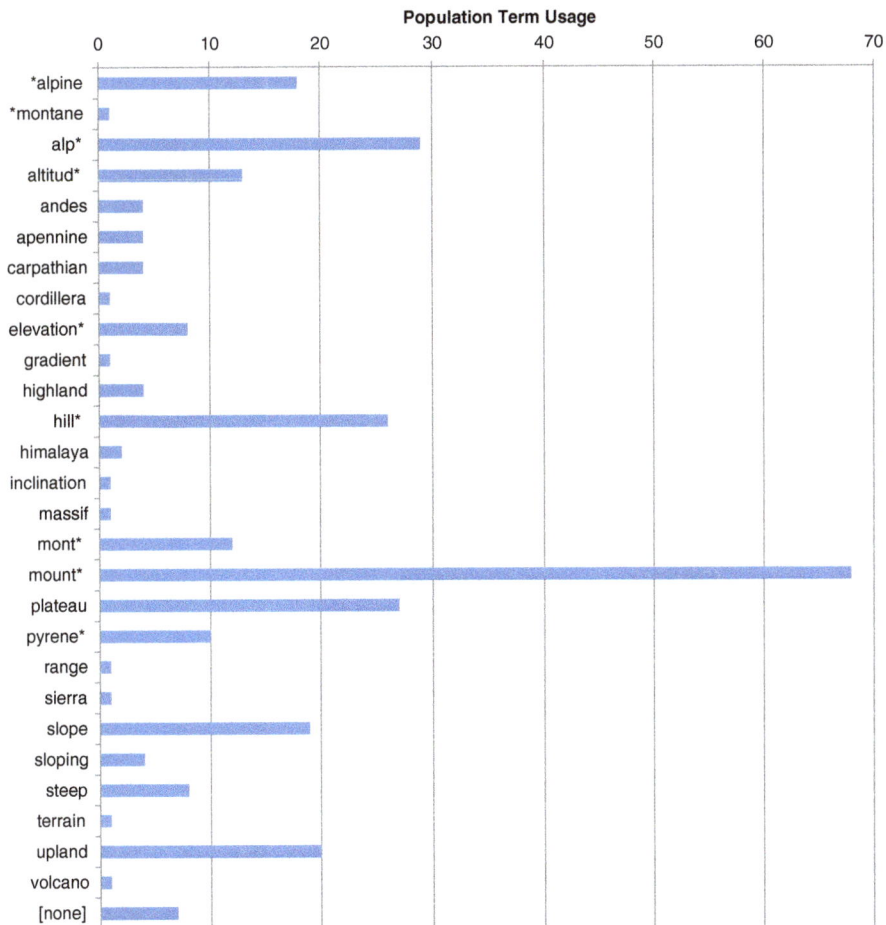

Figure 3 Population descriptors used within the title, abstract and keywords of articles included in the systematic map.

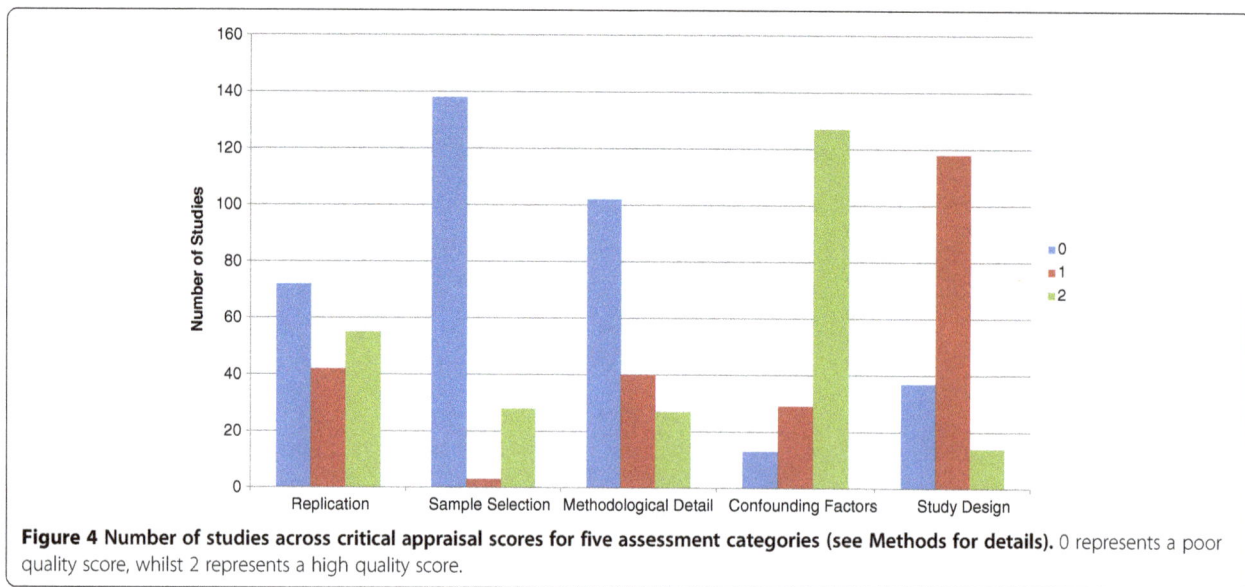

Figure 4 Number of studies across critical appraisal scores for five assessment categories (see Methods for details). 0 represents a poor quality score, whilst 2 represents a high quality score.

The most frequently studied outcomes were biodiversity, soil and vegetation (Figure 11a). Vegetation outcomes predominantly related to all vascular plants, with some named vegetation subgroups (lichen, moss, roots, shrubs and grasses and trees) also reported (Figure 11b) and included such measures as tree density, standing crop, and shoot dry mass. Soil outcomes were predominantly soil chemistry (e.g. pH, solute concentration, elemental concentration), and soil structure (e.g. bulk density, porosity, sand/silt/clay composition) (Figure 11c). Biodiversity

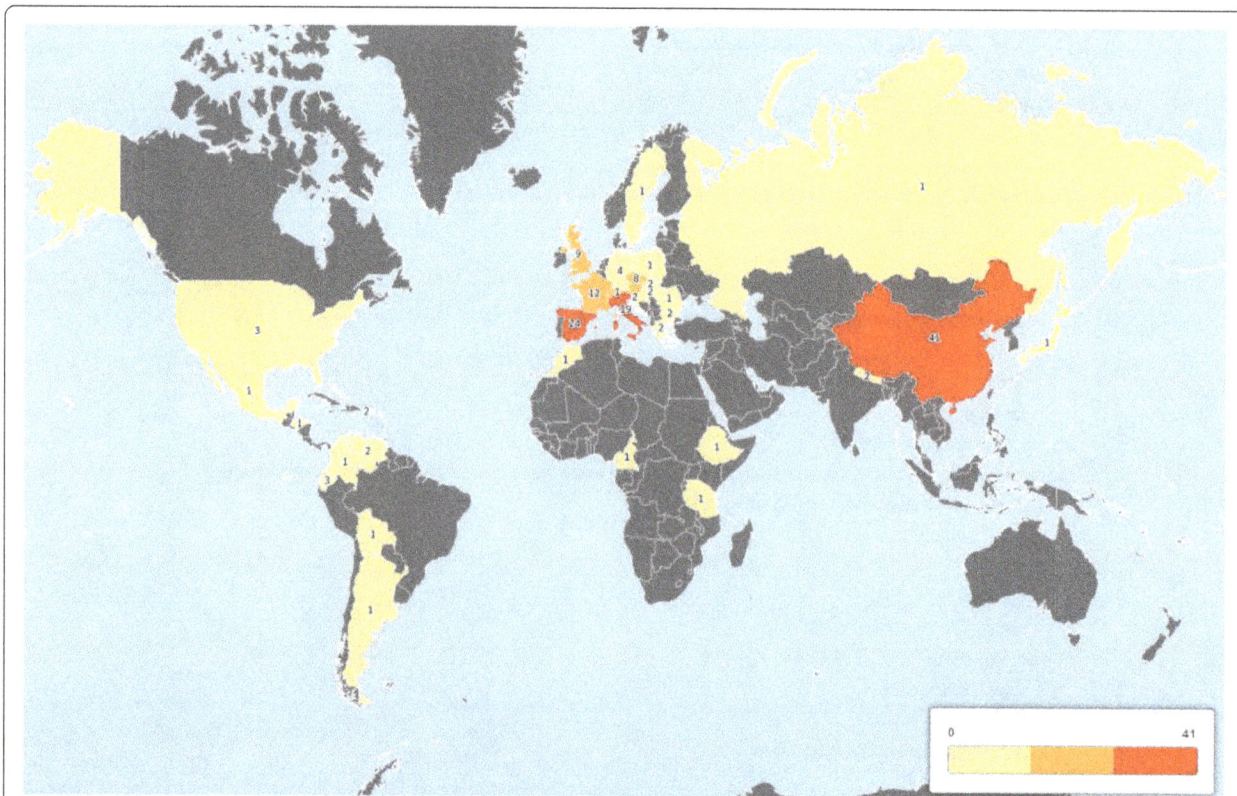

Figure 5 Geographical spread of the studies included in the systematic map, showing the number of studies per country. Studies undertaken across more than one country are counted within each study country. Produced online at www.cartodb.com.

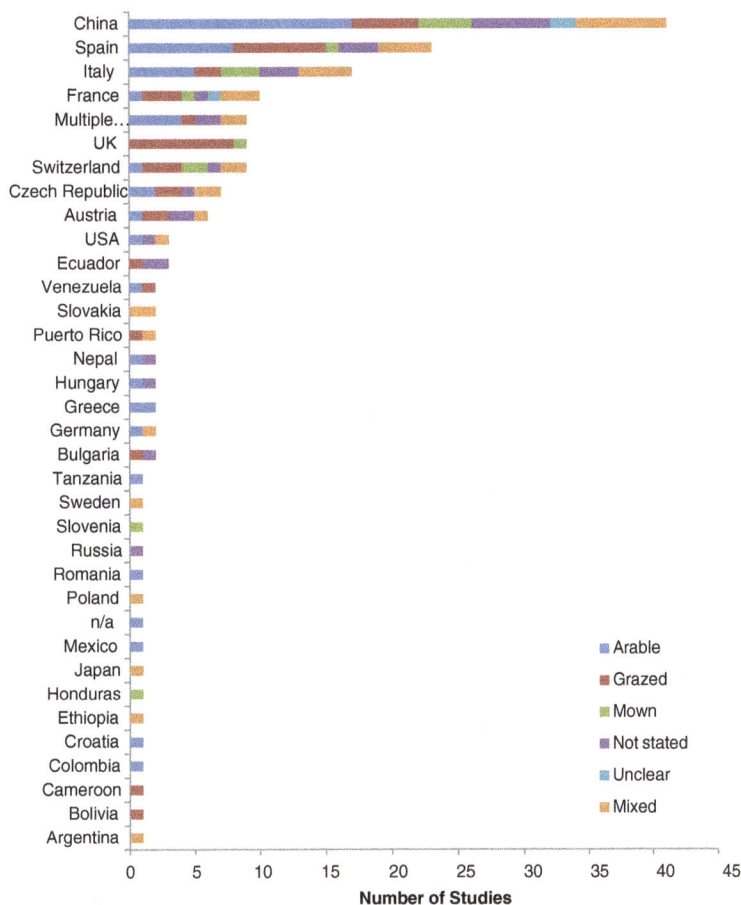

Figure 6 Number of studies performed in different countries separated by farming system. Studies undertaken across more than one country are counted within each country.

outcomes were overwhelmingly measures of vascular plant diversity, including diversity indices, abundance, richness, and evenness.

Additional file 10: Tables S1 and Additional file 11: Table S2 display the systematic map across outcomes and countries enabling an assessment of which bodies of combinable evidence are sufficient to allow synthesis and which areas may represent knowledge gaps. Additional file 10: Table S1 shows the number of studies in each country for the 12 major outcome groups and Additional file 11: Table S2 displays the total critical appraisal score for each of these cells. By highlighting the cells with a high volume and high combined quality of evidence it is apparent that the evidence base in China is substantial, covering a large body of studies on biodiversity, soil and vegetation outcomes, and this evidence has a high total CA score. In general the evidence across Europe is relatively strong (a relatively high total CA score), both in terms of the numbers of studied outcomes and the strength of the evidence. Additional file 12: Table S3 details the mean critical appraisal score, and identifies countries and outcome

groups that have evidence that is in general of particularly high standard. This table demonstrates that, typically, single studies of low susceptibility to bias are present in the evidence base, but that groups of studies that are consistently low susceptibility to bias are not present. Thirty-two of the 91 outcome groups/countries combinations identified by the map were of a mean critical appraisal score of 5 or above.

Key knowledge gaps can be seen across all countries for socioeconomic outcomes (including employment, farming, human attitude) and outcomes relating to environmental hazards (including snow and water). Regions other than Europe and China are underrepresented, with evidence from key mountain ranges missing; for example the Rockies, the Andes, the Caucasus, the Himalaya, the Karakoram, the Great Dividing Range and the Urals.

The subjects above, although gaps, are not necessarily worthy of filling by novel primary research. Some of these gaps are surprising, for example environmental hazards, since avalanche, land slip and flood risk should be particular concerns in mountainous regions. The particular

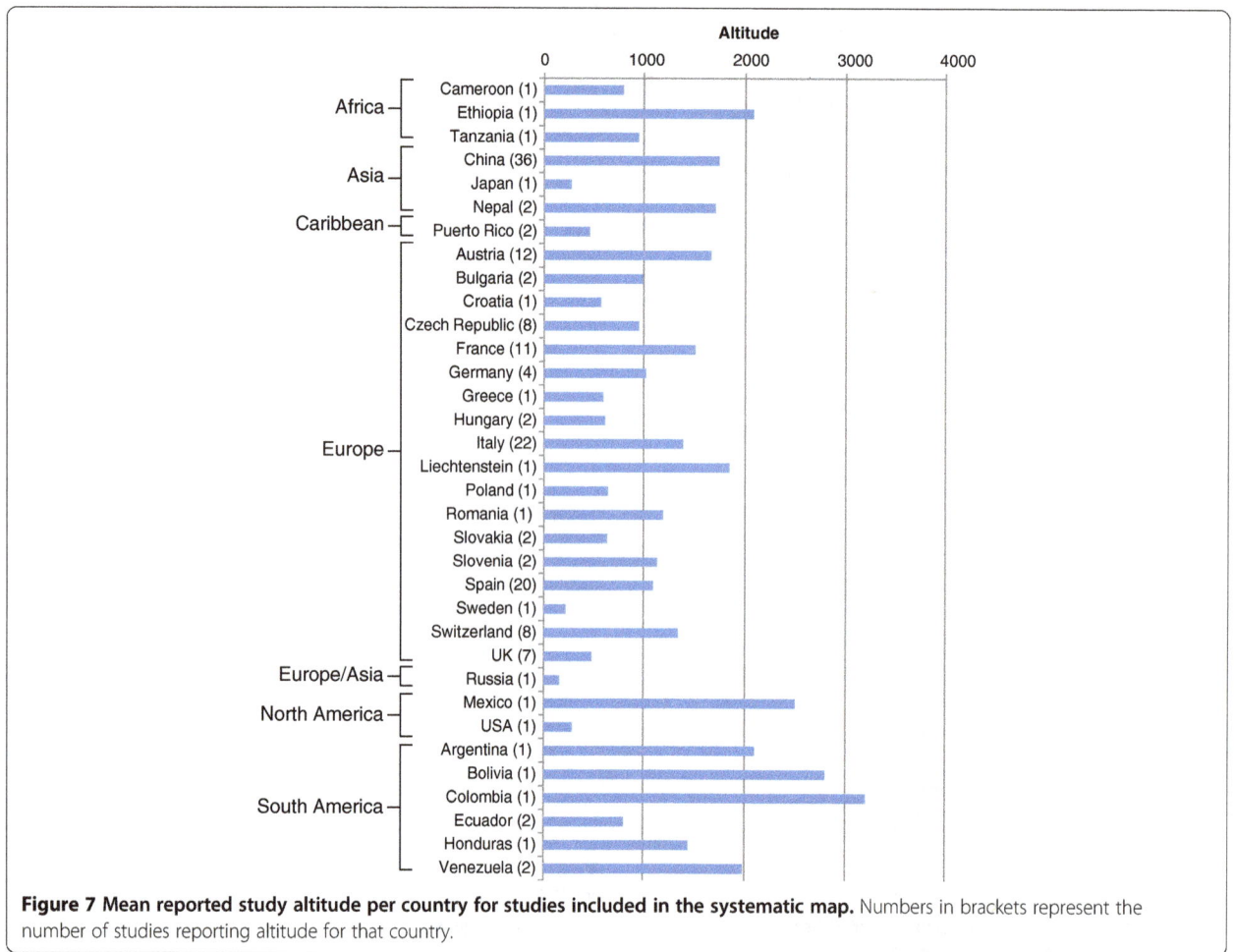

Figure 7 Mean reported study altitude per country for studies included in the systematic map. Numbers in brackets represent the number of studies reporting altitude for that country.

geographical concentration of research is likely to be driven by the availability of research funding and funded demand for knowledge in decision-making.

China and Spain are well-represented by studies reporting soil outcomes. Studies measuring soil outcomes are shown for these two countries in Figures 12 and 13 respectively. Whilst there is a geographical bias in both countries (in the Loess Hilly Plateau in China and the Aisa Valley in the Pyrenees in Spain), there is a moderate spread of a minority of studies across other areas.

Discussion
Key findings

The aim of this map is to document the available evidence on the topic of farmland abandonment in high altitude and mountainous regions, where topography adversely

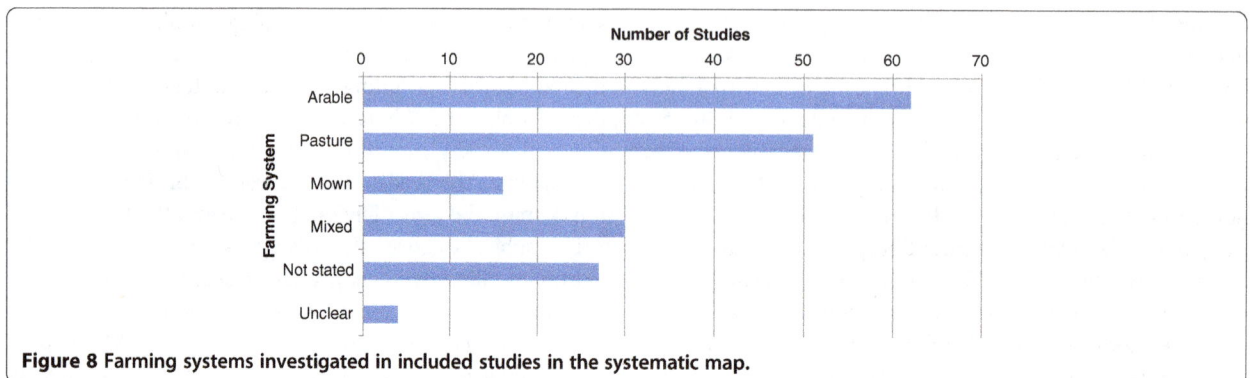

Figure 8 Farming systems investigated in included studies in the systematic map.

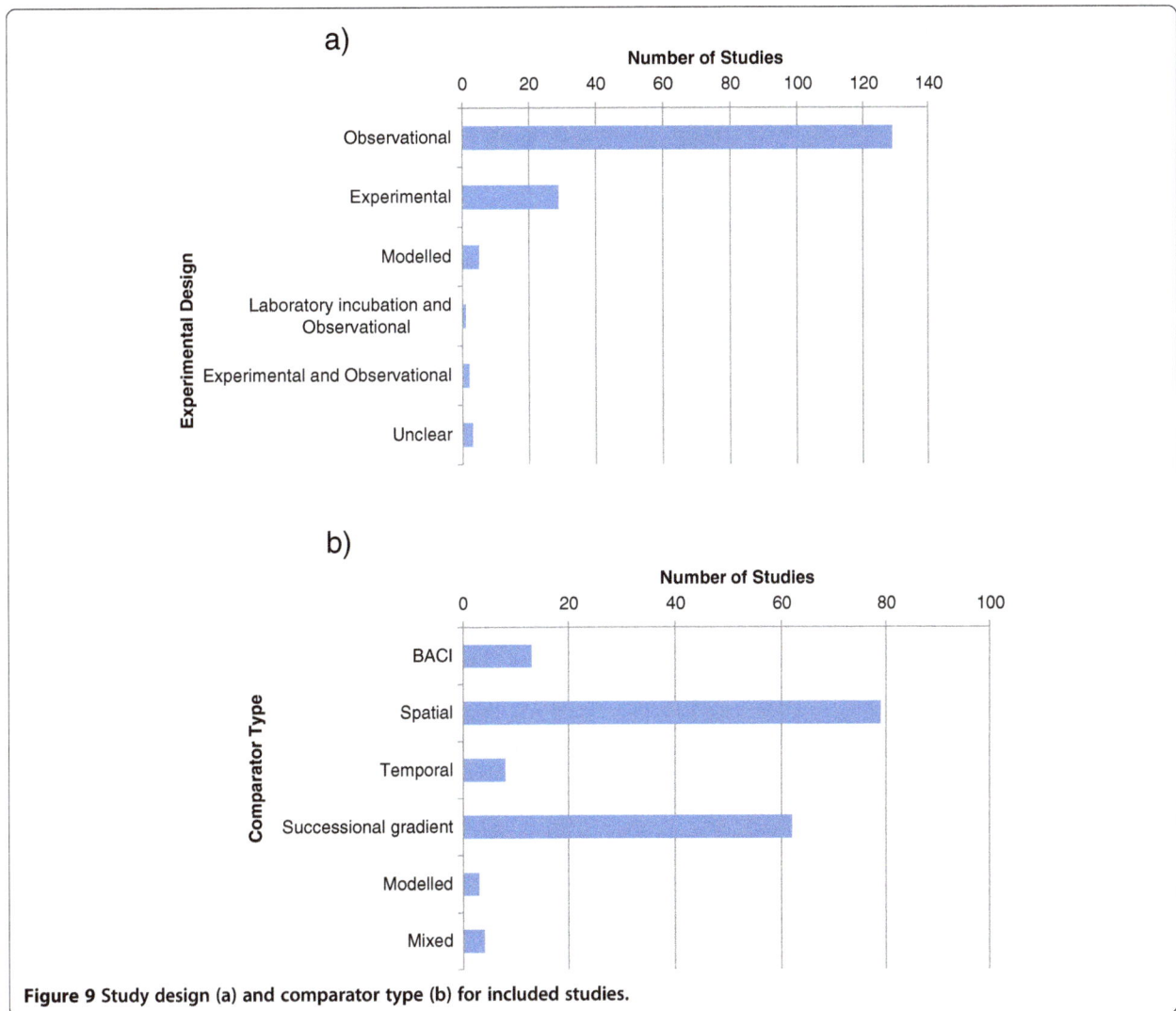

Figure 9 Study design (a) and comparator type (b) for included studies.

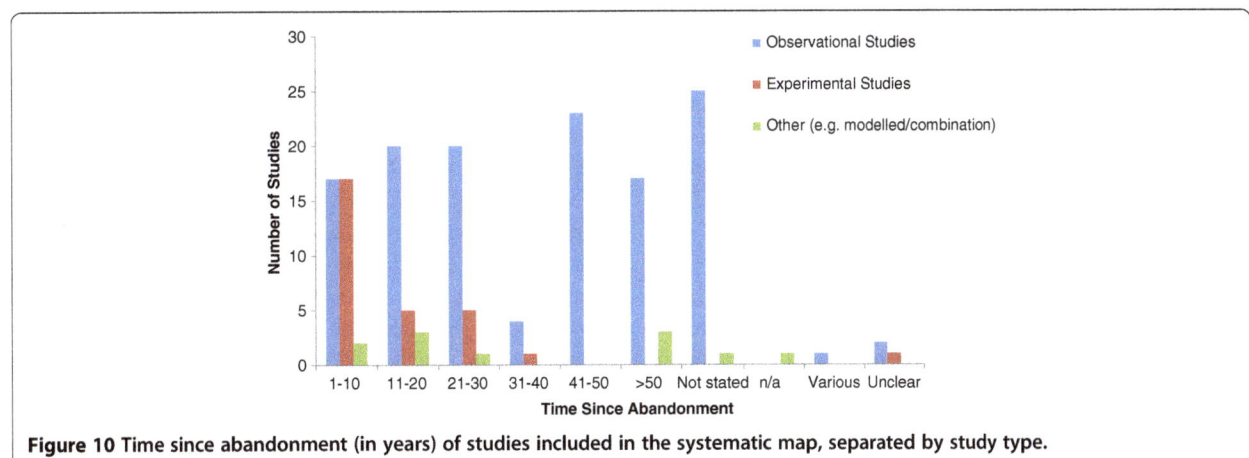

Figure 10 Time since abandonment (in years) of studies included in the systematic map, separated by study type.

a) all outcomes

Number of Studied Outcomes

Categories (top to bottom): animal behaviour, atmosphere, biodiversity, employment, farming, human attitude, light, litter, snow, soil, vegetation, water

b) vegetation outcomes

Number of Studied Outcomes

Categories (top to bottom): lichen, moss, vascular plants, roots, shrubs and grasses, trees

c) soil outcomes

Number of Studied Outcomes

Categories (top to bottom): landslides, runoff, soil chemistry, soil erosion, soil macrofauna, soil microbes, soil productivity, soil structure, soil water

d) biodiversity outcomes

Number of Studied Outcomes

Categories (top to bottom): birds, reptiles, mammals, Coleoptera, Formicidae, Lepidoptera, Orthoptera, Gastropoda, Megadrilacea (earthworms), other Arthropoda, soil macrofauna, soil microbes, mycorrhiza, bryophytes, lichen, trees, vascular plants, woody plants

Figure 11 Included studies' outcome groups for (a) all reported outcomes, (b) vegetation outcomes, (c) soil outcomes, and (d) biodiversity outcomes.

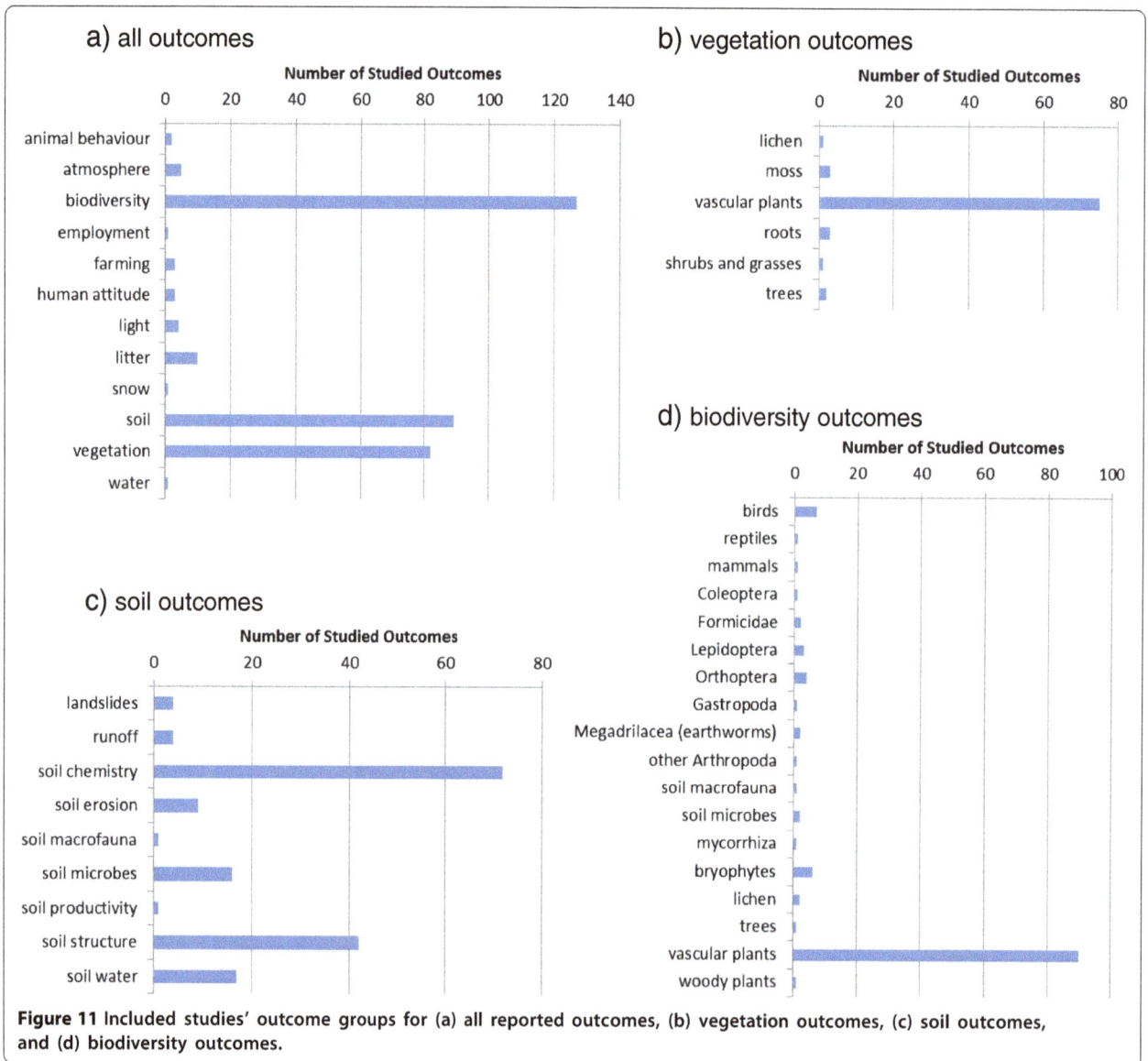

affects farming activities. We have identified a body of 185 articles describing 169 studies of agricultural abandonment. This abandonment spans various farming systems, 35 named countries and 12 major outcome groups. The volume and quality of the various subgroups of evidence (i.e. by country and outcome) vary greatly, and whilst some areas may be represented by high numbers of studies, the average quality of the evidence is not particularly high. There appear to be a number of regions that are well-studied, including the Aisa Valley in Spain and the Loess Plateau in China. Whilst these bodies of evidence may lend themselves well to further synthesis, they may be rather limited in terms of their generalisability because of their limited geographical extent.

Limitations of the systematic map

The following caveats should be highlighted when considering the outputs of this systematic map.

Descriptors of the topography of included studies vary significantly in the evidence base according to researchers and study countries. It is well-known that countries have different official definitions of 'mountainous regions', for example in European Union member states see [13]. We have aimed to include any study that identifies its study region as being high altitude, mountainous, or subject to steep topography using one of the synonyms identified during scoping. By including studies undertaken at high altitude, we have included plateaus in our map. Where these plateaus are extensive, topography may not adversely affect farming practices, and these systems may

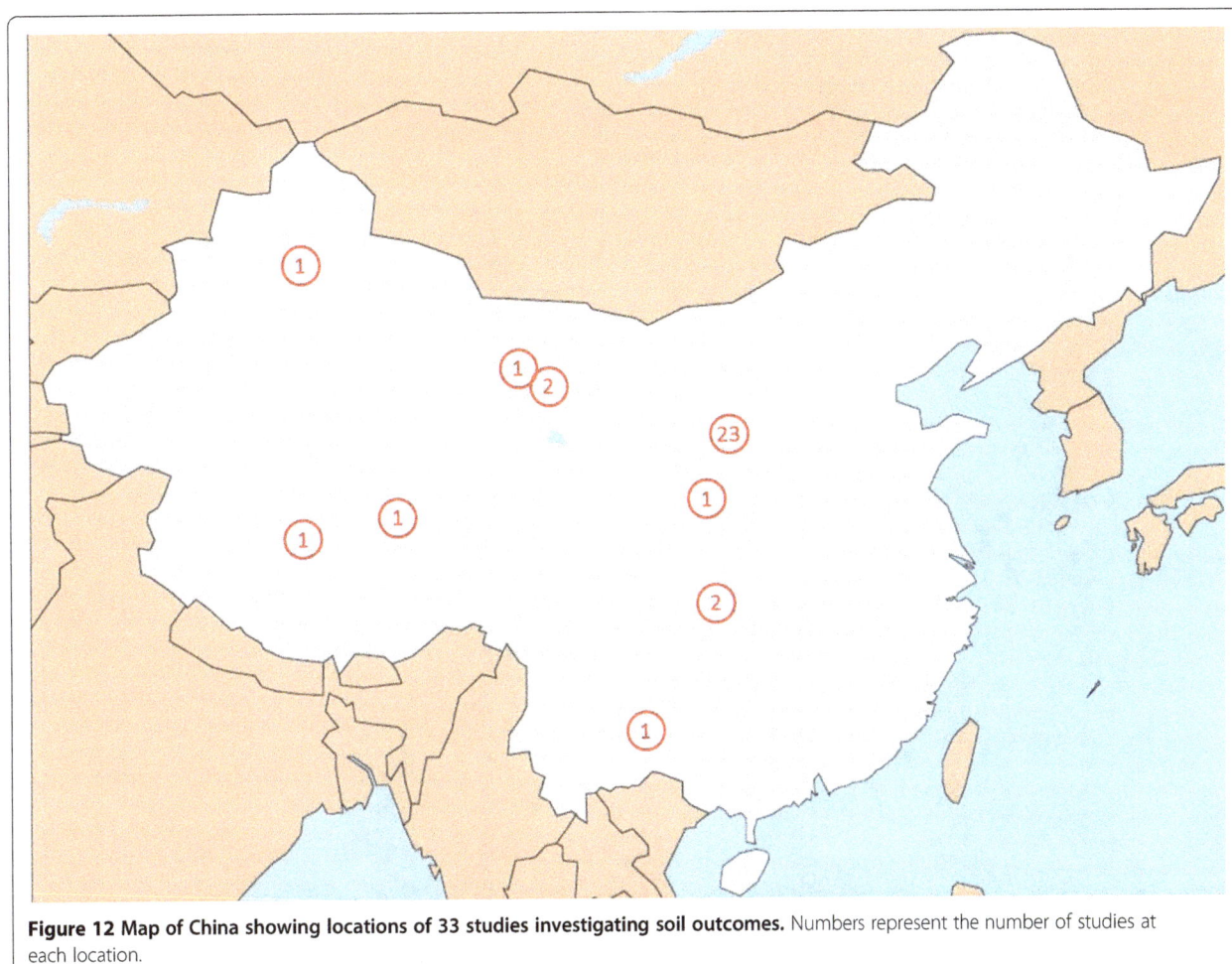

Figure 12 Map of China showing locations of 33 studies investigating soil outcomes. Numbers represent the number of studies at each location.

differ substantially in their causal pathways of abandonment to other high altitude or mountainous regions. This issue is almost exclusively restricted to studies undertaken on the Loess Plateau in China, although we may be overestimating the extent of the issue, since this is also sometimes referred to as the Loess Hilly Plateau. Local expert knowledge would be useful to differentiate between study sites that may and may not be influenced in their agricultural practices by topography.

During the review process we identified 4 descriptors that were not in our original string by iteratively populating a list of descriptors; cordillera, gradient, terrain and inclination. In addition, incline and topography were included in a *post hoc* test in Web of Knowledge (WoK). These 6 synonyms added 894 titles to the original 7,213 hits. Future updates to this map should consider including these additional terms in the search for novel evidence. Given a final inclusion rate of 1.9% relative to the WoK search results we might predict 17 of these articles could be pertinent to the review. To our knowledge this is the first review to employ such post-review synonym relevance assessment. We believe such analysis would be

useful in all systematic reviews and systematic maps in both scoping (currently a common practice) and final reviews/maps. Whilst we performed an assessment of common synonyms during scoping, only a full assessment of the relevant evidence can allow such a complete analysis of synonym completeness.

There is a risk that this map may have failed to identify all of the available literature on the topic, since authors may not have included high altitude/mountain descriptors in their titles, abstracts or keywords. Seven articles were identified that did not contain high altitude/mountain descriptors in these sections but that were, nonetheless, undertaken at altitude. However, it would be unfeasible to screen search results for all abandonment literature in search of relevant high altitude/mountain region studies. This risk can be mitigated by consultation with experts to maximise the inclusion of relevant research, which was a key strength of this review.

Thirteen studies in non-English languages were included in the review (French, 1; Spanish, 1; German, 1; Chinese 10). Articles that could not be translated often appeared to be relevant but could not be clearly translated by the review

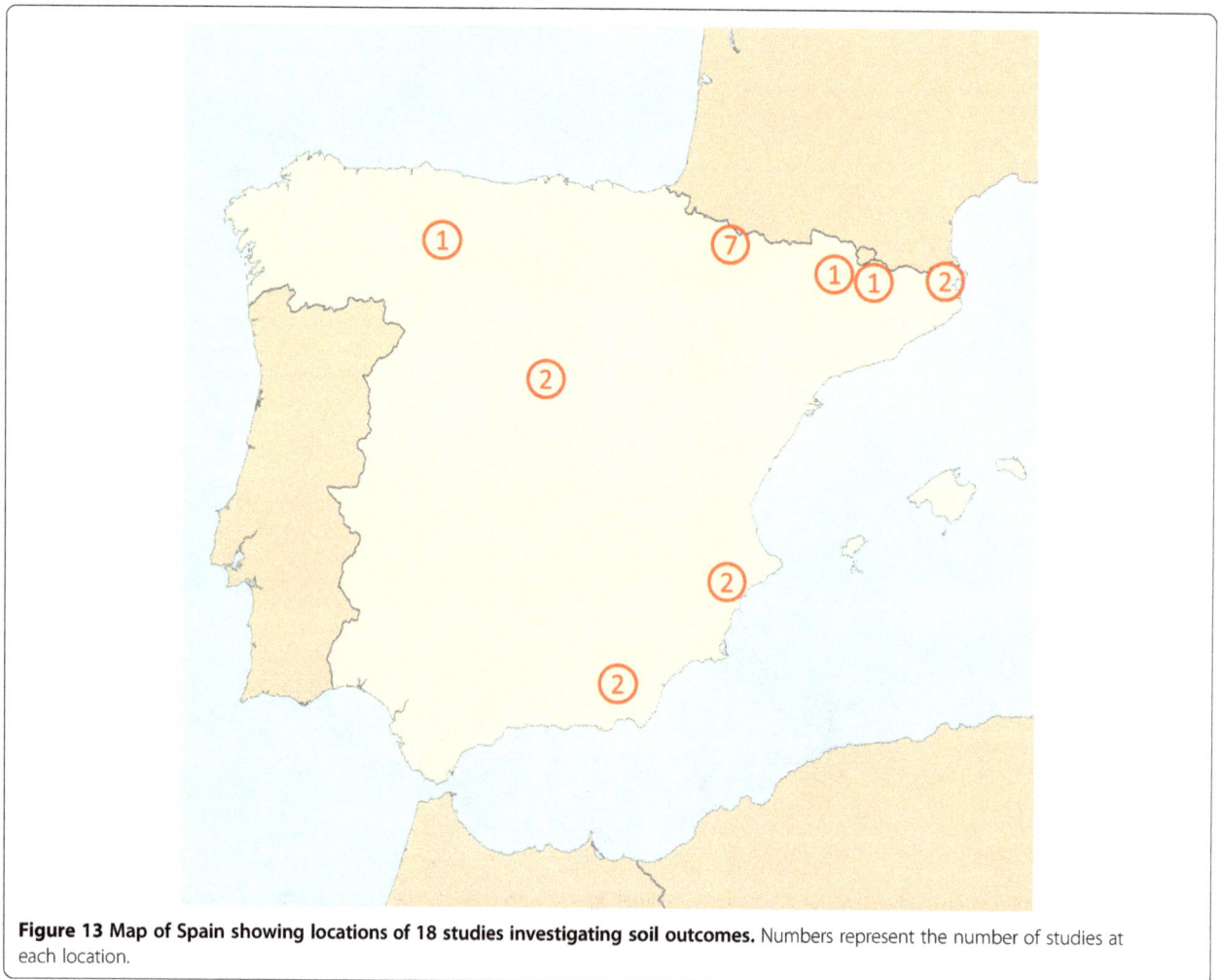

Figure 13 Map of Spain showing locations of 18 studies investigating soil outcomes. Numbers represent the number of studies at each location.

team in sufficient detail to assess the suitability of the comparator, details of the intervention, or the level of methodological detail. In particular, the presence of complete abandonment was difficult to assess. In addition, a number of articles could not be translated using automated translation tools due to restrictions on 'copy-and-paste' functions within digital articles. The ability to include these 26 untranslated articles would add strength to the accuracy of the map and any resultant syntheses.

Some studies examining experimental grazing exclusion were included in this map. These are pertinent to the review question, but they may only represent a small subset of the grazing exclusion literature. We believe that this may be the case since the term 'exclusion' was deliberately not included in our search string. However, some studies were identified as relevant through bibliographic searching. These studies simulated abandonment over a sufficient period for our inclusion criteria (>1 year) but did not refer to abandonment or destocking within their titles, abstracts or keywords and would therefore not have been identified using our search string. Since these studies

may only be a proportion of the available literature on experimental investigation of high altitude and mountain region grazing abandonment we recommend this subset of the literature (experimental, pasture categories) be treated with caution until it can be validated.

Limitations of the evidence base

The following aspects of the evidence base were highlighted during the systematic mapping process.

Many studies were not adequately replicated, with 72 scoring 0 for critical appraisal of replication (studies with 1 to 3 replicates per intervention/comparator group, or those that failed to report replication clearly). Articles were often unclear about the level of replication, and true replication was commonly difficult to discern from pseudoreplication (i.e. within-site sampling).

Few studies described the spatial scale of experiments in sufficient detail. Sample locations were not described, precluding an assessment of the contiguity of samples (i.e. whether closely located samples were pseudoreplicates or true replicates). Furthermore, this lack of detail makes assignment of scale to each study difficult; for example whether a study was undertaken at the plot, field or landscape scale.

One possibly common confounder that is unaccounted for in included studies in this map is the presence of extensive grazing on abandoned arable land. Several articles were excluded from the map due to the confounding effect of arable fields being abandoned and subsequently grazed extensively. This appears to be a relatively common practice in mountainous regions. However, this fully confounds any assessment of the impacts of farmland abandonment since comparator sites aren't equally grazed. Where stated, these studies were excluded, but there is a potential for this to be an undisclosed confounder, as several studies casually mentioned extensive grazing very briefly and sometimes outwith the methodology sections.

Very few studies reported their sample selection procedure. It is assumed that many of these studies selected samples purposefully, and are therefore open to sample selection bias. This may be of particular influence on the studies' findings where sample sizes were low. Twenty-eight studies used random or exhaustive sampling, whilst a further 3 appeared to have used some form of randomisation. The remaining 138 studies used either purposive sampling or failed to report their sampling procedure.

Implications for policy/practice

Systematic maps aim to document and categorise all available evidence on a topic of interest. The outputs therefore represent a first step toward formalising the evidence base and decision makers may find the map useful in initial gauging of the extent of the evidence and extracting relevant evidence on more specific aspects of the subject (e.g. for questions of national or regional importance). The map should help identify cases where there may be sufficient data on a specific policy-relevant question to justify a systematic review and synthesis of effects (see section Implications for synthesis below).

Implications for research

Implications for primary research

This map identifies a number of understudied subtopics that may correspond to knowledge gaps, which could benefit from primary research. In addition, an assessment of susceptibility to bias (critical appraisal) identified areas that have been frequently studied but that typically do not have strong evidence: these areas should be supplemented with high quality research.

Knowledge gaps were identified in the following areas:

1. Africa, Asia (excluding China), North America and South America - all outcomes

2. Europe - natural hazards (fire hazard, avalanche risk and flood risk), socioeconomic outcomes, animal behaviour and atmosphere
3. Global - fire hazard, avalanche risk and flood risk
4. UK - vegetation and soil
5. Czech republic - soil
6. France - soil
7. Key mountain ranges - including the Rockies, the Andes, the Caucasus, the Himalaya, the Karakoram, the Great Dividing Range and the Urals

Frequently studied research areas that were judged to be highly susceptible to bias are as follows:

1. China - overall and for biodiversity, soil and vegetation
2. Italy - overall and vegetation
3. Spain - overall and soil

Implications for synthesis

This review highlights a number of subtopics within the evidence that would be suitable for systematic review. The following questions have suitable numbers of studies to permit synthesis in a full systematic review, although the latter two questions relate to evidence that, in general, has been coarsely coded as highly susceptible to bias:

1. What is the impact of farmland abandonment in high altitude/mountain regions on biodiversity and vegetation in Europe?
2. What is the impact of farmland abandonment in high altitude/mountain regions on biodiversity and vegetation in the European Alps?
3. What is the impact of farmland abandonment in high altitude/mountain regions on soil in Spain?
4. What is the impact of farmland abandonment on the Loess Plateau (China) on biodiversity, soil and vegetation?

The progression from studies within this map to full systematic review is a relatively small task, since the time-consuming stages of full systematic review (namely, searching, screening and full text assessment) have already been undertaken, and all that remains to be done for the above questions is full data extraction (partially completed within this map), full critical appraisal (including external validity assessment) and qualitative/quantitative synthesis where appropriate. In addition we strongly advise the inclusion of stakeholder engagement for these full review questions to ensure that relevant stakeholders are made aware of the synthesis and included in prioritisation and dissemination efforts. Furthermore, as always, this map should be updated to ensure new evidence is included, and calls for evidence pertinent to the review questions above should be made to maximise the likelihood of including all available evidence.

Appendix
How to search the systematic map database
To open the database

- Open the Access database file (.accdb) using Microsoft Access
- Open the Full Text Map table (double click) using the 'Tables' panel on the left

To search for evidence

1. Simple search - e.g. searching for studies undertaken in arable farming systems
   - Navigate to the column titled 'Farming System' using the navigation bar or the right cursor
   - Click the arrow in the right end of the column title
   - Choose only the 'Arable' tick box
   - Click OK
   - The database will filter out only the Arable farming system studies
2. Multi-topic search - e.g. what research exists on arable farming from Switzerland?
   - Proceed as above to filter out only arable farming studies
   - Navigate to the 'Study Country/ies' column and select only 'Switzerland'
3. Searching for a term - e.g. bulk density
   - Proceed to the relevant column
   - Click the arrow in the right end of the column title
   - Select 'Text Filters' > 'Contains'
   - Enter search text and click OK

When finished filtering click the 'Filter' or 'Toggle Filter' buttons in the 'Home' tab to show the full database.

Other included information

- First author
- Email address
- Full reference
- Publication year
- Publication type
- Publication format
- Accessibility notes
- Study timing
- Study length
- Farming system (arable, mown, grazed)
- Altitude
- Experimental design

Additional files

Additional file 1: Search String Development and Database Search Strategy.

Additional file 2: Web Search Engine Search Strategy.

Additional file 3: Organisational Website Search Strategy.

Additional file 4: Bibliographic Checking Strategy.

Additional file 5: Author-submitted Evidence Received after Map Completion.

Additional file 6: Unobtainable and Untranslated Articles.

Additional file 7: Excluded Articles at Full Text Screening.

Additional file 8: Coding Framework.

Additional file 9: Systematic Map (Microsoft Access and Excel versions [CA only in Excel]).

Additional file 10: Table S1. Major outcome groups within studies across different countries arranged by continent.

Additional file 11: Table S2. Evidence strength of studies across different countries arranged by continent.

Additional file 12: Table S3. Average evidence item strength across different countries arranged by continent.

Competing interest
The authors declare that they have no competing interest.

Authors' contributions
NRH managed and planned the conduct of the systematic map and undertook all mapping activities. ASP, DS, and NRH conceived and refined the map question. ASP and DS assisted in interpretation, synthesis and writing. All authors read and approved the final manuscript.

Acknowledgements
The authors thank the European Commission for funding this research. We also thank Nicola Randall from the CEE Systematic Mapping Methods Group for advice on systematic mapping practicalities.

Sources of support
This research is undertaken as part of a project funded by the European Commission's Joint Research Centre through Service Contract Number 153172–2012 A08 GB. All statements/comments within this document belong to the authors and do not necessarily represent the views of the European Commission.

Author details
[1]Centre for Evidence-Based Conservation, School of the Environment and Natural Resources and Geography, Bangor University, Bangor LL57 2UW, UK. [2]School of the Environment and Natural Resources and Geography, Bangor University, Bangor LL57 2UW, UK.

References
1. Pointereau P, Coulon F, Girard P, Lambotte M, Stuczynski T, Sanchez Ortega V, Del Rio A, Anguiano E, Bamps C, Terres J: *Analysis of farmland abandonment and the extent and location of agricultural areas that are actually abandoned or are in risk to be abandoned.* Ispra: European Commission-JRC-Institute for Environment and Sustainability; 2008.
2. MacDonald D, Crabtree J, Wiesinger G, Dax T, Stamou N, Fleury P, Gutierrez Lazpita J, Gibon A: **Agricultural abandonment in mountain areas of Europe: environmental consequences and policy response.** *J Environ Manage* 2000, **59**:47–69.
3. Ramankutty N, Foley JA: **Estimating historical changes in global land cover: Croplands from 1700 to 1992.** *Global Biogeochem Cycles* 1999, **13**:997–1027.
4. Hobbs RJ, Cramer VA: **Why old fields? Socioeconomic and ecological causes and consequences of land abandonment.** In *Old Fields: Dynamics*

and Restoration of Abandoned Farmland. Edited by Cramer VA, Hobbs RJ. Washington DC: Island Press; 2007:1–14.

5. Schuler M, Stucki E, Roque O, Perlik M: **Mountain Areas in Europe: Analysis of mountain areas in EU member states, acceding and other European countries.** In *Final report, January: Nordregio, Nordic Centre for Spatial Development*, European Commission contract No 2002.CE.16.0.AT.136. 2004.

6. Körner C, Paulsen J, Spehn EM: **A definition of mountains and their bioclimatic belts for global comparisons of biodiversity data.** *Alp Bot* 2011, **121**:73–78.

7. Baldock D, Beaufoy G, Brouwer F, Godeschalk F: *Farming at the margins: abandonment or redeployment of agricultural land in Europe.* London: Institute for European Environmental Policy The Hague; 1996.

8. Moravec J, Zemeckis R: **Cross Compliance and Land Abandonment.** In *A research paper of the Cross-Compliance Network (Contract of the European Community's Sixth Framework Programme, SSPE-CT-2005-022727), Deliverable D17 of the Cross-Compliance Network.* 2007.

9. Jodha NS: **Mountain agriculture: the search for sustainability.** *J Farm Syst Res Ext* 1990, **1**:55–75.

10. Rey Benayas J, Martins A, Nicolau JM, Schulz JJ: **Abandonment of agricultural land: an overview of drivers and consequences.** *CAB Rev Perspect Agric Vet Sci Nutr Nat Resour* 2007, **2**:1–14.

11. Plieninger T, Gaertner M, Hui C, Huntsinger L: **Does land abandonment decrease species richness and abundance of plants and animals in Mediterranean pastures, arable lands and permanent croplands?** *Environ Evid* 2013, **2**:1–7.

12. Coppola A: **An economic perspective on land abandonment processes.** In *Working paper n. 1/2004. presented at the AVEC Workshop on "Effects of Land Abandonment and Global Change on Plant and Animal Communities. Anacapri, October 11–13, 2004.* 2004.

13. Haddaway NR, Styles D, Pullin AP: **Environmental impacts of farm land abandonment in high altitude/mountain regions: a systematic map of the evidence.** *Environ Evid* 2013, **2**:18.

14. Cohen J: **A coefficient of agreement for nominal scales.** *Educ Psychol Meas* 1960, **20**:37–46.

8

Are population abundances and biomasses of soil invertebrates changed by *Bt* crops compared with conventional crops? A systematic review protocol

Kaloyan Kostov[1], Christian Frølund Damgaard[2], Niels Bohse Hendriksen[3], Jeremy B Sweet[4] and Paul Henning Krogh[2*]

Abstract

Background: *Bt* crops modified by inserting and expressing the Cry toxin from *Bacillus thuringiensis* have raised environmental concerns over consequences for sustainability of soil biodiversity and ecosystems services in agricultural land. Part of this concern is related to the possible effects of the exposure to Cry toxins of non-target soil invertebrates as a result of *Bt* crops cultivation. Soil invertebrate members of microfauna, mesofauna and macrofauna play significant roles in nutrient cycling and energy flow and thus are crucial for soil ecological functions. In recent years, a number of studies have compared the population abundance and biomass of different members of soil biota in fields planted with genetically modified *Bt* crops and their conventional counterparts. In the present systematic review protocol, we describe the methodology and quality standards to perform a rigorous literature search and a quantitative synthesis of the evidence provided by these studies as required for conducting a Systematic Review.

Methods: The question that the systematic review will ask is whether populations of soil invertebrates differ under *Bt* crops and conventional crops. Relevant research literature will be collected systematically through a comprehensive search strategy. A scoping exercise was performed to identify search terms likely to capture appropriate studies and the results were verified using a list of relevant publications as references. The criteria against which studies will be included in the review are present, as well as the methodology for the quality assessment. To be included the study must contain relevant population abundances or biomass data on soil invertebrates exposed to characterised *Bt* proteins from field studies. The Review Protocol outlines the type of analyses that will be performed to assess bias of the selected studies and if covariables describing the heterogeneity of the studies introduce bias. Comparative effect sizes irrespective of statistical significance of effects will be calculated for individual studies and stored in publicly available databases ready for synthesis of all the studies. These treatment effects on population data will be compared across the studies in a meta-analysis using Hedges' g.

Keywords: *Bt* crops, Non-target organisms, Soil invertebrates, Population changes, Systematic review, Meta-analysis

Background

The technology for genetic modification of plants provides a tool for crop breeding and has been applied for the development of varieties with novel or improved traits. Insect resistance has been achieved via introduction of genes from the bacterium *Bacillus thuringiensis* (Bt), which is the second most distributed GM trait worldwide after the herbicide tolerance. The only GM plant cultivated commercially at large scale in EU is maize designed to produce a *Bt* toxin - Cry1Ab that provides protection against corn borers (Lepidoptera). GM crops producing different types of *Bt* toxins against other pest insects or in combination with other GM traits are also cultivated outside Europe and being considered for EU cultivation [1,2].

Because *Bt* crops contain insecticidal proteins, potential interactions with non-target organisms are of

* Correspondence: phk@dmu.dk
[2]Department of Bioscience, Aarhus University, P.O. Box 314, Vejlsøvej 25, 8600 Silkeborg, Denmark
Full list of author information is available at the end of the article

major concern for the risk assessment [3]. In recent years, many field and laboratory studies have been conducted to evaluate the potential effects of the *Bt* crops on above-ground and soil dwelling non-target organisms.

Soil invertebrates are classified according to their size as microfauna (protozoa, small nematodes), mesofauna (nematodes, Collembola, mites, enchytraeids) and macrofauna (earthworms). They play significant roles in the nutrient cycling and energy flow and are actively involved in physical, chemical and biological processes. It is very likely that any changes in soil properties will affect invertebrate communities, thus their composition and abundance can be used as an indicator of soil quality [4,5].

There are two possible ways that the *Bt* crop can influence soil biota: First, directly through root feeding, root exudates and litter that contain Cry toxin and/or directly through unintended changes in the plant, caused by the genetic modification; and secondly, indirectly through changes in agricultural management practices related to the genetic modification, e.g. changed insecticide regimes. The major concern is that the *Bt* crop cultivation may have an effect on soil invertebrate populations and their communities [6], which could attain a magnitude that would cause undesirable changes in the soil ecosystem functioning according to thresholds set by legislation or environmental authorities.

Systematic review and meta-analysis have been applied routinely for synthesis of data from medical studies assessing the risk or benefits of treatments or drugs. However, the potential to use this methodology for quantitative synthesis of data from impact assessment of GM plants has also been recognized [7]. Meta-analysis of field and laboratory studies assessing the effects of *Bt* crops to non-target organisms have been performed already [8-12]. Although comprehensive literature reviews exists [13-15] a quantitative synthesis of evidence with a quality control as required for a systematic review about the effects of *Bt* crops to the populations of soil invertebrates is still missing. It will contribute firm evidence-based conclusions about the possible impacts on soil biota communities and to the soil ecology in general.

Population abundance and biomass are the major endpoints for monitoring of soil invertebrates in ecological studies, therefore this systematic review will study data from field experiments of these two measurement endpoints in relation to soil-dwelling and surface-dwelling species including protozoa, nematodes, Collembola, mites, enchytraeids, and earthworms.

Objective of the review

This systematic review aims for a synthesis of the field evidence about the effects of *Bt* crops to six groups of soil-dwelling and surface-dwelling invertebrate species, protozoa, nematodes, Collembola, mites, enchytraeids, and earthworms.

The review question (RQ) asked by the present systematic review protocol is:

Are population abundances and biomasses of soil invertebrates changed by *Bt* crops compared with conventional crops?

The question has the following components:

Population: Soil-dwelling and surface-dwelling invertebrate species: protozoa, nematodes, Collembola, mites, enchytraeids, and earthworms.
Exposure: Genetically modified *Bt* crops and the conventional comparator in their concomitant farming practice through the soil environment and in the rhizosphere.
Comparator: Conventional non-GM crops and their concomitant farming practice.
Outcome: Net changes in the population abundances or biomasses contrasted with the comparator.

Methods

The systematic review methodology describes the approach which will be used to find and analyse original articles containing data from field experiments assessing the effects of the *Bt* crop cultivation on soil invertebrates including: search strategy and terms; study inclusion criteria and quality assessment; data extraction and methods for evidence synthesis.

Search strategy

The aim of the search is to find all available studies containing data from field experiments assessing the effect of *Bt* crop cultivation on soil invertebrates. The main approach will be to conduct comprehensive electronic searches in web databases, search engines for scholarly literature and specialized databases. An additional source will be the personal data collected by experts and stakeholders of the GRACE project network. In addition, the reference lists of related review papers and datasets from previously conducted reviews will be checked manually for relevant studies.

Search terms

Search terms defining the population - types of soil invertebrates, the exposure – types of *Bt* proteins, the assessed outcome - population abundances or biomasses, and the method - field studies, will be used for retrieving of relevant studies. The search terms will be organised in strings, which will be modified according to the requirements of each bibliographic database. The used search strings and the results will be documented and presented as an additional file of the systematic review.

The following terms will be used:

Population terms - invertebrate* OR mesofauna OR macrofauna OR arthropod* OR "ground-dwelling arthropod*" OR "surface-dwelling" OR microarthropod* OR microfauna* OR nematod* OR springtail* OR collembola* OR protoz* OR protist* OR earthworm* OR lumbricid* OR enchytraeid* OR oligochaeta OR acar* OR mite*

Exposure terms - cry1* OR cry2* OR cry3* OR "bacillus thuringiensis" OR delta-endotoxin OR Bt OR "cry toxin"

Outcome terms - population* OR abundanc* OR communit* OR diversit* OR biodiversity OR number* OR biomass* OR effect* OR impact*

Method terms (optional) - field

Scoping exercise

Scoping exercise was performed for a preliminary assessment of the availability of relevant research literature and for optimisation of the search strings to be employed in the systematic review. The search for research publications which contained the necessary elements for inclusion in the systematic review was done by simplified and focused search strings in Scopus platform, one string for each of three major groups of soil invertebrates - nematodes, Collembola and earthworms, (population AND exposure AND [method]; e.g. earthworm* AND Bt AND field). In this way, 17 studies (see Additional file 1) were found which subsequently were used as references to evaluate the relevance of the search strings in a pilot search exercise. The aim of pilot search was to determine the exact content of the search strings to be used in the systematic review and was conducted in three literature web databases: Web of knowledge, Scopus and AGRIS.

The content of the search strings was modified by including or excluding terms until all 17 articles (if present in the database) were found among the records and in the same time have not resulted in an excessive amount of irrelevant studies. The search strings which produced manageable numbers of records with all the reference studies among them are shown in Table 1. The search strings thus defined will be used for the systematic review search.

Web databases

Literature databases Databases containing scientific literature including theses, books, abstracts and articles will be searched using the defined strings. The following search service providers and bibliographic databases will be used:

- **Web of knowledge** (webofknowledge.com/) - search service including the following citation databases and platforms:
  Web of Science - platform which consists of nine databases containing scholarly journals, books, book series, reports, conferences, and other articles.
  BIOSIS Citation Index^{SM} - comprehensive reference database for life science research which includes cited references to primary journal literature.
  MEDLINE® - database of the U.S. National Library of Medicine (NLM) contains over 12 million records of journal articles in all areas of the life sciences.
- **CAB Direct** (http://www.cabdirect.org/) -platform for access to all CABI database subscriptions
- **Scopus** (http://www.scopus.com/) - large abstract and citation database of peer-reviewed literature
- **AGRIS** (http://agris.fao.org/agris-search/index.do) - Information system for the agricultural sciences and technology

Specialized databases The Center for Environmental Risk Assessment (CERA) Bibliography Database (http://cera-gmc.org/index.php?action=bibliography_database)

Bibliosafety (http://bibliosafety.icgeb.org/) - The Biosafety Bibliographic Database

Database of the Safety and Benefits of Biotechnology (http://biotechbenefits.croplife.org/)

Biosafety Information Resource Centre (BIRC) (http://bch.cbd.int/database/resources/)

Web search engines

The following search engines will be used:

Google scholar (http://scholar.google.com/)
JSTOR (http://www.jstor.org/)

The first 200 hits will be checked for relevance. The links will be followed once from the original hit.

Personal datasets

Experts and stakeholders from the consultation network created within the GRACE project will be asked to provide data relevant to the topic. The complete list of articles received by personal communication and the source will be recorded and will be included in the additional file of the systematic review.

Manual search

Literature datasets or databases from other reviews related to effects of *Bt* crops on non-target soil invertebrates, as

Table 1 Pilot search results (search conducted on 17.04.2014)

Platform	Search string	Total records	References papers
Web of knowledge	Topic = (invertebrate* OR mesofauna OR macrofauna OR arthropod* OR "ground-dwelling arthropod*" OR "surface-dwelling" OR microarthropod* OR microfauna* OR nematod* OR springtail* OR collembola* OR protoz* OR earthworm* OR lumbricid* OR enchytraeid* OR oligochaetaOR acar* OR mite*) AND Topic = (cry1* OR cry2* OR cry3* OR "bacillus thuringiensis" OR delta-endotoxin OR Bt) AND Topic = (population* OR abundance OR community OR diversity OR biodiversity OR number* OR biomass OR effect* OR impact*) AND Topic = (field)	2657	16 from 17 (one article is not present in the database)
Scopus	(TITLE-ABS-KEY(invertebrate* OR mesofauna OR macrofauna OR arthropod* OR "ground-dwelling arthropod*" OR "surface-dwelling" OR microarthropod* OR microfauna* OR nematod* OR springtail* OR collembola* OR protoz* OR earthworm* OR lumbricid* OR enchytraeid* OR oligochaeta OR acar* OR mite*) AND TITLE-ABS-KEY (cry1* OR cry2* OR cry3* OR "bacillus thuringiensis" OR delta-endotoxin OR bt) AND TITLE-ABS-KEY(population* OR abundance OR community OR diversity OR biodiversity OR number* OR biomass OR effect* OR impact*))	837	17 from 17
AGRIS	+(invertebrate* mesofaunamacrofauna arthropod* "ground-dwelling arthropod*" "surface-dwelling" microarthropod* microfauna* nematod* springtail* collembola* protoz* earthworm* lumbricid*enchytraeid* oligochaetaacar* mite*) + (cry1* cry2* cry3* "bacillus thuringiensis" delta-endotoxin Bt) + (population* abundance community diversity biodiversity number* biomass effect* impact*)	535	12 from 17 (5 articles are not present in the database)

The asterisk (*) is a search query wildcard representing any group of characters, including no character.

well as the reference lists of articles found within the electronic search will be searched manually.

All searches will be performed in English; however, no restriction for language or year of publication will be made. In that way, all studies which have published an abstract in English will be within the scope of our search. Original full texts in either English or German will be included for further analyses. The search results from each of the used sources will be saved and citations will be imported in ENDNOTE® citation manager software. All duplicates will be removed and a list containing the accumulated results will be created and uploaded to the open-access database CADIMA (Central Access Database for Impact Assessment of Crop Genetic Improvement Technologies).

Study inclusion criteria

In order to be included a study needs to fulfil each of the following criteria:

Relevant population(s): Field collected soil invertebrates at species level or at higher taxonomic levels among protozoa, nematodes, Collembola, mites, enchytraeids, and earthworms.
Relevant exposure(s): Field soil and rhizosphere exposure to genetically modified Bt crops and comparator non GM crops and their associated farming practices.
Relevant comparator(s): Non-Bt near-isogenic crop or another non-Bt variety of the same crop species in an

experimental design allowing for any of the comparisons:

- Bt with non-Bt plots, neither of which received any additional insecticide treatments.
- Bt plots not treated with insecticide with non-Bt plots that received insecticides.
- Bt with non-Bt fields when both are subject to insecticide treatments.

Relevant outcomes: Population abundance or biomass.

Appling study inclusion criteria

Two reviewers will independently perform the selection of studies, which fulfil the inclusion criteria at three stages. At first, articles will be selected by their title to remove highly irrelevant studies from the overall search results, followed by the second stage in which the inclusion criteria will be applied against the abstracts of the articles. If there is doubt or lack of enough information from the title and abstract alone to judge whether the article meets the inclusion criteria the full text of the study will be obtained to enable the assessment.

At the beginning of the second stage, after the first 100 publications are processed a test for consistency between two reviewers will be made using Cohen's Kappa (see Additional file 2). The calculation of Cohen's kappa coefficient will represent the agreement between the two reviewers. If the Cohen's kappa coefficient is less than 0.6, the inconsistencies will be discussed and the criteria for inclusion will be adjusted taking into account the main reasons for disagreement.

In the third stage, the articles will be reviewed in full text for the presence of each of the elements needed for the inclusion. The reasons for including or excluding each study at this stage will be recorded and reported.

Study quality assessment

All the studies, which fulfil the inclusion criteria, will be assessed for bias. The aim of this phase is to ensure that the studies are providing evidence, which represent true statistical similarity and level of difference. The internal (design, conduct, and analyses) and external (population, interventions, and outcomes) validity will be assessed by applying four quality domains - selection bias, performance bias, measurement bias and attrition bias and defined as low, high or uncertain and the results will be reported separately for each domain.

One reviewer will perform the assessment using check lists. A random subsample (20%) of the studies will be assessed additionally by second reviewer and the outcomes from both will be compared with Cohen's Kappa. The inconsistencies (Cohen's kappa coefficient is less than 0.6) will be discussed and a third reviewer will be involved in case no agreement could be reached. Records with the evaluation results and the reasons for judgment will be made for each article included at this stage.

Selection bias

Pre-treatment differences between the studied groups and in the baseline characteristics of the study will be assessed in the selection bias domain. The following elements will be assessed:

- **plot location** - low risk if the experimental plots of both the intervention and the control treatment are located in one field with known history. High risk if the plots are located at different fields and there is no information about the history of the field.
- **comparator** - low risk if the comparison is *Bt* crop vs corresponding isogenic line. High risk if another variety is used as the comparator.
- **randomization** - the randomisation is the best way to avoid selection bias; thus, studies which are designed by any block or plot randomization method will be considered as low risk. High risk studies will occur if there is no or poor randomization.
- **replications** - low risk if there are 4 or more replicates per treatment. Moderate risk is if there are 1 to 3 replicates. High risk if there is no replication.

Performance bias

Performance bias arises if the studied groups are influenced by factors different from the intervention, which

may have an effect on the measured outcomes. In the field studies, such influences might come from:

- **plot size** - low risk if the plot size is properly defined, depending on the movement behaviour of the examined taxa and high risk if plot size is excessively small.
- **field management** - low risk if both control and intervention received the same agro-technical management including tillage, fertilizers, fungicides, irrigation, cultivation etc. High risk if there are differences in the management between the treatments.

Measurement bias

The way the measurement of the outcomes is done can influence the true effect estimation if the selected method is not accurate or can be influenced by human subjectivity.

- **sampling of soil invertebrates** - sample collection and extraction procedures are crucial when assessing the abundance and biomass of the soil invertebrates. A common source of errors is the variation in depth and number of soil cores. The techniques differ between taxa, and so a general recommendation is not appropriate. When sampling and extraction are performed using standardized techniques or other recognized methods, the risk of bias will be considered low. If the technique is not suitable for the examined taxa, or is prone to human influence, the risk of bias will be considered high.

Attrition bias

Imbalance in the final set of selected studies will be assessed. In theory, an imbalance may occur if studies with the following properties are excluded:

- **sample size** - low risk if the sample size is equal between the treatments. High risk if it differs due to loss of samples.
- **missing data** - low risk if the amount of data for the measured outcomes is equal for all the treatment. High risk if there is imbalance in the presented outcome data between the treatments.

Data extraction strategy

The aim of the data extraction stage is to retrieve information relevant to the design, performance and measured outcomes, which will be used for the quantitative synthesis and the analysis of the variability between studies. Details about the experimental sites and design will be extracted from the text in the sections describing the materials and methods of the study, as well as the description of sampling and extraction techniques and statistical analysis. Numeric data for the measured outcomes and the corresponding variance will be extracted from tables and figures in the result section of the study. All data will be imported into a standardised Excel table. Each measurement of the population abundance or biomass of the different treatment (e.g. *Bt* crop and comparator) will be included in the table as separate records containing all the defining variables. One review team member will perform the data extraction and checks for errors will be made by another review team member of a random subset (20%) of the data. The discrepancies will be solved by the involvement of a third reviewer. Data to be extracted from each included study are presented in Table 2.

Data analysis

The extracted data will be used to synthesize the evidence provided by the individual studies about the effects of *Bt* crop cultivation on soil invertebrates, as well as to investigate the heterogeneity among the studies.

Only quantitative population data will be used for the meta-analysis of Hedges' g. There will be some aggregation of population abundance or biomass data into taxonomic groups to enable comparison between studies not using the same level of taxonomic resolution.

Assessment of statistical power of included studies

A post-hoc analysis of the power to detect effects with different magnitude small (effect size of 0.2), medium (effect size of 0.5) large (effect size of 0.8) at significance criterion 0.05 of each study will be made (Cohen, 1988).

Measures of treatment effect

Depending on the studied taxon and/or the sampling technique, the population abundance of soil invertebrates can be measured as number of individuals per volume of soil or per surface area, and the biomass as the weight of the population per volume of soil or per surface area. The reported values are usually the mean for the treatment and the calculation of the associated variance. Finally, irrespective of measurement units treatment effect sizes reported in each study will be expressed as Hedges' g.

Dealing with missing data

If measurements of the population abundance or biomass of either the treated plots or the control plots are missing then the study is not included in the meta-analysis. In cases when, for the purpose of reporting the variance, other values than standard deviation or standard error are used, e.g. t, F, p or z-values, an appropriate mathematical method will be used to calculate the pooled standard deviation, if appropriate. If this is not feasible, authors will be contacted to provide the missing data.

Synthesis

Quantitative synthesis Quantitative synthesis will be performed to combine the magnitude of the effects from the individual studies. The meta-analysis will include calculation of the pooled effect size (Hedges'g) for each study accompanied with the corresponding confidence intervals. The results from the meta-analysis will be presented graphically in 'forest plots', where the estimated effect size with the confidence interval of each individual study will be plotted horizontally as the combined effect size and confidence interval will be plotted below them.

The complete dataset will be stored in database for open access after the finalisation of the review. For mixed effects modelling SAS PROC MIXED, PROC GLIMMIX or R ver. 3 will be performed [16,17].

Heterogeneity of the variability across studies is already implicit in the estimate of the standard deviation for Hedges' g.

A range of *effect modifiers* will be extracted from the selected papers and stored in the database. These effect modifiers include comparator properties, pesticide treatment, experimental design, cropping system, crop rotation, tillage date, date of GM experimental cropping system establishment and soil type. When the final dataset allows for assessing the effect of the effect modifiers on the outcome of the meta-analysis and any biases will be reported.

Assessment of heterogeneity The heterogeneity across the studies that may influence the outcome will be assessed. In field studies estimating the effects of *Bt* crop cultivation on soil invertebrates, several sources of heterogeneity may be expected.

Heterogeneity in studied populations The examined taxonomic groups will vary between the studies. The populations under investigation will include species or higher taxonomic groups among the soil invertebrates: protozoa, nematodes, Collembola, mites, enchytraeids, and earthworms.

Heterogeneity in type of exposure A source of heterogeneity related to the type of exposure will be the variability of the GM plant species and the type of *Bt* toxin, i.e. the CRY event, which they produce. Among the most widely studied *Bt* crops are maize producing Cry1Ab (against lepidopteran pests) and Cry3Bb1 (against coleopteran pests), followed by Cry1Ac producing cotton.

Preliminary assessment of the heterogeneity in the studied population and type of exposure among the studies used as references for the pilot search are shown in Table 3.

Table 2 List of variables to be extracted from the papers for the systematic review

Variable name	Definition	Type
Georeference, longitude	GPS coordinate WGS 84 decimal format	decimal
Georeference, latitude	GPS coordinate WGS 84 decimal format	decimal
Location	Location of the experimental site	characters
Crop	Name of the crop	characters
GM Event	Name of the GM event	characters
Treatment property	Type of *Bt* toxin or comparator	characters
Gene stacking	Information about stacked event (0 - no, 1- yes), name of the stacked gene	vector {Binary integer, characters}
Variety	The commercial name of the variety	characters
Comparator properties	Information about the used comparator in relation with the *Bt* variety - isogenic or other.	characters
Insecticide treatment	Information about insecticide treatment (0 - no, 1- yes), the product name, active substance, amount, number, time and method for application.	vector {Binary integer, characters, characters}
Experimental design	RBC = Randomized complete blocks; CR = Completely randomized; Multi location = ML	characters
Plot size	Calculation of the plot size (in square meters)	real number
Plots number	Number of plots per treatment	real number
Cropping system	Conventional; Reduced tillage; Conservation tillage etc.	characters
Crop rotation	Information about the history of the experimental field	characters
Tillage date	Date since last tillage event, including ploughing, harrowing, rotovation etc.	date
Date of cropping system establishment	Date when the *Bt* crop was planted for first time	date
Seeding date	Date of seeding within each growing season	date
Sampling date	Date when the sampling was performed	date
Order	Name of the taxonomic order of the soil invertebrates	characters
Family	Name of the taxonomic family of the soil invertebrates	characters
Species	Name of the taxonomic species of the soil invertebrates	characters
Stage of development	Stage of development of the soil invertebrates	characters
Sample type	e.g. litterbags, bulk soil, rhizosphere	characters
Sample depth	Depth in the soil where the sample was taken	real number
Sample amount	Amount of soil in one sample	real number
Sampling location	Location from which the samples are take, e.g. between rows, within rows, distance from roots	characters
Extraction method	Used technique for extraction of soil invertebrates form the sample	characters
Sample size	Sample size as reported by the author	real number
Measurement endpoint	Type of measured endpoint, e.g. abundance, biomass	characters
Unit	Unit in which the measurement endpoint is presented	characters
Value	Value of the measured endpoint	real number
Variability measure	STD; SEM; CLM; Variance	characters
Quantity	Quantified variability	real number
Soil type	Soil type according to WRB classification	characters
Other fixed and random effects	E.g. fertilization, fungicide use, additional factors and effects of the experimental design	characters
Data origin	Table or figure from which the data originates	characters
Data extraction	How data was extracted, e.g. as exact number from tables or scaled numbers from graphs	characters

Table 2 List of variables to be extracted from the papers for the systematic review (Continued)

Statistical analysis	Description of the statistical methods for analysis of variance	characters
Reference	Bibliographic reference code as found in the GRACE CADIMA database (Central Access Database for Impact Assessment of Crop Genetic Improvement Technologies)	characters
Source of funding	Description of the funding source of the study	characters
Authors affiliation	Type of institution to which the first author belongs	characters
Comments	Any other information which may be relevant	characters
Keywords	Keywords for finding the reference in a systematic review - {list of keywords}	characters

Heterogeneity in methodology The main source of variability between the studies related with the methodology could be expected from differences in the experiment duration and time of sampling. Some authors report one or two years study with one sampling at the end of the season, while others report multi-year studies with several sampling occasions in the beginning, in the middle and in the end of the season. Considering the possible cumulative effect of the Bt- toxin in the soils the measurements carried out in several subsequent years will be compared separately if corresponding data is found.

Other sources of heterogeneity which may have an effect on the outcome of the field studies could be related to differences in plot size, sample size, sampling method and field management, or to be caused by the different factors of the environment such as the soil type and availability of water.

Heterogeneity in comparisons The experimental design of the studies can include one or a combination of the following comparisons: Bt with non-Bt plots, neither of which received any insecticide treatments; Bt- plots not treated with insecticide with non-Bt plots that received insecticides and Bt with non-Bt fields when both are subject to insecticide treatments.

To deal with the above mentioned heterogeneity sub-group meta-analysis will be performed if data suitable to calculate the effect size is found in a minimum of three distinct publications, which contain the same Bt crop

producing and type of Cry toxin) and the same group of studied invertebrates (protozoa, nematodes, Collembola, mites, enchytraeids, and earthworms). Meta–analysis on the finest possible taxa will be made, however if necessary an aggregation of data from lower taxonomic groups will be performed. If heterogeneity in the sampling time is extreme for effect size calculation will be used the peak abundance or biomass of the season.

Statistical analysis of heterogeneity Heterogeneity will be addressed in conventional analyses of correlation, regression and mixed modelling ANOVA to reveal if the heterogeneities have any significant impact on the Bt crop effect estimates. E.g. if the studies can be grouped into soil type categories holding sufficient number of replicates, hypotheses about the effect of soil type on the outcome of effects can be elucidated. This will be one type of sensitivity analysis. In principle all the effect modifiers can be included in sensitivity analyses, i.e. do their inclusion or exclusion affect the assessment of effect levels in terms of Hedges' g. Effects reported by a study will be critically assessed by excluding the risk of confounding between the effect modifiers and the effect measure of Hedges' g.

Sensitivity analysis The validity of the systematic review findings will be verified by sensitivity analysis. Meta-analysis calculations will be undertaken twice using different assumptions related to the quality of experimental performance and reporting of the results, as well as differences of the methods used. Studies will

Table 3 Preliminary heterogeneity assessment of crops, taxonomic groups and types of Bt toxins

Taxonomic group	Arthropods				Earthworms		Nematodes	
Crop/Bt protein	Cry1Ab	Cry1Ac	Cry3Bb1	Cry3Aa	Cry1Ab	Cry3Bb1	Cry1Ab	Cry3Bb1
Maize	6		4		5	3	4	2
Cotton		2						
Rice	1							
Potato				1				

The table is based on the 17 studies identified during the scoping exercise. Digits represent the number of studies. Total number exceed 17 because some studies are examining more than one taxonomic group or use more than one Bt crop.

be included or excluded from the calculations according to the following factors:

- weight of the individual study;
- study quality;
- near isogenic vs. non-isogenic comparator;
- methodological differences;
- missing data;
- outlier studies (results differ significantly from a range of other studies).

Other factors also considered relevant will be applied, described and recorded.

Assessment of publication bias In many research areas it is known that "positive" results are more likely to be published [18]. Publication bias will be assessed using "funnel plot" analysis, where the effect size of an individual study will be plotted on the horizontal axis and the standard error or sample size on the vertical axis. The asymmetry in the funnel plot may indicate publication bias [19].

Review teams
An extraction team at ABI and a review team at Dep. of Bioscience have been established to ensure internal quality assurance of the review process.

Additional files

> **Additional file 1:** List of studies used for the relevance check in the pilot search exercise.
>
> **Additional file 2:** Details on the calculation of Cohen's kappa coefficient.

Competing interests
None of the members of the review team has any financial interest in the outcome of the systematic review nor are they affiliated with any religion or organisation that have expressed a certain position to GMO.

Authors' contributions
KK, PHK, JBS, CFD, NBH prepared this review protocol. KK made the data extraction of the pilot literature search. PHK provided study quality criteria and GRACE project management. CFD performed initial pilot meta-analyses. All authors read and approved the final manuscript.

Acknowledgements
This work is funded within EU-FP7 project GRACE, Grant Agreement KBBE-2011-6-311957. We are grateful for the support and methodological guidance provided by the management and members of the GRACE consortium.

Author details
[1]Agrobioinstitute, 8 Dragan Tzankov Blvd, BG-1164 Sofia, Bulgaria. [2]Department of Bioscience, Aarhus University, P.O. Box 314, Vejlsøvej 25, 8600 Silkeborg, Denmark. [3]Department of Environmental Science, Aarhus University, Frederiksborgvej 399, DK-4000 Roskilde, Denmark. [4]Sweet Environmental Consultants, 6 Green Street, Willingham, Cambridge CB24 5JA, UK.

References
1. Gatehouse JA: **Biotechnological prospects for engineering insect-resistant plants.** *Plant Physiol* 2008, **146**:881–887.
2. Marshall A: **Existing agbiotech traits continue global march.** *Nat Biotechnol* 2012, **30**:207–207.
3. Romeis J, Bartsch D, Bigler F, Candolfi MP, Gielkens MMC, Hartley SE, Hellmich RL, Huesing JE, Jepson PC, Layton R, Quemada H, Raybould A, Rose RI, Schiemann J, Sears MK, Shelton AM, Sweet J, Vaituzis Z, Wolt JD: **Assessment of risk of insect-resistant transgenic crops to nontarget arthropods.** *Nat Biotechnol* 2008, **26**:203–208.
4. Lavelle P, Decäens T, Aubert M, Barot S, Blouin M, Bureau F, Margerie P, Mora P, Rossi JP: **Soil invertebrates and ecosystem services.** *Eur J Soil Biol* 2006, **42**(Supplement 1):S3–S15.
5. De Vries FT, Thébault E, Liiri M, Birkhofer K, Tsiafouli MA, Bjørnlund L, Bracht Jørgensen H, Brady MV, Christensen S, de Ruiter PC, d' Hertefeldt T, Frouz J, Hedlund K, Hemerik L, Hol WHG, Hotes S, Mortimer SR, Setälä H, Sgardelis SP, Uteseny K, van der Putten WH, Wolters V, Bardgett RD: **Soil food web properties explain ecosystem services across European land use systems.** *Proc Natl Acad Sci* 2013, **110**:14296–14301.
6. Krogh PH, Griffiths B, Demšar D, Bohanec M, Debeljak M, Andersen MN, Sausse C, Birch ANE, Caul S, Holmstrup M, Heckmann LH, Cortet J: **Responses by earthworms to reduced tillage in herbicide tolerant maize and Bt maize cropping systems.** *Pedobiologia* 2007, **51**:219–227.
7. Marvier M: **Using meta-analysis to inform risk assessment and risk management.** *J Verbr Lebensm* 2011, **6**:113–118.
8. Duan JJ, Lundgren JG, Naranjo S, Marvier M: **Extrapolating non-target risk of Bt crops from laboratory to field.** *Biol Lett* 2010, **6**:74–77.
9. Duan JJ, Marvier M, Huesing J, Dively G, Huang ZY: **A meta-analysis of effects of Bt crops on honey bees (Hymenoptera: Apidae).** *PLoS One* 2008, **3**:e1415.
10. Marvier M, McCreedy C, Regetz J, Kareiva P: **A meta-analysis of effects of Bt cotton and maize on nontarget invertebrates.** *Science* 2007, **316**:1475–1477.
11. Naranjo SE: **Impacts of Bt crops on non-target invertebrates and insecticide use patterns.** *CAB Rev: Perspect Agric, Vet Sci, Nutr Nat Resour* 2009, **4**:1–11.
12. Wolfenbarger LL, Naranjo SE, Lundgren JG, Bitzer RJ, Watrud LS: **Bt crop effects on functional guilds of non-target arthropods: a meta-analysis.** *PLoS ONE* 2008, **3**:e2118.
13. Carpenter JE: **Impact of GM crops on biodiversity.** *GM Crops* 2011, **2**:7–23.
14. Icoz I, Stotzky G: **Fate and effects of insect-resistant Bt crops in soil ecosystems.** *Soil Biol Biochem* 2008, **40**:559–586.
15. O'Callaghan M, Glare TR, Burgess EPJ, Malone LA: **Effects of plants genetically modified for insect resistance on nontarget organisms.** *Annu Rev Entomol* 2005, **50**:271–292.
16. A programming environment for data analysis and graphics version 3.0.1 (2013-05-16). http://www.r-project.org/.
17. SAS Institute Inc: *SAS/STAT® 9.3 User's Guide.* Cary, NC: SAS Institute Inc; 2011.
18. Walker E, Hernandez AV, Kattan MW: **Meta-analysis: its strengths and limitations.** *Cleve Clin J Med* 2008, **75**:431–439.
19. Sterne JAC, Egger M: **Funnel plots for detecting bias in meta-analysis: guidelines on choice of axis.** *J Clin Epidemiol* 2001, **54**:1046–1055.

What is the effect of phasing out long-chain per- and polyfluoroalkyl substances on the concentrations of perfluoroalkyl acids and their precursors in the environment? A systematic review protocol

Magnus Land[1*], Cynthia A de Wit[2], Ian T Cousins[2], Dorte Herzke[3], Jana Johansson[2] and Jonathan W Martin[4]

Abstract

Background: There is a growing concern in Sweden and elsewhere that continued emissions of per- and polyfluoroalkyl substances (PFASs) may cause environmental as well as human health effects. PFASs are a broad class of man-made substances that have been produced and used in both commercial products and industrial processes for more than 60 years. Although the production and use of some PFASs has been phased-out in some parts of the world, it is not known what effect these actions to date have had on PFAS concentrations in the environment. Owing to the wide diversity of PFASs, it is difficult to generalize their properties, environmental fate and production histories. However, the strength and stability of the C-F bond renders the perfluoroalkyl moieties resistant to heat and environmental degradation. Several PFASs are now occurring even in very remote areas in large parts of the world, but the environmental transport and fate of substances within this group is not well understood. A systematic review may be able to determine whether the concentrations of these substances in different environments are changing in any particular direction with time, and whether the phase-outs have had any effects on the concentration trends.

Methods: Searches for primary research studies reporting on temporal variations of PFAS concentrations in the environment will be performed in the scientific literature as well as in other reports. Relevant samples include both abiotic and biological samples including humans. No particular time, document type, language or geographical constraints will be applied. Two authors will screen all retrieved articles. Double screening of about 10% of the articles will be performed by all authors at both title/abstract and full-text levels. Kappa tests will be used to test if the screening is consistent. Relevant articles will be critically appraised by four authors (double checking of 25% of the articles). Quality assessment will focus on selection bias, dating of samples, sample integrity and analytical procedures. Data synthesis will be based on statistical analysis of temporal concentration trends.

Keywords: Perfluoroalkyl acids, Perfluoroalkane acids, PFOA, PFOS, Temporal trends, Phase-out, Source, Emission, Environmental fate, Regulation, Concentration

* Correspondence: magnus.land@eviem.se
[1]Mistra EviEM, The Royal Swedish Academy of Sciences, Box 50005, SE-104 05 Stockholm, Sweden
Full list of author information is available at the end of the article

Background

Per- and polyfluoroalkyl substances (PFASs) are a broad class of man-made substances that have been produced and used in commercial products and industrial processes for over 60 years [1]. For example, PFASs have been used in water-, soil-, and stain-resistant coatings for clothing, leather, upholstery, and carpets; oil-resistant coatings for food contact paper; aviation hydraulic fluids; fire-fighting foams; paints, adhesives, waxes, polishes, and other products; and industrially as surfactants, emulsifiers, wetting agents, additives, and coatings [2-7]. The perfluoroalkyl moieties (C_nF_{2n+1} −) of PFAS molecules are both hydrophobic and lipophobic [2], and the extreme strength and stability of the C-F bond [8] renders the perfluoroalkyl moieties resistant to heat and environmental degradation processes. Nevertheless, owing to the wide diversity of PFASs (i.e. chain-lengths, molecular weight, degree and pattern of fluorination, presence of polar functional groups), it is difficult to generalize their properties, environmental fate, and production histories [9]. For purposes of this document we have separated the discussion into two broad categories, which will be the focus of our study:

1) Perfluoroalkyl carboxylic acids (PFCAs) and their precursors
2) Perfluoroalkane sulfonic acids (PFSAs) and their precursors

We decided to focus on PFCAs and PFSAs because these are by far the two most widely studied PFAS classes, and multiple temporal trend datasets were known to be available. PFCAs and PFSAs are well studied due to their persistence and ubiquity in the environment and biota as well as their potentially harmful effects (see brief review below). We use the terminology recommended by Buck *et al.* [9] throughout this document. A list of abbreviations used in this protocol is provided in Additional file 1: List of abbreviations.

Perfluoroalkyl carboxylic acids (PFCAs) and their precursors

PFCAs occur in the environment due to emission from intentional manufacturing, as impurities in commercial products containing other PFASs, or as environmental degradation products of other PFASs [3-5]. The PFCA manufactured in the largest quantity, perfluorooctanoic acid (PFOA), was produced mainly by electrochemical fluorination (ECF) by 3M until 2002 (3M had >80% of global market) and was primarily used as a processing aid (emulsifier) in the manufacture of polytetrafluoroethylene (PTFE). The ECF manufacturing process produces a mixture of linear (70%) and branched (30%) isomers. After 3M phased-out the production of PFOA

in 2002, other companies continued to manufacture PFOA mainly through the telomerization process which produces only linear isomers. For a thorough description of the sources of PFCA homologues and their precursors see the reviews of Prevedouros *et al.* [3], Wang *et al.* [4] and Wang *et al.* [5].

The most common PFCA measured in the abiotic environment is PFOA. Once present in the environment, the PFCAs have no significant known mode of environmental degradation and are thus highly persistent. For example, in wastewater treatment plants PFOA did not degrade, but actually increased in the outflow owing to degradation of unidentified PFCA precursors [10]. Engineering solutions using oxidants and catalysts have been developed to degrade PFCAs on a small scale in the lab. For example, UV photo-oxidation with indium oxide (In_2O_3) as the catalyst is effective for PFOA degradation [11], but these processes will not occur in the natural environment.

PFCAs have an acid dissociation constant (pKa) in the range 0–1 and are thus completely dissociated anions in environmental media which typically have pH above 4 [12-15]. Once discharged into the environment, unlike typical persistent organic substances, the majority of PFCAs do not sorb appreciably on particles but are instead present mainly in the dissolved phase in surface waters. Sorption to the organic fraction of particles increases with the length of the perfluoroalkyl chain [16], but the organic-carbon water partition coefficient (K_{OC}) of a long-chain PFCA (C8 and higher) such as PFOA [16] is still orders of magnitude lower than for typical persistent organic substances. Their low sorption and high prevalence in surface waters is evident from field investigations demonstrating that PFOA in biosolids applied to agricultural land can be re-mobilized by rainfall [17], and it has been shown that the sediment/water partition coefficient is low (PFOA K_{OC} = 2.4 ± 0.12 cm^3 g^{-1}) [18]. As a result, PFCAs can be transported long distances by rivers and ocean currents and now occur in the open marine environment, even in the remote Northern Atlantic, Pacific, and Arctic Oceans [19,20]. PFCAs can also be detected at low concentration in the ambient atmosphere, where they may be directly emitted [21], and/or formed in situ by oxidation of semivolatile PFCA-precursors such as the fluorotelomer alcohols (FTOH) [22].

Overall, the direct and indirect sources of PFCAs to the environment are various, and the relative importance of each source is temporally variable, PFCA-specific and not well quantified. For a given environmental medium in a given region, it is challenging to determine the source contribution to the PFCA contamination profile because in addition to uncertainties in the relative importance of multiple source types, the contamination profile will be transformed as a result of differences in the fate and transport of the individual PFCAs after release. The relative importance of atmospheric versus marine transport of PFCAs, of direct atmospheric emission versus atmospheric oxidation of PFCA-precursors, or of telomerization versus electrochemical manufacturing of PFCAs and their precursors is the subject of much recent and ongoing research. For PFCAs and their precursors, the perfluoroalkyl chain-length can have a considerable effect on their environmental fate and partitioning [16,23].

Occurring to a large extent in the dissolved phase in surface waters, PFCAs are relatively bioavailable compared to typical hydrophobic organic substances. PFCAs have been detected in numerous biological samples, e.g., fish [24], bird eggs [25-27], invertebrates, reptiles and marine mammals including polar bear [28], as well as humans [29,30].

Bioaccumulation has been shown to occur in mammals and birds, increasing with perfluoroalkyl chain length [31]. A limited number of field-Biomagnification Factors (BMFs) and Trophic Magnification Factors (TMFs) are available for long-chain PFCAs (and perfluorooctane sulfonate, PFOS) and they provide evidence that biomagnification of these substances takes place. Tomy *et al.* [32] reported bioaccumulation of perfluoroalkyl acids (PFAAs, including PFCAs and PFSAs) occurring in Arctic marine food webs. Kelly *et al.* [33] compared different parts of the marine food web, showing that the TMF is below one in the case of piscivorous food webs if air breathing organisms are excluded but becomes larger than one if air breathing organisms are taken into account. Loi *et al.* [34] observed trophic magnification for long-chain PFCAs (perfluorodecanoic acid (PFDcA), perfluoroundecanoic acid (PFUnA), perfluorododecanoic acid (PFDoA)) and PFOS in a subtropical food web. In freshwater systems, Martin *et al.* [35] found BMFs of long-chain PFCAs to vary from 0.4 to 3.4 between lake trout and prey organisms in Canada (PFOA had the lowest BMF and PFUnA the highest). In terrestrial samples, Muller *et al.* [36] reported BMFs and TMFs from caribou to lichen for PFAAs. Highest BMFs of 75 and 46 were found for PFDcA and PFUnA, respectively. TMFs of the food chain wolf – caribou – lichen varied between 2.4 and 7.1 for all long-chain PFAAs with PFDcA and PFOS showing the highest TMFs of 7.1 and 6.7 respectively. In summary, BMFs and TMFs above 1 indicate trophic biomagnification for PFAAs with a perfluoroakyl chain length containing 8 or more perfluorinated carbons in the terrestrial and freshwater ecosystems studied. Some short-chain (e.g. butane- and hexane-based) alternatives also appear to be persistent but to not bioaccumulate to the same extent, as they are excreted rapidly from the organisms studied [37].

PFCAs are not acutely toxic based on standard toxicity endpoints. However, they have been reported to have endocrine disrupting properties [38-40]. PFAAs are structurally similar to natural long-chain fatty acids and may displace them in biochemical processes and at receptors, such as peroxisome proliferator-activated receptor alpha (PPARα) and the liver-fatty acid binding protein (L-FABP). PFCAs, particularly the long-chain PFOA, PFNA (perfluorononanoic acid) and PFDcA but not the short-chain PFHxA (perfluorohexanoic acid), are highly potent peroxisome proliferators in rodent livers and affect mitochondrial, microsomal, and cytosolic enzymes and proteins involved in lipid metabolism [41-44]. PFCAs cause hepatomegaly in rodents [42] which is an indicator for hepatotoxicity. Perfluorobutanoic acid (PFBA) has a less pronounced effect on indicators of peroxisome proliferation [41].

Starting in 2000, various actions were undertaken by industry and regulators to reduce the release of PFCAs and precursors. In 2000, 3M announced a global phase-out by 2002 of its production of products based on perfluoroalkyl chains containing 6, 8 and 10 carbons, including PFOA [45]. In 2006, eight major PFCA, fluoropolymer and fluorotelomer manufacturers joined the US EPA 2010/15 Stewardship Program to work towards the elimination of long-chain PFCAs and their precursors from emissions and products by 2015 [46]. On the regulatory front, PFOA, its ammonium salt ammonium perfluorooctanoate (APFO), and C11–C14 PFCAs were included in the Candidate List of Substances of Very High Concern under the European chemicals regulation, REACH [47]. Although long-chain PFCAs are being stepwise phased out by the major manufacturers and heavily regulated in Japan, Western Europe and the United States (US) [45,46,48], new manufacturers (largely in continental Asia) have begun to produce long-chain PFCAs and their precursors. As a result of the phase out of many long-chain PFASs in recent years, many alternative fluorinated products have been introduced [49] and these new, alternative industrial processes and products have resulted in new sources of PFCAs and other fluorinated substances. For example, there has likely been increasing emissions of PFHxA due to the increasing use of side-chain polymers based on 6:2 FTOHs in surface treatment products [49]. A time-series of key production events and regulatory actions for PFCAs is shown in Figure 1.

Perfluoroalkane sulfonic acids (PFSAs) and their precursors

Similar to PFCAs, sources of PFSAs include release during manufacture and use of the PFSAs as well as from the degradation of various precursor substances [6,50]. Commercial scale manufacture of perfluorooctane sulfonyl fluoride (POSF)-based products began by 3M in

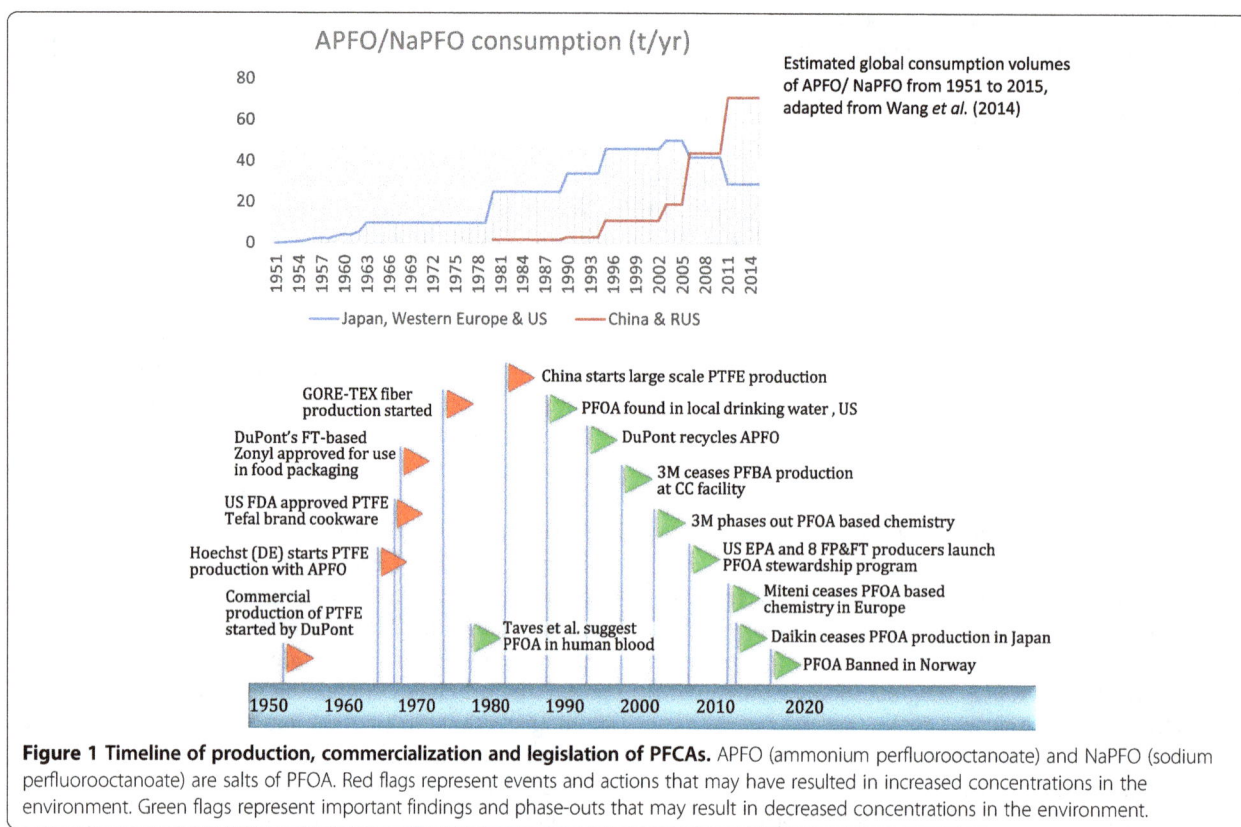

Figure 1 Timeline of production, commercialization and legislation of PFCAs. APFO (ammonium perfluorooctanoate) and NaPFO (sodium perfluorooctanoate) are salts of PFOA. Red flags represent events and actions that may have resulted in increased concentrations in the environment. Green flags represent important findings and phase-outs that may result in decreased concentrations in the environment.

the late 1950s with product lines based on N-methyl perfluorooctane sulfonamido ethanols (N-MeFOSEs) used in surface treatment applications (e.g., carpets, upholstery and textiles) and in the late 1960s product lines based on N-ethyl perfluorooctane sulfonamido ethanols (NEtFOSEs) were introduced for use in paper and board packaging applications. PFOS and various salts were manufactured for direct use in a variety of products (e.g. aqueous film-forming foams (AFFFs) for firefighting and mist suppressants in acids baths used for metal plating (for a more complete list of uses of PFOS and its salts see Paul *et al.* [6])). Commercial use of PFOS and its salts first started around 1970 [6]. Uses and emissions of PFSAs and their precursors have been estimated to have steadily risen after manufacture started until a maximum usage was reached at the end of the 20th century [6]. PFSAs and precursors have been manufactured by ECF which produced a mixture of linear (70%) and branched (30%) isomers [51]. 3M also historically made products based on perfluorohexane and perfluorodecane sulfonyl fluoride [50]. For a more thorough description of sources of PFSA homologues and their precursors see the reviews of Paul *et al.* [6] and Armitage *et al.* [50].

PFOS and other PFSAs are widely distributed in the global environment [52-56], biota [57-62] and humans [29,63-70]. Due to the dominance of POSF-based products

historically, PFOS is usually found to be the most abundant PFSA. PFSAs are stronger acids than PFCAs with pKa's < 0 and are thus fully dissociated anions in environmental media [15]. Although properties vary with chain-length, the environmental fate and bioaccumulation behavior of PFSAs is broadly similar to that of PFCAs; PFSAs are persistent, are mostly distributed to surface waters [50], bind weakly to organic phases [16] compared to other persistent organic substances, are shown to bioaccumulate in the laboratory [71,72] and biomagnify in food webs [73,74]. Also similar to PFCAs, the global distribution of PFSAs is governed by a combination of direct release and transport as well as release and transport of precursors that subsequently degrade to PFSAs [50]. One difference in behavior is that PFSAs with perfluoroalkyl chains of the same length tend to sorb [16] and bioaccumulate [31,75,76] more strongly than PFCAs, [31,75,77] which is an effect of the different anionic head groups. Consequently, PFSAs with perfluoroalkyl chain lengths of C6 (i.e. perfluorohexane sulfonate) and higher are considered to be long-chain (http://www.oecd.org/ehs/pfc/), whereas for PFCAs those with perfluoroalkyl chain lengths of C8 (i.e. PFOA) and higher are considered to be long-chain.

In 2000 the major manufacturer of PFOS in the US (3M) announced their plan to cease the production of

C_6, C_8 and C_{10} perfluoroalkane sulfonyl fluoride (PASF)-based products and completed the phase out in 2002 [78]. In 2006 the EU adopted a Marketing and Use Directive (2006/122/EC) that bans the use of PFOS in semi-finished products (maximum content of PFOS: 0.005% by weight) as of summer 2008. In 2009, PFOS (and related substances derived from the POSF) were listed under Annex B (restriction of production and use) of the Stockholm Convention on Persistent Organic Pollutants. After 2000, China filled the gap created between global supply and demand of PFOS (and related products) caused by the 3M phase out. According to Zhang et al. [79], the production volume of PFOS increased from 30 t in 2002 to 247 t in 2006. Since then, caused by international legislation to restrict or eliminate PFOS production, the production volume of PFOS has declined to about 100 t/y in 2008.

Since 2003, 3M has commercialized new surface treatment products based on perfluorobutane sulfonate (PFBS, C4 sulfonate) [80]. Although likely to be as environmentally persistent as PFOS, PFBS is thought to be less bioaccumulative and toxic. It is likely that PFBS is released during manufacture and use of 3M's new surface treatment products and that increasing usage will lead to increasing levels in the environment. Evidence for the environmental release of PFBS can be found from its reported presence in the aquatic environment [81,82], biota [83] and humans [84]. A time-series of key production events and regulatory actions for PFSAs is shown in Figure 2.

Genesis of the systematic review

A systematic review on PFASs in the environment was originally suggested by the Swedish Chemicals Agency at a general stakeholder meeting arranged by Mistra EviEM in January 2012, to which a broad spectrum of organizations were invited and encouraged to suggest topics for systematic reviews. The Swedish Chemicals Agency is responsible for the Swedish environmental quality objective "A non-toxic environment" [85]. There is a growing concern in Sweden that continued discharges of PFASs may cause environmental as well as human health effects. Borg and Håkansson [86] performed a risk assessment consisting of an exposure assessment based on Swedish biomonitoring data, a hazard assessment with toxicological data from studies on mammals, birds and fish, and a risk characterization. The result of the environmental risk characterization indicated a cause for concern for seals and otters for hepatotoxicity and reproductive toxicity. For reproductive toxicity in birds, a cause for concern was indicated for PFOS where the highest level in peregrine falcons eggs (sampled in 2006) exceeded the toxic effect level identified in a study by Molina et al. [87]. Regarding human health effects, at least three different municipal drinking water plants in Sweden have been shut down due to high concentrations

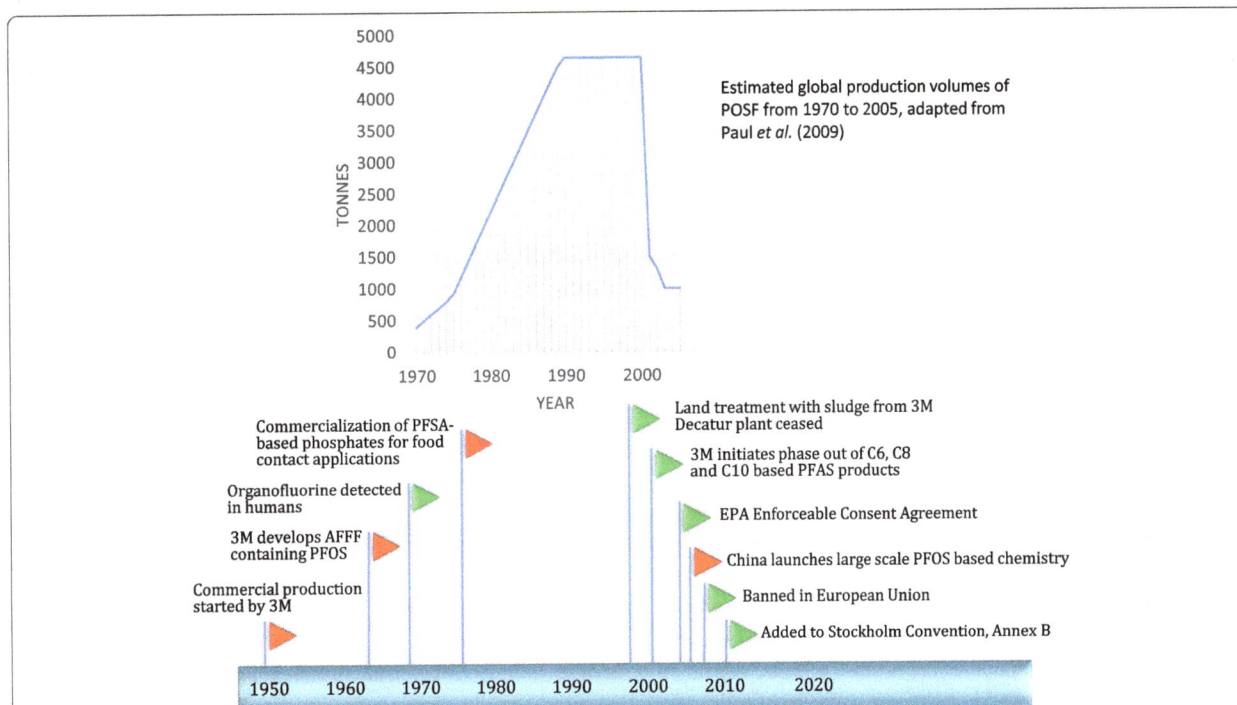

Figure 2 Timeline of production, commercialization and legislation of PFSAs. POSF (perfluorooctane sulfonyl fluoride) is the major raw material used to manufacture PFOS. Red flags represent events and actions that may have resulted in increased concentrations in the environment. Green flags represent important findings and phase-outs that may result in decreased concentrations in the environment.

of PFOS in the groundwater source. It is believed that the high concentrations are caused by spills of fire-fighting foam at adjacent airports or firefighting training sites.

A specific question regarding PFASs was not formulated at the general stakeholder meeting arranged by Mistra EviEM, but it emerged in later discussions with the Swedish Chemical Agency that increasing concentrations of short-chain substances such as PFBS have been observed [84], and that such trends are of great concern. There is a need to find out whether this is a local or global trend and whether similar trends have been observed for additional PFASs with recently increased production volumes.

Another stakeholder that Mistra EviEM has consulted is the Swedish Agency for Marine and Water Management (SWAM), which is responsible for the Swedish environmental quality objectives "Flourishing Lakes and Streams" and "A Balanced Marine Environment, Flourishing Coastal Areas and Archipelagos". Even though SWAM's major interest in this topic is the toxic effects, the Agency is also interested in more information on sources, transportation processes and environmental fate of PFASs in general.

Since the consultations resulted in several conceivable review questions, the main stakeholders and a few scientists were invited to a discussion where the goal was to find a common ground and agree on the most relevant question. It was concluded that there is little or no controversy regarding the properties of PFCAs and PFSAs; they are in many cases persistent, bioaccumulative and toxic (PBT) and some of them are classified and treated as such. It is however less clear how new compounds with shorter carbon chains should be treated. Even though they may be toxic and extremely persistent they may not be bio-accumulative enough to be classified as PBT-compounds or vPvB compounds. The Swedish Chemicals Agency needs more information on the environmental fate of both phased-out and replacement compounds. To compile and evaluate that information is not necessarily an easy task since, e.g., the results in monitoring studies can appear to be somewhat contradictory. In some cases one particular compound has shown opposite temporal trends in different sample types even within the same region.

In order to make the review as relevant as possible to as many people as possible, a wider group of Swedish stakeholders were invited to discuss what the review should focus on. This group of stakeholders included governmental agencies, municipal drinking water producers, environmental consultants and NGOs. Some of the key points put forward at the meeting included

- PFOS and PFOA are the two most important phased-out substances, but their precursors are also

important. Among replacement substances, the most toxic (if that is known) should be prioritized.
- Time trends in both biotic and abiotic samples should be included
- Contaminated areas as point sources are important to study. In what way will contaminated areas influence future time trends of concentrations in the surrounding environment?
- If time trends of replacement substances are scarce, the mere information on occurrences in the environment is also interesting.

The last point is out of the scope of this review and will not be considered. Additional questions that the stakeholders were seeking more information about included

- Are PFASs distributed differently between environmental media due to varying properties?
- How far should mitigation of contaminated areas be pushed? Is it possible to establish global or regional baseline or background concentrations?
- Can important knowledge gaps be identified?

One additional stakeholder that Mistra EviEM has consulted is the FluoroCouncil, which is a global organization representing a range of different fluorotechnology companies that manufacture, formulate or process fluoropolymer products, fluorotelomer-based products, fluoro-surfactants, and fluoro-surface property modification agents. The FluoroCouncil is involved in the 2010/2015 PFOA Stewardship Program, a global partnership between U.S. EPA and industry based on voluntary goals to eliminate PFOA from facility emissions and product content by 2015.

Objective of the review

Although compounds such as PFOS and PFOA have been phased out in some parts of the world, for many reasons it cannot be taken for granted that this will lead to swift declines in environmental or human exposure. If new emissions of a PFAS were to cease, the high persistence of PFCAs and PFSAs may lead to delayed and very slow declines in exposure. In addition, existing environmental burdens of precursor substances (including fluorinated polymers) might continue to act as an indirect pseudo-source of PFCAs and PFSAs in the future, depending on their stability. Furthermore, continued production and use of these compounds in parts of the world where phase-out policies have yet to be implemented could influence human and environmental exposure in regions far from this production, owing to long-range transport or exporting of commercial products containing PFASs. An additional complication is the vast spatial heterogeneity of current PFCA and PFSA burdens in the world's oceans. For example, despite all recent regulatory actions (Figures 1 and 2), remote marine ecosystems in the Arctic are predicted to receive increased exposure to PFOS and PFOA in the future as ocean currents slowly transport relatively contaminated ocean water from mid-latitudes northwards to remote regions where current ocean concentrations are lower [50,88,89]. With all these complexities, uncertainties, and global spatial heterogeneity, it is difficult to predict the future of human and environmental exposure to PFASs. For this reason, an objective and systematic review of temporal trends reported in the literature was deemed a high priority.

In summary, PFCAs and PFSAs have unique environmental chemistry and much of the environmental fate of substances within these groups is still poorly understood. The objective of the proposed systematic review is to find out whether the concentrations of these substances in the environment are changing in any particular direction, and whether any spatial differences or changes in temporal concentration trends can be related to the implemented phase-outs. In addition, any temporal changes in the distribution of linear vs branched perfluoroalkyl chains for individual PFCAs and PFSAs will be explored to determine if isomer pattern changes can be associated with the phase-outs.

The review team has phrased the review question as *"What is the effect of phasing out long-chain per- and polyfluoroalkyl substances on the concentrations of perfluoroalkyl acids and their precursors in the environment?"* The question builds on the following elements:

- Population/Subject: abiotic and biological samples including general human populations.
- Intervention: Legislative or voluntary phase-out of production and use of long-chain PFASs
- Comparator: Before intervention
- Outcome: Change of concentrations of the phased-out substances and their precursors and substitutes.

Methods
Searches
Searches in literature databases will be made using the search terms displayed in Table 1. Using the Boolean operators indicated this translates into the search string below, where * is a wildcard that can be any number of characters, and a question mark is exactly one arbitrary character.

(perfluor* OR polyfluor* OR fluorotelomer* OR PF?S OR PF?A OR PFC OR PFT OR PFHxS OR FOSE OR FOSA OR PAPS) AND (((trend OR variation) NEAR (time OR temporal)) OR ((change OR increase OR decrease) NEAR/5 (level or concentration)) OR "time series" OR

Table 1 Suggested search strategy

	Term 1	AND	Term 2
	perfluor*		((trend OR variation) NEAR (time OR temporal))
OR	polyfluor*	OR	((change OR increase OR decrease) NEAR/5 (level or concentration))
OR	fluorotelomer*	OR	"time series"
OR	PF?S	OR	((snow OR ice OR sediment) NEAR (core OR column OR cap))
OR	PF?A	OR	archive*
OR	PFC	OR	"specimen bank"
OR	PFT	OR	"long-term monitoring"
OR	PFHxS	OR	"repeated measurements"
OR	FOSE	OR	historic*
OR	FOSA		
OR	PAPS		

Search terms are connected with the Boolean operator AND, and words within each term is connected by OR. Truncation is denoted by *. Question mark is a wildcard that represents exactly one character.

((snow OR ice OR sediment) NEAR (core OR column OR cap)) OR archive* OR "specimen bank" OR "long-term monitoring" OR "repeated measurements" OR historic*)

The search string shown above and in Table 1 is designed for Web of Science. Other literature databases may use other wildcards or require less complex search strings. Adjusted search strings used in individual databases are shown in Additional file 2: Search strings used in individual literature databases. No particular constraints regarding time, document type, or language will be applied when searching for literature. At a later stage some languages may however be excluded due to limitations in translation resources. Literature databases that will be used for searching are listed in Table 2. The table also indicates the fields that will be searched and number of hits obtained in preliminary searches.

The comprehensiveness of the searches will be tested by cross-checking the hits with 1) a list of papers that the review team *a priori* think should be found by the searches, and 2) bibliographies in review articles. The list of papers that should be found is shown in Additional file 3: List of relevant articles that should be found in searches for literature.

In addition to data in the scientific literature it is anticipated that data will be found also in the grey literature. Such data will be searched for using the search engines Google and Google Scholar on the internet. For these search engines the search string below will be used. The same search string translated to Scandinavian languages and to German will also be used.

(pfas or pfaa or pfos or pfoa) (change or trend or temporal or increase or decrease).

Table 2 Electronic databases used for searching

Database	Searched field	No of hits[3]	Date
ISI Web of Science[1]	topic	3938 (3891)	2014-05-13
Science Direct	title, abstract and keywords	3272 (2869)	2014-05-13
Engineering Village[2]	subject/title/ abstract	1217 (728)	2014-05-14
Scopus	title, abstract and keywords	2502 (639)	2014-05-14
Academic search premier	title/abstract/ subject/keyword	851 (815)	2014-05-14
Wiley Online Library	abstract or title or keywords	641 (548)	2014-05-14
Directory of Open Access Journals	all fields	42 (39)	2014-05-14

[1]Including Web of Science™ Core Collection, Medline® and SciELO Citation Index.
[2]Including Geobase and Georef.
[3]Values in parentheses indicate number of items remaining after removal of duplicates when all databases are combined.

In addition, websites of relevant specialist organisations (listed below) will be searched. In this case the search strings will be website specific. All searches will however be reported in the systematic review. If no reports are found on the website, but the organisation still is thought to have relevant results, a letter will be sent to the appropriate person to try to obtain the information.

- Swedish Environmental Protection Agency (SEPA)
- US Environmental Protection Agency (USEPA)
- FluoroCouncil
- Society of the Plastics Industry
- Arctic Monitoring and Assessment Programme (AMAP)
- Danish Environmental Protection Agency
- Norwegian Environment Agency
- Finnish Environment Institute (SYKE)
- German Umweltsbundesamt
- UK Environment Agency
- HELCOM
- EEA
- Stockholm Convention GMP report to COP
- Netherlands RIVM
- National food agencies
- Environment Canada
- Canadian Northern Contaminants Program
- US CDC, National Health and Nutrition Examination Survey (NHANES)
- Other relevant organisations we identify

The reviewers will also send a standardized email to their scientific networks asking for information on possible studies that are soon to be available or in the grey literature.

Study inclusion criteria

The interventions relevant for the systematic review (phase-outs of long-chain PFASs) are intended to have a global rather than a local impact, and therefore the concentrations at one location may be impacted by a combination of more than one phase-out. Consequently, measured concentrations are in most studies not directly related to any particular phase-out. In fact, useful time trends may even be found in studies where the purpose of the study was completely unrelated to any phase-out of PFASs at all. This means that studies qualified for the systematic review will not be restricted to intervention studies, but will potentially include any study reporting on at least one temporal PFAS concentration trend. Since it is known when and where the interventions have been implemented it may be possible to relate these to the changes in concentrations, and hence, intervention should still be a justified question element.

Relevant temporal trends may be obtained by recurring measurements at a given location (monitoring) or by means of environmental archives (e.g. dated sediment cores, ice cores) or specimen banks (e.g. biota, human diet, human samples). There will be no geographic limitations. However, to be included in the systematic review, the articles must pass each of the following relevance criteria.

- Relevant population or subject: abiotic and biological samples, including general human populations, exposed to ambient loads of PFASs and their precursors. Populations with occupational exposures related to manufacturing of PFASs, or with deliberate exposures in controlled trials, will be excluded.
- Types of outcome: time trends in concentrations of PFASs covering at least two years. Ideally the studies should provide concentration data from both before and after an intervention. In that case it is possible to compare the concentrations before and after the intervention. This is however not a prerequisite for inclusion in the systematic review since there are other means of evaluating temporal trends (see section Data synthesis and presentation).

At the title and abstract level all retrieved articles will be screened by two reviewers. To check that the screening is consistent and complies with the agreed inclusion/ exclusion criteria, a small subset (10%) of the retrieved articles will also be double screened by the other reviewers. In case it cannot be decided whether the article should be included or excluded on the title and abstract level, the article will pass to the full text level. To evaluate the consistency of the screening, Kappa tests will be used. When screening at the full-text level the articles will be screened in the same manner as at the title and abstract level. A subset of at least 10% of the articles will be double screened. Again, Kappa tests will be used to test the consistency between the reviewers. If any Kappa test shows unacceptable discrepancies ($\kappa < 0.6$) the inclusion/exclusion criteria will be revisited by all reviewers and defined in a more unambiguous way. At both title/abstract and full text levels, excluded articles will be coded with a reason for exclusion. A list of all articles excluded at full text screening, with reasons for exclusion, will be provided in the systematic review.

Potential effect modifiers and reasons for heterogeneity

The outcome may to a large extent be a function of time, and one obvious factor influencing the outcome is the timing of the study relative to the interventions. A study with data from just a few years directly after the intervention is likely to show a smaller effect than a more recent study covering a longer study period.

Although the interventions are intended to have a global impact, the outcome may for several reasons vary considerably depending on the location of the studies. Since the interventions are not yet implemented globally, it is reasonable to assume that the effect will be smaller in regions close to present sources compared to regions close to past sources where the interventions have been implemented.

Contaminated areas may also be a significant factor. In areas where the phase-out has been implemented, contaminated areas may still be present and leaching the phased-out substances to the surrounding environments.

In more remote regions, such as the Arctic, the outcome may depend on the predominating mode of transport. In terrestrial or high-altitude areas where the PFAS source is dominated by long-range transport of volatile precursors, the response to the interventions may be quicker compared to coastal areas where the PFAS source is dominated by direct long-range transport in the aquatic environment. The measured outcome may also depend on sample type.

Food web dynamics, e.g. temporal shifts in diet and/or growth rates over the time-scale of decades, can influence long-term temporal trends of persistent organic contaminants. For example, in the northern Baltic Sea, herring have been shown over the long term to grow more slowly which is related to shift in feeding at one trophic level higher than they did in the past. Bioenergetics modelling undertaken by Peltonen et al. [90] suggests that changes in the feeding behaviour of herring can explain the "levelling off" of the downward temporal trends of polychlorinated dibenzo-p-dioxins and polychlorinated dibenzofurans (PCDD/Fs) in herring from the northern Baltic Sea [91]. Among studies of human samples, the results may vary between cross-sectional studies, where different people are studied every year, and prospective/longitudinal studies, where the same individuals are followed over time.

Analytical quality may also be a reason for heterogeneity. In the early years standards were not always available for accurate calibration of the analytical instruments. Both accuracy and precision have been improved in recent years, and the detection limits for PFASs were in general 100 times higher in 2000 than in 2014. The ability to distinguish between linear and branched perfluroalkyl chains and to analyse a wider range of substances have helped to improve our understanding of heterogeneities between studies.

In summary, potential effect modifiers and reasons for heterogeneity may include:

Timing of the study relative to the interventions
Proximity to past and present sources
Geographical differences
Mode of predominant transport
Type of sample e.g. sampling method, matrix, individual vs pooled
For human data, design of study (e.g. cross-sectional, prospective, age, sex)
Analytical quality in relation to time of the study (e.g. availability of standards)
Food web changes over time
Species differences

Study quality assessment

During critical appraisal of relevant studies, information that directly affects the internal validity of each study will be recorded in a pre-designed Excel spreadsheet. The assessment of the internal validity will focus on the following aspects:

- Selection bias. In this case the comparison groups are formed by samples taken at different times. The information about samples and what they represent must be sufficient to determine whether samples from different sampling occasions are comparable.
- Dating of samples. In most cases it is known when samples are collected, but samples from environmental archives such as sediment or ice cores are more complicated to date accurately. Dating by means of e.g. isotope techniques that give an absolute age to each individual sample may be regarded as high quality dating. Dating by means of historical markers (e.g. peak concentrations of other contaminants) providing relative ages may be regarded as acceptable dating. Post-depositional perturbations caused by for instance thawing-freezing cycles in snow or bioturbation in sediments should be discussed and assessed.

- Sample integrity. Are sample pre-treatment, sample preservation, prevention of contamination and storage methods suitable for the sample type?
- Analytical quality. Are analytical procedures appropriate? Did instrumentation, internal standards, procedural blanks, LOD/LOQ or field blanks change over time? If so, what is the risk that these parameters have influenced the measured temporal trends?

Study length and sampling frequency will be recorded but will generally not be part of the regular quality assessment since these parameters presumably will be reflected by the standard errors and confidence intervals of the effect sizes in each study. Study length and sampling frequency will thus affect the weighting of each study in the quantitative synthesis. The number of individual or pooled samples will also be recorded. Standard errors are likely to be smaller in studies with multiple and individually analysed samples at each sampling occasion than studies with just one sample from each time point. A high number of pooled samples at each sampling occasion may increase the chance of obtaining results representative for these time points and may therefore decrease the scatter along a regression line. In this way the number of pooled samples is also likely to affect the weighting of each study in the quantitative synthesis.

Information important for the assessment of external validity, or how transferable the studies are to the context of the question, will be recorded as well. As indicated by the study inclusion criteria, relatively few restrictions regarding populations/subjects and geographical locations will be applied. The external validity is therefore mainly related to subgroup analysis that will be performed. Information about environmental setting and important sample covariates should be reported to such an extent that each study confidently can be grouped with other studies or be judged to stand alone. Also, the study should cover a time period that is possible to relate to any known relevant phase-out.

Relevant studies that are judged to have low risk for selection bias, have accurately dated samples, ensured good sample integrity, have used appropriate analytical procedures, and provide sufficient information regarding external validity will meet the quality criteria and hence be included in the quantitative synthesis. Any reason for not including a study based on quality will be recorded in the data spreadsheet and a list of all articles excluded during critical appraisal, with reasons for exclusion, will be presented in the systematic review. However, studies that fail to qualify for the quantitative synthesis will be used in sensitivity analyses where it will be tested if the inclusion or exclusion of these studies influence the overall results.

Critical appraisal of relevant studies will be carried out by four of the reviewers. To ensure a high consistency between the reviewers at least 25% of the relevant articles will be double-checked and Kappa tests will be used to test the consistency. In addition to this, any articles that for some reason are difficult to critically appraise will be discussed by the entire review team before any decisions are made.

Data extraction strategy

Data (concentrations and other important study information) will be extracted from relevant studies and recorded in pre-designed Excel data sheets. Information directly relevant to the internal validity of the studies will be extracted and recorded by the reviewers that carry out the critical appraisal, while the rest of the data will be extracted and recorded by two other reviewers. However, the design of the data extraction sheets and how they should be filled will be approved in advance by all reviewers, and double-checking by all reviewers will be done for at least 25% of the articles. Data will always be recorded as reported in the primary studies. All necessary transformations and calculations will be performed at the analysis stage.

The outcome data that will be recorded is the PFAS concentrations in the studied samples at different times and/or slopes of trends during specified time periods. The location of these data within each article, as well as an indication whether the data have been graphically extracted from figures, will be recorded. The types of data that will be compiled include sample type (air, sediment, species etc.), matrix (filter, 2 cm slice, liver, dissolved vs total water etc.), number of pooled or individual samples, covariates (sampling time, age, sex etc.), statistical variates (concentrations, slope of trend, linear-non-linear etc.), geographical location, time period covered, number of years covered. Data extraction may include contact with individual scientists for complementary information or for raw data or unpublished data.

Data synthesis and presentation

Meta-analyses for each PFAS will be performed using random effect models [92]. One possible effect size will be based on average concentrations during specified periods before and after the implementation of an intervention, respectively. All studies do however not use the same scale. Different units for concentration are used for different sample types (e.g. water and sediment), and different sample types show different concentration levels depending on the partition coefficient between the sample types. Different units for concentration may also be used for a certain sample type. For example, the concentrations in sediment samples are sometimes given as ng/g bulk sediment and sometimes as ng/g TOC (Total

Organic Carbon). All of this means that raw differences in average concentrations cannot be used as effect size. Therefore response ratios will be used as the effect size. The response ratio R is calculated according to equation 1, where $\bar{C}_0$ and $\bar{C}_I$ are the average concentration during a specified period before and after the implementation of the intervention, respectively.

$$R = \frac{\bar{C}_I}{\bar{C}_0} \quad (1)$$

Another effect size that will be explored is the normalized rate of change in concentration. It is possible that even though there is no significant difference in average concentrations, there can be a significant trend in some direction after the intervention. Moreover, although the evidence for a causal relationship between the outcome and the intervention will be weaker, the rate of change in concentration can be evaluated when pre-intervention data is missing or when more than one intervention influence the outcome. This means that also non-intervention studies can be included. The rate of change will be calculated by means of regression analyses. Again, since different scales are used in different studies, the rate of change needs to be normalized. If the rate of change is normalized to the average concentration before the intervention the effect size C' is calculated as

$$C' = \frac{1}{\bar{C}_0} \cdot \frac{\partial C}{\partial t} \quad (2)$$

where C_0 is concentration at time t = 0. It may also be possible to compare the normalised rates of change in concentration before and after the intervention, respectively, if that data exist.

Obviously, with such a wide range of subjects or populations that potentially can be included in the systematic review it will be necessary to perform subgroup analyses. It is envisaged that for each PFAS analysed, the splitting into subgroups will be based on e.g. sample type, species and geographical region. A narrative synthesis will be prepared using tables of study characteristics and results. Visualisation of results may be done using forest plots and other graphical representations.

Additional files

Additional file 1: List of abbreviations.

Additional file 2: Search strings used in individual literature databases.

Additional file 3: List of relevant articles that should be found in searches for literature.

Competing interests
The authors declare that they have no competing interests.

Authors' contributions
This systematic review protocol is based on a draft written by ML. All authors discussed the draft and suggested improvements. The major part of the background section is written by JWM, ITC and DH. All authors read and approved the final manuscript.

Acknowledgements
The preparation of this protocol and the forthcoming review is financed by the Mistra Council for Evidence-based Environmental Management (Mistra EviEM). EviEM is funded by the Swedish Foundation for Strategic Environmental Research (Mistra) and hosted by the Royal Swedish Academy of Sciences. The authors wish to thank two anonymous reviewers for their insightful comments. We are also grateful for the comments made by stakeholders during a public review of an earlier version of this protocol.

Author details
[1]Mistra EviEM, The Royal Swedish Academy of Sciences, Box 50005, SE-104 05 Stockholm, Sweden. [2]Department of Environmental Science and Analytical Chemistry (ACES), Stockholm University, SE-106 91 Stockholm, Sweden. [3]Norwegian Inst Air Res, FRAM High North Res Ctr Climate & Environm, Tromso, Norway. [4]Division of Analytical and Environmental Toxicology, 10-102C Clinical Sciences, Edmonton, Alberta T6G 2G3, Canada.

References
1. Lindstrom AB, Strynar MJ, Libelo EL. Polyfluorinated compounds: past, present, and future. Environ Sci Technol. 2011;45:7954–61.
2. Kissa E. Fluorinated Surfactants and Repellents. 2nd ed. New York (NY): Marcel Dekker, Inc.; 2001.
3. Prevedouros K, Cousins IT, Buck RC, Korzeniowski SH. Sources, fate and transport of perfluorocarboxylates. Environ Sci Technol. 2006;40:32–44.
4. Wang Z, Cousins IT, Scheringer M, Buck RC, Hungerbühler K. Global emission inventories for C4–C14 perfluoroalkyl carboxylic acid (PFCA) homologues from 1951 to 2030, Part I: production and emissions from quantifiable sources. Environ Int. 2014;70:62–75.
5. Wang Z, Cousins IT, Scheringer M, Buck RC, Hungerbühler K. Global emission inventories for C4–C14 perfluoroalkyl carboxylic acid (PFCA) homologues from 1951 to 2030, part II: the remaining pieces of the puzzle. Environ Int. 2014;69:166–76.
6. Paul AG, Jones KC, Sweetman AJ. A first global production, emission, and environmental inventory for perfluorooctane sulfonate. Environ Sci Technol. 2009;43:386–92.
7. Armitage JM, MacLeod M, Cousins IT. Modeling the Global Fate and Transport of Perfluorooctanoic Acid (PFOA) and Perfluorooctanoate (PFO) Emitted from direct sources using a multispecies mass balance model. Environ Sci Technol. 2009;43:1134–40.
8. Smart BE, Dixon DA. Bond-energies and stabilities of poly(perfluoroethers). Abstr Pap Am Chem Soc. 1994;207:31-FLUO.
9. Buck RC, Franklin J, Berger U, Conder JM, Cousins IT, de Voogt P, et al. Perfluoroalkyl and polyfluoroalkyl substances in the environment: terminology, classification, and origins. Integr Environ Assess Manag. 2011;7:513–41.
10. Becker AM, Gerstmann S, Frank H. Perfluorooctane surfactants in waste waters, the major source of river pollution. Chemosphere. 2008;72:115–21.
11. Li XY, Zhang PY, Jin L, Shao T, Li ZM, Cao JJ. Efficient photocatalytic decomposition of perfluorooctanoic acid by indium oxide and its mechanism. Environ Sci Technol. 2012;46:5528–34.
12. Igarashi S, Yotsuyanagi T. Homogeneous liquid-liquid-extraction by ph dependent phase-separation with a fluorocarbon ionic surfactant and its application to the preconcentration of porphyrin compounds. Mikrochim Acta. 1992;106:37–44.
13. Lopez-Fontan JL, Sarmiento F, Schulz PC. The aggregation of sodium perfluorooctanoate in water. Colloid Polym Sci. 2005;283:862–71.
14. Wang ZY, MacLeod M, Cousins IT, Scheringer M, Hungerbuhler K. Using COSMOtherm to predict physicochemical properties of poly- and perfluorinated alkyl substances (PFASs). Environ Chem. 2011;8:389–98.
15. Vierke L, Berger U, Cousins IT. Estimation of the acid dissociation constant of perfluoroalkyl carboxylic acids through an experimental investigation of their water-to-air transport. Environ Sci Technol. 2013;47:11032–9.

16. Higgins CP, Luthy RG. Sorption of perfluorinated surfactants on sediments. Environ Sci Technol. 2006;40:7251–6.

17. Gottschall N, Topp E, Edwards M, Russell P, Payne M, Kleywegt S, et al. Polybrominated diphenyl ethers, perfluorinated alkylated substances, and metals in tile drainage and groundwater following applications of municipal biosolids to agricultural fields. Sci Total Environ. 2010;408:873–83.

18. Ahrens L, Yeung LWY, Taniyasu S, Lam PKS, Yamashita N. Partitioning of perfluorooctanoate (PFOA), perfluorooctane sulfonate (PFOS) and perfluorooctane sulfonamide (PFOSA) between water and sediment. Chemosphere. 2011;85:731–7.

19. Benskin JP, Ahrens L, Muir DCG, Scott BF, Spencer C, Rosenberg B, et al. Manufacturing Origin of Perfluorooctanoate (PFOA) in Atlantic and Canadian Arctic Seawater. Environ Sci Technol. 2012;46:677–85.

20. Yamashita N, Taniyasu S, Petrick G, Wei S, Gamo T, Lam PKS, et al. Perfluorinated acids as novel chemical tracers of global circulation of ocean waters. Chemosphere. 2008;70:1247–55.

21. Barton CA, Zarzecki CJ, Russell MH. A site-specific screening comparison of modeled and monitored air dispersion and deposition for perfluorooctanoate. J Air Waste Manage Assoc. 2010;60:402–11.

22. Ellis DA, Martin JW, De Silva AO, Mabury SA, Hurley MD, Sulbaek Andersen MP, et al. Degradation of fluorotelomer alcohols: a likely atmospheric source of perfluorinated carboxylic acids. Environ Sci Technol. 2004;38:3316–21.

23. Armitage JM, MacLeod M, Cousins IT. Comparative Assessment of the Global Fate and Transport Pathways of Long-Chain Perfluorocarboxylic Acids (PFCAs) and Perfluorocarboxylates (PFCs) Emitted from Direct Sources. Environ Sci Technol. 2009;43:5830–6.

24. EFSA. Opinion of the Scientific Panel on Contaminants in the Food chain on Perfluorooctane sulfonate (PFOS), perfluorooctanoic acid (PFOA) and their salts. FSA J. 2008;653:1–131.

25. Ahrens L, Herzke D, Huber S, Bustnes JO, Bangjord G, Ebinghaus R. Temporal trends and pattern of polyfluoroalkyl compounds in tawny owl (Strix aluco) eggs from Norway, 1986–2009. Environ Sci Technol. 2011;45:8090–7.

26. Holmstrom KE, Johansson AK, Bignert A, Lindberg P, Berger U. Temporal trends of perfluorinated surfactants in swedish peregrine falcon eggs (Falco peregrinus), 1974–2007. Environ Sci Technol. 2010;44:4083–8.

27. Vicente J, Bertolero A, Meyer J, Viana P, Lacorte S. Distribution of perfluorinated compounds in Yellow-legged gull eggs (Larus michahellis) from the Iberian Peninsula. Sci Total Environ. 2012;416:468–75.

28. Houde M, De Silva AO, Muir DCG, Letcher RJ. Monitoring of perfluorinated compounds in aquatic biota: an updated review PFCs in aquatic biota. Environ Sci Technol. 2011;45:7962–73.

29. Lindh CH, Rylander L, Toft G, Axmon A, Rignell-Hydbom A, Giwercman A, et al. Blood serum concentrations of perfluorinated compounds in men from Greenlandic Inuit and European populations. Chemosphere. 2012;88:1269–75.

30. Sturm R, Ahrens L. Trends of polyfluoroalkyl compounds in marine biota and in humans. Environ Chem. 2010;7:457–84.

31. Conder JM, Hoke RA, De Wolf W, Russell MH, Buck RC. Are PFCAs bioaccumulative? A critical review and comparison with regulatory lipophilic compounds. Environ Sci Technol. 2008;42:995–1003.

32. Tomy GT, Pleskach K, Ferguson SH, Hare J, Stern G, Macinnis G, et al. Trophodynamics of some PFCs and BFRs in a Western Canadian Arctic Marine Food Web. Environ Sci Technol. 2009;43:4076–81.

33. Kelly BC, Ikonomou MG, Blair JD, Surridge B, Hoover D, Grace R, et al. Perfluoroalkyl contaminants in an arctic marine food web: trophic magnification and wildlife exposure. Environ Sci Technol. 2009;43:4037–43.

34. Loi EIH, Yeung LWY, Taniyasu S, Lam PKS, Kannan K, Yamashita N. Trophic magnification of poly- and perfluorinated compounds in a subtropical food web. Environ Sci Technol. 2011;45:5506–13.

35. Martin JW, Whittle DM, Muir DCG, Mabury SA. Perfluoroalkyl contaminants in a food web from lake Ontario. Environ Sci Technol. 2004;38:5379–85.

36. Muller CE, De Silva AO, Small J, Williamson M, Wang XW, Morris A, et al. Biomagnification of perfluorinated compounds in a remote terrestrial food chain: lichen-caribou-wolf. Environ Sci Technol. 2011;45:8665–73.

37. Russell MH, Nilsson H, Buck RC. Elimination kinetics of perfluorohexanoic acid in humans and comparison with mouse, rat and monkey. Chemosphere. 2013;93:2419–25.

38. Du GZ, Huang HY, Hu JL, Qin YF, Wu D, Song L, et al. Endocrine-related effects of perfluorooctanoic acid (PFOA) in zebrafish, H295R steroidogenesis and receptor reporter gene assays. Chemosphere. 2013;91:1099–106.

39. Joensen UN, Veyrand B, Antignac JP, Jensen MB, Petersen JH, Marchand P, et al. PFOS (perfluorooctanesulfonate) in serum is negatively associated with testosterone levels, but not with semen quality, in healthy men. Hum Reprod. 2013;28:599–608.

40. Kjeldsen LS, Bonefeld-Jorgensen EC. Perfluorinated compounds affect the function of sex hormone receptors. Environ Sci Pollut Res. 2013;20:8031–44.

41. Ikeda T, Aiba K, Fukuda K, Tanaka M. The induction of peroxisome proliferation in rat-liver by perfluorinated fatty-acids, metabolically inert derivatives of fatty-acids. J Biochem. 1985;98:475–82.

42. Kudo N, Bandai N, Suzuki E, Katakura M, Kawashima Y. Induction by perfluorinated fatty acids with different carbon chain length of peroxisomal beta-oxidation in the liver of rats. Chem Biol Interact. 2000;124:119–32.

43. Upham BL, Deocampo ND, Wurl B, Trosko JE. Inhibition of gap junctional intercellular communication by perfluorinated fatty acids is dependent on the chain length of the fluorinated tail. Int J Cancer. 1998;78:491–5.

44. Vanden Heuvel JP. Perfluorodecanoic acid as a useful pharmacologic tool for the study of peroxisome proliferation. Gen Pharmacol Vasc Syst. 1996;27:1123–9.

45. 3M. Letter to US EPA Re: phase-out plan for POSF-based products (226–0600). US EPA Adm Rec. 2000;226:1–11.

46. 2010/2015 PFOA Stewardship Programme. http://www.epa.gov/oppt/pfoa/pubs/stewardship/index.html.

47. ECHA. Candidate list of substances of very high concern for authorisation. 2014; http://echa.europa.eu/web/guest/candidate-list-table.

48. Ritter S. Fluorochemicals go short. Chem Eng News. 2010;88:12–7.

49. Wang ZY, Cousins IT, Scheringer M, Hungerbuhler K. Fluorinated alternatives to long-chain perfluoroalkyl carboxylic acids (PFCAs), perfluoroalkane sulfonic acids (PFSAs) and their potential precursors. Environ Int. 2013;60:242–8.

50. Armitage JM, Schenker U, Scheringer M, Martin JW, MacLeod M, Cousins IT. Modeling the global fate and transport of perfluorooctane sulfonate (PFOS) and precursor compounds in relation to temporal trends in wildlife exposure. Environ Sci Technol. 2009;43:9274–80.

51. Martin JW, Asher BJ, Beesoon S, Benskin JP, Ross MS. PFOS or PreFOS? Are perfluorooctane sulfonate precursors (PreFOS) important determinants of human and environmental perfluorooctane sulfonate (PFOS) exposure? J Environ Monit. 2010;12:1979–2004.

52. Benskin JP, Muir DCG, Scott BF, Spencer C, De Silva AO, Kylin H, et al. Perfluoroalkyl acids in the Atlantic and Canadian Arctic Oceans. Environ Sci Technol. 2012;46:5815–23.

53. Filipovic M, Berger U, McLachlan MS. Mass balance of perfluoroalkyl acids in the Baltic sea. Environ Sci Technol. 2013;47:4088–95.

54. Hu JY, Yu J, Tanaka S, Fujii S. Perfluorooctane Sulfonate (PFOS) and Perfluorooctanoic Acid (PFOA) in water environment of Singapore. Water Air Soil Pollut. 2011;216:179–91.

55. Kwok KY, Yamazaki E, Yamashita N, Taniyasu S, Murphy MB, Horii Y, et al. Transport of Perfluoroalkyl substances (PFAS) from an arctic glacier to downstream locations: implications for sources. Sci Total Environ. 2013;447:46–55.

56. Wang X, Halsall C, Codling G, Xie Z, Xu B, Zhao Z, et al. Accumulation of perfluoroalkyl compounds in tibetan mountain snow: temporal patterns from 1980 to 2010. Environ Sci Technol. 2014;48:173–81.

57. Ahrens L. Polyfluoroalkyl compounds in the aquatic environment: a review of their occurrence and fate. J Environ Monit. 2011;13:20–31.

58. Braune BM, Letcher RJ. Perfluorinated sulfonate and carboxylate compounds in eggs of seabirds breeding in the Canadian Arctic: temporal trends (1975–2011) and interspecies comparison. Environ Sci Technol. 2013;47:616–24.

59. Giesy JP, Kannan K. Global distribution of perfluorooctane sulfonate in wildlife. Environ Sci Technol. 2001;35:1339–42.

60. Hart K, Gill VA, Kannan K. Temporal trends (1992–2007) of perfluorinated chemicals in northern sea otters (Enhydra lutris kenyoni) from South-Central Alaska. Arch Environ Contam Toxicol. 2009;56:607–14.

61. Lofstrand K, Jorundsdottir H, Tomy G, Svavarsson J, Weihe P, Nygard T, et al. Spatial trends of polyfluorinated compounds in guillemot (Uria aalge) eggs from North-Western Europe. Chemosphere. 2008;72:1475–80.

62. Shi YL, Pan YY, Yang RQ, Wang YW, Cai YQ. Occurrence of perfluorinated compounds in fish from Qinghai-Tibetan Plateau. Environ Int. 2010;36:46–50.

63. Calafat AM, Kuklenyik Z, Caudill SP, Reidy JA, Needham LL. Perfluorochemicals in pooled serum samples from United States residents in 2001 and 2002. Environ Sci Technol. 2006;40:2128–34.

64. Chen CL, Lu YL, Zhang X, Geng J, Wang TY, Shi YJ, et al. A review of spatial and temporal assessment of PFOS and PFOA contamination in China. Chem Ecol. 2009;25:163–77.

65. Ericson I, Gomez M, Nadal M, van Bavel B, Lindstrom G, Domingo JL. Perfluorinated chemicals in blood of residents in Catalonia (Spain) in relation to age and gender: A pilot study. Environ Int. 2007;33:616–23.

66. Fromme H, Midasch O, Twardella D, Angerer J, Boehmer S, Liebl B. Occurrence of perfluorinated substances in an adult German population in southern Bavaria. Int Arch Occup Environ Health. 2007;80:313–9.

67. Ji K, Kim S, Kho Y, Paek D, Sakong J, Ha J, et al. Serum concentrations of major perfluorinated compounds among the general population in Korea: dietary sources and potential impact on thyroid hormones. Environ Int. 2012;45:78–85.

68. Karrman A, Domingo JL, Llebaria X, Nadal M, Bigas E, van Bavel B, et al. Biomonitoring perfluorinated compounds in Catalonia, Spain: concentrations and trends in human liver and milk samples. Environ Sci Pollut Res. 2010;17:750–8.

69. Kato K, Calafat AM, Wong LY, Wanigatunga AA, Caudill SP, Needham LL. Polyfluoroalkyl compounds in pooled sera from children participating in the national health and nutrition examination survey 2001–2002. Environ Sci Technol. 2009;43:2641–7.

70. Volkel W, Genzel-Boroviczeny O, Demmelmair H, Gebauer C, Koletzko B, Twardella D, et al. Perfluorooctane sulphonate (PFOS) and perfluorooctanoic acid (PFOA) in human breast milk: results of a pilot study. Int J Hyg Environ Health. 2008;211:440–6.

71. Fernandez-Sanjuan M, Faria M, Lacorte S, Barata C. Bioaccumulation and effects of perfluorinated compounds (PFCs) in zebra mussels (Dreissena polymorpha). Environ Sci Pollut Res. 2013;20:2661–9.

72. Lasier PJ, Washington JW, Hassan SM, Jenkins TM. Perfluorinated chemicals in surface waters and sediments from northwest georgia, usa, and their bioaccumulation in lumbriculus variegatus. Environ Toxicol Chem. 2011;30:2194–201.

73. Fang S, Chen X, Zhao S, Zhang Y, Jiang W, Yang L, et al. Trophic magnification and isomer fractionation of perfluoroalkyl substances in the food web of Taihu Lake, China. Environ Sci Technol. 2014;48:2173–82.

74. Kannan K, Tao L, Sinclair E, Pastva SD, Jude DJ, Giesy JP. Perfluorinated compounds in aquatic organisms at various trophic levels in a Great Lakes food chain. Arch Environ Contam Toxicol. 2005;48:559–66.

75. Martin JW, Mabury SA, Solomon KR, Muir DCG. Bioconcentration and tissue distribution of perfluorinated acids in rainbow trout (Oncorhynchus mykiss). Environ Toxicol Chem. 2003;22:196–204.

76. Martin JW, Mabury SA, Solomon KR, Muir DCG. Dietary accumulation of perfluorinated acids in juvenile rainbow trout (Oncorhynchus mykiss). Environ Toxicol Chem. 2003;22:189–95.

77. Zhao S, Zhu L, Liu L, Liu Z, Zhang Y. Bioaccumulation of perfluoroalkyl carboxylates (PFCAs) and perfluoroalkane sulfonates (PFSAs) by earthworms (Eisenia fetida) in soil. Environ Pollut. 2013;179:45–52.

78. Olsen GW, Mair DC, Church TR, Ellefson ME, Reagen WK, Boyd TM, et al. Decline in perfluorooctanesulfonate and other polyfluoroalkyl chemicals in American Red Cross adult blood donors, 2000–2006. Environ Sci Technol. 2008;42:4989–95.

79. Zhang L, Liu JG, Hu JX, Liu C, Guo WG, Wang Q, et al. The inventory of sources, environmental releases and risk assessment for perfluorooctane sulfonate in China. Environ Pollut. 2012;165:193–8.

80. Renner R. The long and the short of perfluorinated replacements. Environ Sci Technol. 2006;40:12–3.

81. Kirchgeorg T, Weinberg I, Dreyer A, Ebinghaus R. Perfluorinated compounds in marine surface waters: data from the Baltic Sea and methodological challenges for future studies. Environ Chem. 2010;7:429–34.

82. Zhou Z, Liang Y, Shi Y, Xu L, Cai Y. Occurrence and transport of perfluoroalkyl acids (PFAAs), including short-chain PFAAs in Tangxun Lake, China. Environ Sci Technol. 2013;47:9249–57.

83. Persson S, Rotander A, Karrman A, van Bavel B, Magnusson U. Perfluoroalkyl acids in subarctic wild male mink (Neovison vison) in relation to age, season and geographical area. Environ Int. 2013;59:425–30.

84. Glynn A, Berger U, Bignert A, Ullah S, Aune M, Lignell S, et al. Perfluorinated alkyl acids in blood serum from primiparous women in Sweden: serial sampling during pregnancy and nursing, and temporal trends 1996–2010. Environ Sci Technol. 2012;46:9071–9.

85. The environmental objectives portal. http://www.miljomal.se/Environmental-Objectives-Portal/.

86. Borg D, Håkansson H. Environmental and Health Risk Assessment of Perfluoroalkylated and Polyfluoroalkylated Substances (PFASs) in Sweden. In: Book Environmental and Health Risk Assessment of Perfluoroalkylated and Polyfluoroalkylated Substances (PFASs) in Sweden (Editor ed.^eds.), vol. 6513. City: Swedish Environmental Protection Agency; 2012.

87. Molina ED, Balander R, Fitzgerald SD, Giesy JP, Kannan K, Mitchell R, et al. Effects of air cell injection of perfluorooctane sulfonate before incubation on development of the white leghorn chicken (Gallus domesticus) embryo. Environ Toxicol Chem. 2006;25:227–32.

88. Dietz R, Bossi R, Riget FF, Sonne C, Born EW. Increasing perfluoroalkyl contaminants in east Greenland polar bears (Ursus maritimus): a new toxic threat to the Arctic bears. Environ Sci Technol. 2008;42:2701–7.

89. Wania F. A global mass balance analysis of the source of perfluorocarboxylic acids in the Arctic ocean. Environ Sci Technol. 2007;41:4529–35.

90. Peltonen H, Kiljunen M, Kiviranta H, Vuorinen PJ, Verta M, Karjalainen J. Predicting effects of exploitation rate on weight-at-age, population dynamics, and bioaccumulation of PCDD/Fs and PCBs in herring (Clupea harengus L.) in the Northern Baltic Sea. Environ Sci Technol. 2007;41:1849–55.

91. Miller A, Hedman JE, Nyberg E, Haglund P, Cousins IT, Wiberg K, et al. Temporal trends in dioxins (polychlorinated dibenzo-p-dioxin and dibenzofurans) and dioxin-like polychlorinated biphenyls in Baltic herring (Clupea harengus). Mar Pollut Bull. 2013;73:220–30.

92. Borenstein M, Hedges LV, Higgins JPT, Rothstein HR. Introduction to meta-analysis. John Wiley & Sons, Ltd; 2009.

10

What are the impacts of urban agriculture programs on food security in low and middle-income countries: a systematic review

Marcel Korth[1*], Ruth Stewart[1,2], Laurenz Langer[1], Nolizwe Madinga[1], Natalie Rebelo Da Silva[1], Hazel Zaranyika[1], Carina van Rooyen[1] and Thea de Wet[1]

Abstract

Background: Urban Agriculture is considered to contribute to improved food security among the income poor in urban contexts across developing countries. Much literature exists on the topic assuming a positive relationship. The aim of this review was to collect and analyse available evidence on the impact of urban agriculture in low and middle-income countries.

Methods: We employed systematic review methods to identify all relevant and reliable research on UA's impact on food security and nutrition. Only impact evaluations that set out to measure the effectiveness of UA interventions on food security, as compared to the effects of not engaging in UA, qualified for inclusion. Studies had to have a comparison group and at least two data points.

Results: Systematic searches resulted in 8142 hits, and screening of abstracts resulted in 198 full texts identified. No studies met the review's inclusion criteria. Therefore, the review found no available evidence that supports or refutes the suggestion that urban agriculture positively impacts on individual or household food security in low and middle-income countries. The largest proportion of studies at full text stage was excluded based on study design, as they were not impact evaluations, i.e. they did not have a comparison group and at least data points. Two observations were made: Firstly, searches yielded a range of studies that consider *associations* between UA and certain aspects of food security. Secondly, there is a large pool of cross-sectional studies on UA's potential to contribute to increased food security, particularly from west and east Africa.

Conclusions: The research currently available does not allow for any conclusions to be made on whether or not urban agriculture initiatives contribute to food security. The fact that impact evaluations are absent from the current evidence-base calls for increased efforts to measure the impact of urban agriculture on food security in low and middle-income countries through rigorous impact evaluations. With regard to systematic review methodology, this review alludes to the value of compiling a systematic map prior to engaging in a full systematic review.

Keywords: Urban agriculture, Food security, Nutrition, Impact, Systematic review, Urbanisation

* Correspondence: mkorth@uj.ac.za
[1]CEE Johannesburg, Centre for Anthropological Research, University of Johannesburg, Johannesburg, South Africa
Full list of author information is available at the end of the article

Background

The emergence of urban agriculture

The twenty-first century has often been described as 'the first urban century'. Unprecedented rural–urban migration has led to rapid urban growth. Whilst in 1900 a mere 13 per cent of the world's population lived in urban areas, the UN-Habitat [1] estimates that by 2030 this level will have risen to 60 per cent. Furthermore, virtually all of this population growth over the next few decades will be absorbed by cities in low and middle-income countries and thus increase pressure on often already exhausted urban resources and administrations.

Among the most pressing needs of any urban agglomeration is to achieve food security. Urban populations depend on reliable and stable availability of food products as well as affordable and convenient access to them. High levels of urban income poverty paired with rising food prices, however, often make the formal urban food supply system unaffordable to the urban poor. An informal supply system, consisting of street vendors, informal markets, home-based enterprises as well as Urban Agriculture (UA), since exists alongside formal interventions. These informal networks predominantly satisfy the urban poor's demand for cheap and easily accessible foodstuffs.

Approaches to urban agriculture

Whilst urban decision makers and academics alike have identified UA as the most beneficial and promising pillar of informal food supply systems [2-7], the evidence for such claims is unclear. Although UA has been an integral part of urban livelihoods[a] throughout human history the concept came only to the fore in the late 1980s/early 1990s, evoking interest among international donors and development practitioners [6]. A United Nations Development Program (UNDP) report compiled by Smit and colleagues in 1996 [7] estimated UA to be reaching some 800 million urban dwellers who used UA as a livelihood strategy in the early 1990s. A number of studies with promising titles such as 'hunger-proof cities', 'Agropolis' and 'Cities feeding people' [8] indicate the potential generally ascribed to UA. Critics of UA quickly pointed to the weak empirical evidence of some of these studies and the low overall scale of UA amongst urban poor [9]. During the first years of the urban century, UA had therefore slipped past the focus of the international development community. Yet the peak of global food prices in 2008 shed new light on the idea of locally produced food products and household subsistence production. Urban agriculture subsequently once more was considered as a major intervention to improve urban food security [10].

For example, the City of Johannesburg in its Growth and Development Strategy (GDS) 2040, identifies UA as its main intervention to address food security within the city [11]. On a global scale, the UN High Level Task Force on the Global Food Crisis [12] identified UA as an important strategy to alleviate urban food insecurity and to build cities that are more resilient to crises. A joint FAO/World Bank paper of the same year [13] also expressed that "the World Bank and FAO (...) will promote [UA] related programs and projects in the context of the MDGs and more specifically MDG1 'Eradicate extreme poverty and hunger' and MDG7 'Ensure environmental sustainability'." The FAO published the 'Urban Producers Resource Book' [14] as an outcome of its 'Food for the city' programme. FAO's 'Food for the cities' program forms part of a wider network of organisations, consisting of the UNDP/UN-Habitat 'Sustainable Cities Program', the IDRC's 'Urban Poverty and Environment program', and the Resource Center on Urban Agriculture and Food Security (RUAF), which all strongly advocate UA as a tool to address risks associated with urban food insecurity. In the light of this strong support for UA one must note, however, that some fundamental questions regarding UA remain unanswered. Renewed interest in the topic did not necessarily converge with new knowledge about UA. Little is known about the true extent and impact of UA in urban livelihoods.

Definitions of urban agriculture

Urban agriculture is a complex concept. A large variety of urban farming systems exists internationally, with varying characteristics depending on local socio-economic, geographic and political conditions, further complicating a universally applicable definition. Luc Mougeot developed the currently most widely used definition of UA in 2001 [15]. Using technical criteria of UA he explained that, 'urban agriculture is an industry located within (intra-urban) or on the fringe (peri-urban) of a town, a city or a metropolis, which grows and raises, processes and distributes a diversity of food and non-food products, (re-)using largely human and material resources, products and services found in and around that urban area, and in turn supplying human and material resources, products and services largely to that urban area' [15]. For the purpose of our review, this translates into an understanding of UA as a social intervention, geographically constrained to urban and/or peri-urban areas, involving any form of agriculture with the aim of improving the food security of actors involved.

A wide variety of produce results from UA and can best be classified according to their respective methods of production. Horticulture, animal husbandry, aqua culture and forestry can all be found in urban locations and generate products ranging from, inter alia, fruits and vegetables, dairy products, meat, fish, herbs and fire wood. In terms of end-points, UA's products can either be used for consumption, surplus sale or trade and commercial activities. Actors in UA display a similar diversity. Whilst early literature assumed UA to be a livelihood strategy almost exclusively used by low-income groups and recent migrants to the city with the aim to increase household levels of food security, most scholars today identify a wider range of actors in UA [16]. The popularity of roof top gardens in many high-income countries for example shows the increased reach of UA.

Urban agriculture and food security

Urban Agriculture is thought to increase food security through two main pathways: improved access to food, and increased income [6]. This relates to a broad understanding of food security, follwing the 2009 Declaration of the World Summit on Food Security, which defines the state of 'food security to exist when all people, at all times, have physical and economic access to sufficient, safe, nutritious food to meet their dietary needs and food preferences for an active life' [17].

We will briefly outline these two presumed paths. The first of these assumes that home-grown foodstuffs increase the total amount of food available to a household and thus can prevent hunger and malnutrition. At the same time the availability of fresh, home-grown food products, in particular fruits and vegetables, advances the nutritional status of household members and thereby impacts positively on health outcomes. Direct access to food allows households to consume a more diverse diet that is richer in valuable micronutrients. Especially animal husbandry is believed to provide an important source of animal protein, which is commonly limited in poor households' diets. Studies on UA and its impact on nutrition focus on dietary diversity and kilocalorie consumption [18]. In his 1998 analysis of child nutrition and UA in Kampala, Maxwell [5] also connected the aspect of maternal care to UA, arguing that mothers who

engaged in UA, as opposed to other forms of non-farm employment away from home, have an increased ability to take care of their children. This was in turn considered to positively impact levels of child nutrition. However, even proponents of UA highlight the fact that there is currently no detailed empirical evidence for UA's impact on nutrition levels [18,19].

Secondly, UA is considered to increase household cash income. Domestic producers can either save income, as the household limits its need to purchase food, and/or increase income by selling or trading their products. Higher cash income at household level, in return is considered to be positively linked with food security, as households are believed to have greater access to food products both in terms of quantity and quality. This relationship however depends to a large extent on the calorie elasticity of income, ie the relation between a change in income and a change in calorie consumption [20]. Households with a low elasticity might not experience improved levels of nutrition linked to the increase in income when income is not spent on more nutritious food. Given the low input costs of UA, most scholars nevertheless believe it to have high potential in addressing urban poverty and food insecurity [21]. Figure 1 below depicts these two pathways.

Critics of urban agriculture

Luc Mougeot [15] in 2001 has observed whilst "little could be found in the academic literature which would condemn UA at large ... opposition has tended to come ... from urban planning, public health and environmental circles." The absence of easily accessible empirical data on UA's scale and impact may explain urban planners' reluctance to embrace the concept. Jac Smit, an outspoken proponent of UA, estimates that about 800 million people are involved in UA worldwide. His

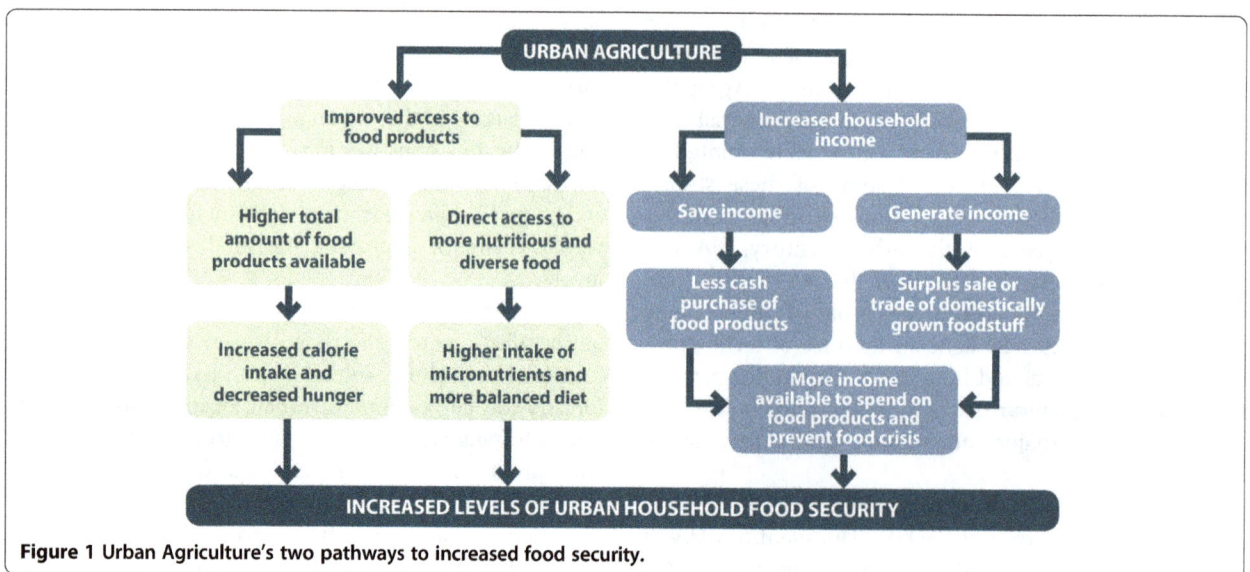

Figure 1 Urban Agriculture's two pathways to increased food security.

estimations came as a result of a number of visits to major urban centers in developing countries, funded by UNDP in the early 1990s, and a range of national surveys. Accurate data on the scale of UA is further limited by different survey designs and differences in conceptual definitions. While scholars [22,23] once identified Lusaka as the urban agriculture capital, a recent baseline study by the Africa Food Security Urban Network [8] found that only three per cent of households in Lusaka currently use UA as a livelihood strategy. The same problem arises with regard to assumed impact of UA on urban livelihoods. Ellis and Sumberg [9] point to the absence of control groups in research regarding UA and criticise that "(UA) claims too much by equating all food production in towns with improved food security for poor people".

Urban agriculture and the need for a systematic review

Hence, the need for a systematic approach to gather and synthesise available data on the impact of urban agriculture is evident. Urban planners and decision makers cannot be expected to base their policy recommendation on what Zezza and Tasciotti call "anecdotal evidence" [18]. Systematic review methodology provides means of identifying, synthesising and analysing the findings of a range of rigorous studies to answer a focused question. Petrosino and colleagues describe systematic reviews as "the most reliable and comprehensive statement about what works" [24].

A number of existing systematic reviews have touched on the issue of UA and food security. Berti et al. [25] reviewed the effectiveness of agriculture interventions in improving nutrition outcomes. Even though they did not make specific reference to the urban sector, Berti and colleagues' findings might indicate trends applicable to urban households. The review of nutritional outcomes of 30 agricultural interventions found that the majority of interventions increase food production, but that this did not present a direct link to improved nutrition. Of the interventions that fostered levels of households' nutritional status, most used a multiple approach focusing on nutrition education, amongst others, in combination with increased food production. Home gardening was found to be the most successful agricultural intervention in increasing levels of household nutrition. This finding is likely to resonate with UA interventions, as home gardening presents a common feature in various types of UA. Limitations in this review include the fact that conclusions are made which are based on studies with very high risk of bias.

Masset et al. [26] conducted a similar systematic review in 2011, which focused on agriculture interventions and levels of nutrition for children under the age of five. This review was registered with the EPPI-Centre at the University of London. Again, this review observed that agricultural interventions did increase levels of food production. However, this was not positively linked to improved levels

of child nutrition either. The review further found that household diets do change on account of the presence of agricultural interventions, but that this change was not related to improved child nutrition as no improvement in levels of micronutrients consumed by the participants could be identified. The review's conclusions are limited by the fact that the report does not aggregate its analysis by the level of included studies' risk of bias, despite the acknowledgement that included studies are heterogeneous in terms of rigour of study design.

Lastly, a systematic review by the Dutch Ministry of Foreign Affairs' Policy and Operations Evaluation Department [27] focused on the question which agriculture interventions improve food security. It identified increased production, improved value chains, market regulations and safe legal tenure as the four main pillars. Although the review claims that it considers urban food security, it neither differentiates between rural and urban food security impacts in its synthesis, nor does it account for such distinction in the reviewed impact studies. According to the review, improved irrigation and the use of genetic crop modification have been found to increase agricultural production in significant ways. Value chain interventions benefited farmers through improved income from the sale of cash crop. Market deregulations had an ambivalent impact on food security. Whilst the reduction of monopolies and the lowering of government involvement were believed to increase food security during crisis situations, its long-term impact differed greatly amongst the studies. Finally, improved land tenure benefited food security in all reviewed cases. This finding might be important in the context of UA, as it supports proponents' calls to give urban farmers access to land and change discriminative urban policies to legalise and support agriculture interventions in urban areas. Since only a minority of urban farmers engages in UA for commercial reasons, interventions that focus on value chains and market regulations have limited impact on UA. The usage of genetically modified crops and better irrigation techniques might also be seen as an attribute of large scale farming projects. Individual and household farmers which present the majority of urban farmers and practice agriculture on mainly small plots within the city might not be able to make use the above techniques to increase their agricultural output.

As a result, the impact of UA on food security and nutrition is currently unanswered. Much literature has been published on the topic assuming a positive relationship, and the concept enjoys the outspoken support of international development agencies such as the IDRC and UNDP. Most research, however, lacks empirical evidence and few studies have generated reliable facts about the scale and impact of UA. Bearing in mind UA's potential to make a meaningful contribution to securing food security among urban populations, this systematic review was developed in response to perceived gaps in the current evidence-base.

Objectives of this review

The aim of this review was to collect and analyse available evidence on the impact of urban agriculture in low and middle-income countries. We sought to provide a solid evidence-base for policy-makers, practitioners and members of the international donor community on the feasibility, benefits and cost of urban food cultivation. Specifically, this review attempted to answer the question: *What are the impacts of urban agriculture programs on food security in low and middle-income countries?* In addressing this question, we focussed on outcomes that measure levels of food security at the individual, household and/or community levels.

According to the definition adopted at the 1996 World Food Summit in Rome "food security exists when all people, at all times, have physical and economic access to sufficient, safe and nutritious food to meet their dietary needs and food preferences for an active and healthy life" [28]. This reflects a broad approach to food security, including the elements of food availability, food accessibility, food reliability, food quality and food preference [8]. Food security is most commonly defined in terms of the three pillars of: availability (including consistency of that availability), access (with the specification of not just access to food, but access to sufficient food for a nutritious diet), and use [13]. The FAO add a fourth pillar, that of stability and apply it to all three of the others i.e. the stability of availability, of access and of use [29].

Levels of food security and nutrition are interdependent with households' socio-economic status. The economic impact of the usage of UA as an income-generating scheme in order to purchase more or different food is therefore also considered in this review. Sales of domestically produced food stuffs can be measured either in terms of quantity or with regard to their monetary value.

Methods

We employed systematic review methods as promoted by the Collaboration for Environmental Evidence. In doing so, we sought all the relevant and reliable research on UA and its impact on food security and nutrition, in order to synthesise it in meaningful ways.

This systematic review process includes searching comprehensively for all available potentially relevant evidence, and then filtering it, firstly for relevance and secondly for risk of bias. In doing so, we used structured approaches to describe and critique the available research. The review data is recorded on specialist systematic review software (EPPI-Reviewer 4) to enable transparent and accurate analysis. Details of the protocol employed in this review were published by Stewart et al. [30].

Searches

We searched for relevant literature using the search strategy presented below. Searches were run between July and September 2013 and conducted in English and Spanish. Title and abstracts of any literature identified in other languages were translated using Google Translate, and any studies that appeared to be relevant from the translation of their abstract were translated in full and considered for inclusion in the review.

Our search strategy combined the key concepts of UA and impact evaluations. Search 1, 2 and 3 were run independently, and combined as per 4 below.

1. 'Urban agriculture' (("Urban farming") OR ("Food supply") OR ("Food planning") OR ("Sustainable agriculture") OR ("Food aid") OR ("Urban agriculture") OR ("Food processing") OR ("Food distribution") OR ("urban food production"))
2. 'Urban' AND 'agriculture' ((urban OR city OR peri-urban OR periurban OR metropolis OR town) AND (agriculture OR farming OR farm OR crop OR livestock OR smallholding OR small-holding OR chickens OR poultry))
3. 'impact evaluation' ((impact OR outcome OR evaluation OR effectiveness OR trial OR comparison study OR comparison study OR non-comparison study OR social performance assessment OR impact OR effects OR randomised controlled trial OR controlled clinical trial OR randomised OR placebo OR clinical trials OR randomly OR program evaluation OR controlled OR control group OR comparison group OR control groups OR comparison groups OR controls OR control OR intervention OR evaluate OR evaluations OR RCT OR experiment* OR (evaluation OR program evaluation OR economic evaluation OR (clinical trials OR trials OR randomised controlled trials) OR (experiments OR "controls (experimental)" OR trials)))
4. (1 OR 2) AND 3.

The above terms have been identified and tested using Science Direct. Searches were limited using specific 'human' filters, and exclusively included literature published since 1980 as only isolated UA initiatives existed before then. Search terms were adapted as necessary and applied to the following electronic databases and websites:

Online databases and repositories:

African Journals Online: http://ajol.info/index.php/index/browse/alpha/index.
AGRICOLA: http://agricola.nal.usda.gov/.
AGRIS: http://agris.fao.org/.
Asia Journals online: http://asiajol.info/index.php/index.
Bioline international: http://www.bioline.org.br/.

Campbell Collaboration Library: http://www.
campbellcollaboration.org/library.php/.
Cochrane Collaboration Library: http://www.
thecochranelibrary.com/.
Collaboration for Environmental Evidence Library:
http://www.environmentalevidence.org/Library.htm.
DAC Evaluation Abstracts: http://www.oecd.org/pages/
0,3417,en_35038640_35039563_1_1_1_1_1,00.html.
Database of impact evaluations (3ie): http://
www.3ieimpact.org/database_of_impact_evaluations.html.
Dialnet Database (Spanish Searches): http://dialnet.
unirioja.es/servlet/buscador.
Emergency Events Database: http://www.emdat.be/
external-outputs.
EPPI-Centre Library: http://www.eppi.ioe.ac.uk/.
Green File: http://www.greeninfoonline.com.
IDRC (Canada): http://www.idrc.ca/EN/Resources/
ResearchDBs/Pages/default.aspx.
Internet library sub-Saharan Africa: http://www.ilissa-
frica.de/en/.
ISI Web of Science: http://www.webofknowledge.com.
Isidore: Open Access Portal (French): http://www.
rechercheisidore.fr/index.
Online Access to Research in the Environment: http://
www.oaresciences.org/en/.
PubMed: http://www.ncbi.nlm.nih.gov/pubmed/.
Research4 DFID: http://www.research4development.info/.
Resource Centres on Urban Agriculture & Food
Security RUAF: http://www.ruaf.org/ruaf_bieb/appflow/
bieb_search.asp.
Sabinet: http://www.sabinet.co.za.
Science Direct: http://www.sciencedirect.com/.
Social Science Research Network: http://www.ssrn.com/.
UNESDOC: http://www.unesco.org/new/en/unesco/
resources/online-materials/publications/unesdoc-
database/.
United Nations Economic Commission for Africa
(UNECA) institutional repository
http://repository.uneca.org.

Websites:

Centre for Agricultural Bioscience International:
http://www.cabi.org/default.aspx?
site=170&page=1016&pid=2196.
J-PAL: www.povertyactionlab.org/.

Search engines:
Google Books: www.google.co.za (screened first 100 en-
tries on title and content page).
Citation searches for the following key publications were
conducted using Google Scholar [3,5,9,10,18,22,23,31-39].
In addition, reference lists of all full texts were checked
for further papers of relevance. Where previous reviews

were available, their reference lists were searched and relevant studies were screened as individual studies.

Inclusion and exclusion criteria/screening search results
We employed a two stage screening process, where all
search hits' abstracts and titles were screened according to
the criteria listed below. Full texts of all potentially relevant studies were then obtained. Where in doubt of an abstract's relevance, we collected full texts. In the second
stage, all full texts were screened according to the same
criteria as below. Two reviewers worked independently on
this task and any disagreements were discussed and resolved with a third reviewer.

Population
The review focused on people in urban and peri-urban
contexts in developing countries, who use forms of UA.
We did not exclude any group of people on age or socio-
economic group. Only research conducted in countries
classified as low or middle-income countries by the world
Bank qualified to be included in the review (see Additional
file 1: Appendix 1). Though not applicable to the review
in the end, research that focused on both developing high
income countries (HICs) would have been considered as
long as it was possible to isolate the impact of UA on the
former.

Intervention
This review searched for UA in all its forms when used as
a livelihood strategy. This included growing plants to eat
or sell (for example, herbs, fruit, vegetables or flowers)
and animal husbandry. Urban Agriculture when purely
used as a leisure activity, such as home or roof top gardens
that are not intended to contribute to either food or
income in the household, did not qualify for inclusion.

Outcomes
This review focused on food security outcomes as
described in Figure 1, including changes in access to, and
quality of, food. Studies that did not include either one of
the two were excluded. Nutrition refers to both access to
and quality of food; we therefore also considered studies
that assess the impact of UA on nutrition levels.

As UA can lead to a change in income levels, which in
turn can have an effect on food security, we considered
studies that assess impacts of UA on income when the
study also related to food security. Studies that address
impacts on income with no link made to food security,
as studies that only focus on the environmental and social aspects of UA, were excluded.

Study design
Only impact evaluations which set out to measure the
effectiveness of UA interventions on urban food security

at the household, individual and/or community level, as compared to the effects of not engaging in UA, qualified for inclusion in this review. Studies that did not provide a comparison group were excluded. This was defined as a second group of participants who did not receive the UA intervention. Similarly, studies that did not measure change over time (i.e. included at least two data points at any point before, during and after intervention roll-out) were excluded.

Language

Studies were not excluded from this review on the basis of language, except that searches were conducted in English (and Spanish to a limited extent). The review team had the scope to translate studies published in English, French, Spanish, Portuguese, German, Dutch, Afrikaans, Zulu and Sotho languages. Abstracts of identified papers in other languages were first translated using Google Translate and, if deemed relevant for inclusion in the review, were assessed by a team member fluent in the language of the report. We assessed full texts in English, French and Spanish.

Describing studies

As the full-text screening yielded no studies that met the inclusion criteria, no further descriptive data were extracted from studies. Additional file 1: Appendix 2, drawing on the review's protocol [30] outlines which data the review had set out to collect from individual studies, how the risk of bias was going to be assessed and which data synthesis strategy was going to be applied to the included studies.

Results

Results from searching and screening

Searches were conducted between July and September 2013 and resulted in over 8100 hits. By far the largest number of hits was generated through CAB Abstracts (2030), followed by those resulting from searching through reference lists and citations (1236). Specialised databases such as the International Information System for the Agricultural Science and Technology (AGRIS) Database and the Resource Centre for Urban Agriculture and Food Security's (RUAF) online library proved very helpful with 518 and 428 hits respectively.

All abstracts were screened, leaving 256 abstracts that appeared initially to meet our inclusion criteria. After removal of 30 duplicates from these 256, full texts of the articles were sought. One hundred and ninety-eight full texts were obtained; 28 reports could not be located[b].

During the second stage screening, all 198 full texts were screened against our predetermined inclusion criteria, including geographical location, focus on urban agriculture as opposed to rural cultivation, the study's focus on the impact of urban agriculture on food security, and the study's

methodological criteria pertaining to basic principles of impact evaluations (see above for detailed inclusion criteria). Impact evaluations are "studies which can attribute changes in selected outcomes ... to a specific intervention" [40]. Hence, two key characteristics are inherent in impact evaluations, i.e. they observe changes over time, and they compare changes in selected outcomes between at least two groups. In our methodology screening we therefore looked for, firstly, the presence of intervention and comparison groups, and secondly the existence of two data points over time.

The screening of the 198 full texts yielded no studies that met the review's inclusion criteria. These articles are listed in Additional file 1: Appendix 3. Three were not assessing food security, four were not set in developing countries, and 12 were not about urban agriculture but about rural farming. The remaining 173 studies were excluded on grounds of study design, as they did not meet the criteria of having at least one intervention and one control group *and* at least two data points. Therefore this review did not find evidence to support or refute the theory that UA interventions increase food security in low and middle-income countries. Figure 2 below summarises the results of our searching and screening process.

Discussion

As discussed earlier in this report, Urban Agriculture is thought to increase food security either by directly improving access to food or by increasing household income, which in return enhances access to food products. In response to the gaps in the evidence-base and the general assumption that urban agriculture initiatives contribute positively to food security, this review set out to address the question of what impact urban agriculture initiatives have, positive or negative, on food security outcomes.

The review found no available evidence, which supports or refutes the suggestion that urban agriculture positively impacts on individual or household food security in low and middle-income countries.

Nonetheless, we made two important observations during searching and screening. Our searches yielded a range of studies that study the *association* between UA and certain aspects of food security. For instance, Maxwell analysed the determinants of the nutritional status of children less than five years of age in Kampala, Uganda [5]. The author's results indicate that nutritional status of children, measured by height for age, is significantly higher among farming households than in non-farming households. Gallagher and colleagues [41] found no significant difference between farming and non-farming households in their analysis of food diversity and vegetable consumption among farming and non-farming households in Nairobi's

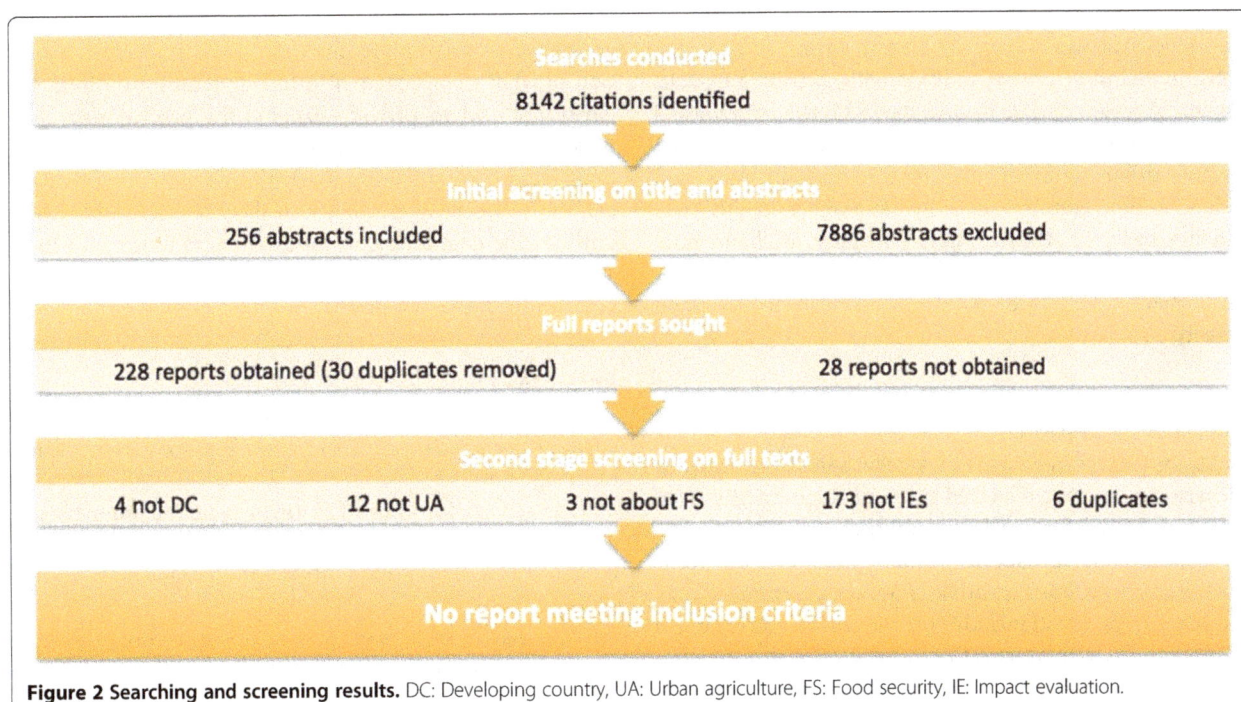

Figure 2 Searching and screening results. DC: Developing country, UA: Urban agriculture, FS: Food security, IE: Impact evaluation.

informal settlements of Kibera. Further systematic analysis of association studies will be useful in the design of impact evaluations by providing important information on the relationship between UA and food security.

It is important to note that there is a wealth of studies discussing urban agriculture's potential to contribute to either food security or economic standing of households and communities (see Additional file 1: Appendix 3). This is particularly true for many parts of eastern and western Africa, south-east Asia, and Latin America. The majority of these, however, consist of descriptive studies with cross-sectional designs that typically take stock of UA characteristics in a specific community at one specific point in time. Other studies are qualitative in nature and therefore provide little measurable data on the impact of urban agriculture. Further examination of this pool of literature will be valuable in developing a theory of change for how UA may impact, positively or negatively, on food security. Future trials could then measure UA initiatives' effectiveness against that theory of change.

In some instances, systematic reviews that yield no included studies, are considered as a reflection of a review question that is either concerned with too narrow a population, addresses an overly focussed intervention, or seeks outcomes that are too specific. The strength of our review was that it set out to consider urban agriculture in all its forms, so long as it is used as a livelihood strategy as opposed to leisure activity. Food security outcomes were to be considered in terms of both quantity and quality of food. Our search criteria were sufficiently

comprehensive and the search was widely spread across 28 databases, as well as additional websites and citations searches.

The question of this review remains topical and important. The fact that no studies met our inclusion criteria underlines, however, the importance of critically interrogating how the impact of urban agriculture is measured in future impact evaluations. Moreover, our Spanish language searches were limited to one database. A systematic map of Spanish and Portuguese language impact evaluations based on searches in a wide range of Spanish and Portuguese language databases will be helpful in establishing the evidence-base for studies published in those languages, which we may have missed in this current review.

Review conclusions
Implications for policy and management
This review had set out to address shortcomings in the evidence-base of urban agriculture's contribution to urban food security, by providing an evidence-based review of all high-quality impact evaluations on the topic[c]. It was expected that this would assist in making policy recommendations by informing urban decision makers on the question if and/or how agriculture interventions are best used to improve levels of food security in urban areas in developing countries.

However, our extensive searching and screening did not reveal any impact evaluations that met our inclusion criteria. Notwithstanding the fact that policy-making takes place irrespective of whether or not reliable research evidence exists, drawing inferences and recommendations regarding what works and what doesn't from studies that met our inclusion criteria only partially, would introduce bias to the review and be inconsistent with the principles of systematic reviews. The available evidence suggests that current urban intervention programmes in low and middle-income countries have no supporting effectiveness evidence-base, i.e. the research currently available does not allow for any conclusions to be made on whether or not urban agriculture initiatives contribute to food security.

Implications for research

As all systematic reviews, this review relied on rigorous primary studies that collected relevant, high-quality data. Hence, the need for improved emphasis on rigorous impact evaluations measuring urban agriculture interventions' impact on food security as a core outcome, stands out. Ideally, these will include control groups and measure developments over a period of time at two, it not more, data points. Funding for these should be made available. Established researchers who have published broadly on the topic, some in a range of cities around the world, are encouraged to collaborate with impact evaluation experts in order to set up rigorous impact evaluations that answer key questions around urban agriculture and its food security impact on households and communities. In addition, existing and new qualitative data should be drawn upon in developing theories of change, which in return will aid the conceptualisation and analysis of impact evaluations.

In the systematic review community worldwide, systematic maps have become increasingly common in recent years [42-44]. This review alludes to the value of conducting a systematic map prior to engaging in a full systematic review. A map lays out the available evidence and provides an overview of the characteristics of the evidence-base. It can provide a summary of methodologies typically employed by researchers in the topic area and guide the exact focus of any subsequent full systematic review. This is particularly valuable in a topic area that draws on research from various disciplines, or interventions and outcomes that have seen little consideration by systematic reviewers. Members of the systematic review community, particularly those who conduct reviews in areas not historically covered by systematic reviews, are encouraged to consider the value of compiling a systematic map for the review question at hand prior to developing their full review protocol. Producing a systematic map for this project prior to engaging in a full systematic review would have outlined the characteristics of the existing evidence-base. In return, it would have presented the review team with suggestions for changes to the inclusion and exclusion criteria of the full review.

Lastly, we suggest that this review be updated in due course so as to provide answers to the important question of what the impacts of urban agriculture on food security are and to provide insights for practitioners and policy makers. Meanwhile, compiling a systematic map that details the currently available research on the *association* between agriculture and food security and which outlines the characteristics of urban agriculture and food security research, will provide a valuable tool for both practitioners and researchers.

Endnotes

[a]Livelihood is commonly defined as comprising "the capabilities, assets ... and activities required for a means of living" [45].

[b]Our search for these full texts included inter-library requests, searches on a range of international academic databases, and searches through google and google scholar.

[c]This refers to all high-quality impact evaluations that would be identified through the review's search strategy as outlined above.

Additional file

Additional file 1: Appendix 1. List of included countries. **Appendix 2.** Proposed data extraction and quality appraisal. **Appendix 3.** List of studies screened at full-text stage and primary reasons for exclusion.

Competing interests
We have no competing interests in the outcomes of this review, financial or otherwise. The review was supported by a grant from the University of Johannesburg's Research Council. Some of the authors are involved in a related systematic review on smallholder farming, supported by the Canadian Ministry for Foreign Affairs, Trade and Development, and registered with the Campbell Collaboration under the title 'What is the effectiveness of agriculture interventions on agricultural investment, yields, and income for smallholder farmers in Africa?'

Authors' contributions
LL conceived the idea for this review. RS developed the protocol, with input from MK, led searches, screened full texts, and led the project. LL led the literature review, contributed to the protocol, conducted searches and screened full texts. MK contributed to the protocol, carried out searches, screened full texts, conducted analysis, with input from RS, and drafted final report. NM conducted searches and screened full texts. HZ contributed to the literature review. TdW secured funding. All authors contributed to drafts, read and approved the final manuscript. Core funding for this review was made available by the central research committee of the University of Johannesburg. Additional thanks go to Dr Nicola Randall, Director of the Centre for Evidence Based Agriculture (CEBA) at Harper Adams University, UK, who assisted with some of the searches, and an anonymous reviewer who provided valuable input into the final manuscript.

Author details
[1]CEE Johannesburg, Centre for Anthropological Research, University of Johannesburg, Johannesburg, South Africa. [2]EPPI-Centre, Social Science Research Unit, Institute of Education, University of London, London, UK.

References

1. UN-Habitat: *The Locus of Poverty is Shifting to the Cities.* ; 2006. Available from: http://mirror.unhabitat.org/documents/media_centre/APMC/THE%20BAD%20NEWS.pdf (Accessed: 14-05-2012).

2. Barker N, Dubbeling S, Guendel U, Koschella S, De Zeeuw H: *Growing Cities, Growing Food. Urban Agriculture on the Policy Agenda: A Reader on Urban Agriculture.* Feldafing, Germany: German Foundation for International Development (DSE); 2001.

3. Cofie OO, Van Veenhuizen R, Drechsel P: *Contribution of Urban and Peri-Urban Agriculture to Food Security in Sub Saharan Africa.* Kyoto: RUAF paper presented at the 3rd WWF; 2003.

4. FAO: *Urban Agriculture: An Oxymoron? The State of Urban Agriculture.* Rome: FAO; 1996.

5. Maxwell DG: **Does urban agriculture help prevent malnutrition? Evidence from Kampala.** *Food Policy* 1998, **23**(5):411–424.

6. Mougeot LJA: *Agropolis. The Social, Political and Environmental Dimensions of Urban Agriculture.* London: IDRC; 2005.

7. UNDP (United Nations Development Programme): *Urban Agriculture: Food, Jobs and Sustainable Cities.* New York: UNDP; 1996.

8. African Food Security Urban Network (AFSUN), Crush J, Hovorka A, Tevera D: *Urban Food Production and Household Food Security in Southern African Cities*, Urban Food Security, Volume 4. Kingston and Cape Town: Queen's University and AFSUN; 2010.

9. Ellis F, Sumberg J: **Food production, urban areas and policy responses.** *World Dev* 1998, **26**(2):213–225.

10. Simatele DM, Binns T: **Motivation and marginalization in African urban agriculture: The case of Lusaka, Zambia.** *Urban Forum* 2008, **19**(2):1–21.

11. City of Johannesburg: *Growth and Development Strategy.* ; 2011. Available from: http://www.joburg.org.za/gds2040/gds2040_strategy.php (Accessed: 12-06-2012).

12. UN High Level Task Force on the Global Food Crisis: *Comprehensive Framework for Action.* ; 2008. Available from: http://un-foodsecurity.org/sites/default/files/OUTLINE_Summary%20UCFA_EN.pdf (Accessed: 14-06-2012).

13. WB/FAO (World Bank/Food and Agriculture Organisation): *Urban Agriculture for Sustainable Poverty Alleviation and Food Security*, WB/FAO publication. ; 2012. Available from: http://www.fao.org/fileadmin/templates/FCIT/PDF/UPA_-WBpaper-Final_October_2008.pdf (Accessed: 14-06-13).

14. Food and Agriculture Organization of the United Nations (FAO): *Urban Producers Resource Book*, FAO publication. ; 2007. Available from: http://www.fao.org/docrep/010/a1177e/a1177e00.HTM (Accessed: 17-03-14).

15. Mougeot LJA: **Urban Agriculture: Definitions, Presence, Potentials and Risks.** In *Growing Cities, Growing Food. Urban Agriculture on the Policy Agenda: A Reader on Urban Agriculture.* Edited by Barker *et al.* Feldafing, Germany: German Foundation for International Development (DSE); 2001.

16. ETC: *Annotated Bibliography on Urban Agriculture.* Leusden, Netherlands: ETC; 2003.

17. FAO, IFAD and WFP: *The State of Food Insecurity in the World 2013. The Multiple Dimensions of Food Security.* Rome: FAO; 2013.

18. Zezza A, Tascotti L: **Urban agriculture, poverty and food security: Empirical evidence from a sample of developing countries.** *Food Policy* 2010, **35**:265–273.

19. Amar Klemesu M: **Urban Agriculture and Food Security, Nutrition and Health.** In *Growing Cities, Growing Food. Urban Agriculture on the Policy Agenda: A Reader on Urban Agriculture.* Edited by Barker, *et al.* Feldafing, Germany: German Foundation for International Development (DSE); 2001.

20. Dawson PJ, Sanjuán AI: **Calorie consumption and income: panel cointegration and causality.** *Appl Econ Lett* 2011, **18**(15):1455–1461.

21. Frayne B: **Survival of the Poorest: Migration and Food Security in Namibia.** In *Agropolis. The Social, Political and Environmental Dimensions of Urban Agriculture.* Edited by Mougeot N. London: IDRC; 2005.

22. Bishwapriya S: **Urban agriculture: Who cultivates and why? A case-study of Lusaka, Zambia.** *Food Nutr Bull* 1985, **7**(3):15–24.

23. Rakodi C: **Self-reliance or survival? Food production in African cities with particular reference to Zambia.** *Afr Urban Stud* 1985, **21**:53–63.

24. Petrosino A, Boruch RF, Soydan H, Duggan L, Sanchez-Meca J: **Meeting the challenges of evidence-based policy: the Campbell collaboration.** *Ann Am Acad Politic Soc Sci* 2001, **578**(1):14–34.

25. Berti PR, Krasevec J, FitzGerald S: **A review of the effectiveness of agriculture interventions in improving nutrition outcomes.** *Public Health Nutr* 2004, **7**(5):599–609.

26. Masset E, Haddad L, Cornelius A, Isaza-Castro J: *A Systematic Review of Agricultural Interventions That aim to Improve Nutritional Status of Children.* London: EPPI-Centre, Social Science Research Unit, Institute of Education, University of London; 2011.

27. IOB: *Improving Food Security. A Systematic Review of the Impact of Interventions in Agricultural Production, Value Chains, Market Regulation, and Land Security,* IOB Study, No. 363. Netherlands: Ministry of Foreign Affairs; 2011.

28. FAO: *The Rome Declaration on World Food Security.* Rome: FAO; 1996. Available from: http://www.fao.org/docrep/003/w3613e/w3613e00.HTM.

29. FAO: *The Urban Producer's Resource Book: A Practical Guide for Working with Low Income Urban and Peri-Urban Producer.* Rome: FAO; 2007. Available from: ftp://ftp.fao.org/docrep/fao/010/a1177e/a1177e.pdf (Accessed: 14-06-2012).

30. Stewart R, Korth M, Langer L, Rafferty S, Rebelo Da Silva N, van Rooyen C: **What are the impacts of urban agriculture on food security in low and middle-income country?** *Environ Evid* 2012, **2**(7):1–13.

31. Atkinson SJ: *Food for the Cities: Urban Nutrition in Developing Countries, Public Health and Policy Publication No. 5.* London: London School of Hygiene and Tropical Medicine; 1992.

32. Binns T, Lynch K: **Feeding Africa's growing cities into the 21st century: The potential of urban agriculture.** *J Int Dev* 1998, **10**:777–793.

33. Crush J, Hovorka A, Tevera D: **Food security in Southern African cities: The place of urban agriculture.** *Progress Dev Stud* 2011, **11**(4):285–305.

34. Drescher AW: **Food for the Cities: Urban Agriculture in Developing Countries.** In *SHS Acta Horticulturae.* ; 2004:643. International Conference on Urban Horticulture.

35. Maxwell DG: **Alternative food security strategy: A household analysis of urban agriculture in Kampala.** *World Dev* 1995, **23**:1669–1681.

36. Mbiaba B: **Classification and description of urban agriculture in Harare.** *Dev South Africa* 1995, **12**(1):75–86.

37. Rogerson CM: **Feeding Africa's cities: The role and potential for urban agriculture.** *Afr Insight* 1992, **22**(4):220–234.

38. Sawio C: *Feeding the Urban Masses? Towards an Understanding of the Dynamics of Urban Agriculture in Dar es Salaam, Tanzania*, PhD Thesis. Worcester: Clark University; 1995.

39. Tevera D: **Urban agriculture in Africa: A comparative analysis of findings from Zimbabwe, Kenya and Zambia.** *Afr Urban Quart* 1999, **11**(2/3):181–187.

40. White H: **Theory-based impact evaluation: principles and practice.** *J Dev Effective* 2009, **1**(3):271–284.

41. Gallaher CM, Kerr JM, Njenga M, Karanja NK, WinklerPrins AMGA: **Urban agriculture, social capital, and food security in the Kibera slums of Nairobi, Kenya.** *Agric Human Values* 2013, **30**(1):389–404.

42. Stewart R, Erasmus Y, Zaranyika H, Rebelo Da Silva N, Korth M, Langer L, Randall N, Madinga N, de Wet T: **The size and nature of the evidence-base for smallholder farming in Africa: a systematic map.** *J Dev Effective* 2014, **6**(1):58–68.

43. Gough D, Thomas J, Oliver S: **Clarifying differences between review designs and methods.** *Systematic Rev* 2012, **1**:28.

44. Snilstveit B, Vojtkova A, Bhavsar A, Gaarder M: *Evidence Gap Maps - A Tool for Promoting Evidence-Informed Policy and Prioritizing Future Research*, Policy Research Working Paper, 6725. Washington, DC: World Bank; 2013. Availabe from: http://www-wds.worldbank.org/external/default/WDSContentServer/IW3P/IB/2013/12/13/000158349_20131213135609/Rendered/PDF/WPS6725.pdf (Accessed: 17-03-14).

45. Carney D (Ed): *Sustainable Rural Livelihoods. What Contribution can we Make?.* London: UK: Department for International Development; 1998.

11

Does the growing of Bt maize change abundance or ecological function of non-target animals compared to the growing of non-GM maize? A systematic review protocol

Michael Meissle[1*], Steven E Naranjo[2], Christian Kohl[3], Judith Riedel[1] and Jörg Romeis[1]

Abstract

Background: Since 1996, genetically modified (GM) crops have been grown on an ever increasing area worldwide. Maize producing a Cry protein from the bacterium *Bacillus thuringiensis* (Bt) was among the first GM crops released for commercial production and it is the only GM crop currently cultivated in Europe. A major part of the regulatory process that precedes the commercial release of GM crops is the environmental risk assessment. Because Bt maize is modified to produce insecticidal proteins, potential interactions with non-target organisms are a major area of concern to be addressed in the risk assessment. In particular, beneficial arthropods that provide important agro-ecological services, such as pollination, decomposition, and biological control are the focus. This systematic review will evaluate if the growing of Bt maize changes abundance or ecological function of non-target animals compared to the growing of conventional, non-GM maize. The review will be limited to plot or field level data including field margins. Potential cropping system effects and off-field effects will not be addressed. Bt maize will be compared to conventional maize either untreated or treated with chemical insecticides.

Methods: Stakeholders from academia, competent authorities, industry, and civil society organizations were given the opportunity to comment on the review question and an earlier draft of this review protocol.
Keyword searches will be conducted in a range of abstracting and full text literature databases. Retrieved records will be screened against a set of inclusion criteria, first on title and abstract level, then on full text level. Selected studies will be evaluated for risks of bias (quality assessment). Data on field/plot characteristics, maize cultivars, insecticide treatments, non-target animal taxa, sampling methods, and response variables of populations and ecological functions will be extracted. Meta-analysis will be conducted using the effect size estimator Hedge's d on a range of comparisons and including sensitivity analysis. The review process will be fully documented in CADIMA, an open access online data portal for evidence synthesis.

Keywords: Agriculture, Biotechnology, Bt crops, Corn, Cry proteins, Fauna, Genetically modified plants, Insect-resistant plants, Meta-analysis, Systematic review

* Correspondence: michael.meissle@agroscope.admin.ch
[1]Agroscope, Institute for Sustainability Sciences ISS, Reckenholzstrasse 191, 8046 Zurich, Switzerland
Full list of author information is available at the end of the article

Background

Since 1996, genetically modified (GM) crops have been grown on an ever increasing area worldwide, reaching 170 million hectares in 2012 [1]. This area represents more than 10% of the global cultivation area of arable crops [2]. The GM maize event MON810, targeting corn borers (Lepidoptera) and producing Cry1Ab from the bacterium *Bacillus thuringiensis* (Bt), has been cultivated in Europe on significant areas for more than a decade. In 2012, it was planted on a total area of 129,000 ha in 5 European countries led by Spain [1,3]. Before new GM plants can be released into the environment, regulatory approval needs to be obtained. While MON810 is the only transformation event currently approved for commercial cultivation in the EU, applications for events expressing Cry3 proteins targeting the corn rootworm (Coleoptera) as well as stacked events (producing several Bt proteins simultaneously) have been submitted [4]. A major part of the regulatory process is the environmental risk assessment. Typical environmental protection goals are farmland biodiversity conservation and sustainable land use by maintaining agro-ecological functions (functional biodiversity) [5-7]. Typical environmental concerns related to those protection goals are a decline in biodiversity and a disruption of agro-ecological functions. This may ultimately lead to yield reduction (Figure 1). For

Europe, the European Food Safety Authority (EFSA) lists the following specific areas of risk to be addressed in the environmental risk assessment [8]: (1) Persistence and invasiveness, including plant-to-plant gene flow; (2) Potential for plant to micro-organisms gene transfer; (3) Interaction of the GM plant with target organisms; (4) Interaction of the GM plant with non-target organisms; (5) Impacts of the specific cultivation, management and harvesting techniques; (6) Effects on biochemical processes; and (7) Effects on human and animal health (Figure 1).

Because Bt maize is modified to produce insecticidal proteins, potential interactions with non-target organisms are a major area of concern to be addressed in the risk assessment. One focus of the assessment is on beneficial arthropods that provide important agro-ecological services, such as pollination, decomposition, and biological control (Figure 1). There are three plausible pathways how those organisms may be harmed: (i) the production of Bt proteins may lead to direct adverse effects on non-target animals; (ii) transformation-related, unintended changes in plant composition of nutrients or toxicants may lead to adverse effects on non-target animals; and (iii) Bt proteins and/or transformation-related effects may influence herbivorous species which triggers adverse food web effects. Common to all pathways is that changes of field populations of non-target species are a prerequisite for harm.

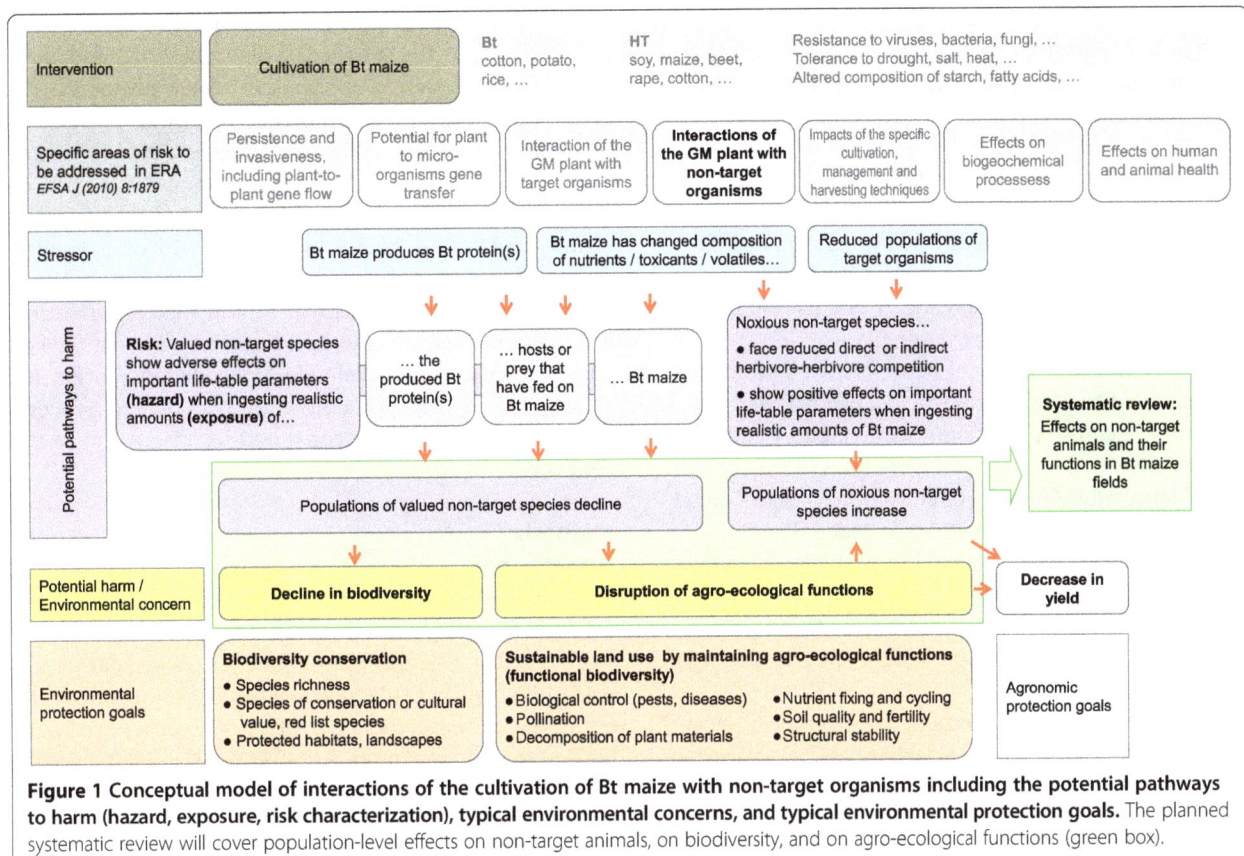

Figure 1 Conceptual model of interactions of the cultivation of Bt maize with non-target organisms including the potential pathways to harm (hazard, exposure, risk characterization), typical environmental concerns, and typical environmental protection goals. The planned systematic review will cover population-level effects on non-target animals, on biodiversity, and on agro-ecological functions (green box).

This systematic review thus focuses on populations of non-target animals in Bt maize fields (Figure 1). Bt maize often replaces conventional pest control systems that include the application of chemical insecticides. Therefore, pest control with Bt maize will be compared to both insecticide-treated and untreated conventional fields. However, the assessment of impacts of cultivation, management, and harvesting techniques is not the focus of this review and will not be addressed in more depth (even though they might need to be addressed in the environmental risk assessment in some jurisdictions including the EU, see Figure 1).

A range of biosafety research programmes have been conducted in European countries such as Spain, Germany, Czech Republic, Hungary, Poland, and Romania. Usually, those programmes included experimental field studies on potential effects of Bt maize on non-target arthropods [9] and many data have become available in the last few years. Those studies include data on Lepidoptera-active and Coleoptera-active Bt maize. Scientists in the USA have conducted meta-analyses on global field data on the impact of Bt crops on non-target invertebrates published between 1992 and early 2008 [10-13]. The current project builds on the previous meta-analyses and incorporates datasets published after 2008. With a broader scope, the planned review will also consider datasets not included in previous meta-analyses. The objective of the review is to answer the following question:

Does the growing of Bt maize change abundance or ecological function of non-target animals compared to the growing of conventional, non-GM maize?

The following components are contained in the question:

- *Population:* Animals (invertebrates and vertebrates) recorded in maize crops and their margins except the targets of the expressed Bt proteins: *Ostrinia nubilalis, Sesamia nonagrioides,* and *Helicoverpa zea* for Cry1, *S. nonagrioides* and *H. zea* for Vip3, and *Diabrotica* spp. for Cry3 proteins.
- *Intervention:* Growing of maize cultivars genetically modified to produce insecticidal proteins derived from *Bacillus thuringiensis,* including Cry and Vip proteins.
- *Comparator:* Growing of conventional maize either untreated or treated with insecticides. The comparison of conventional to Bt maize is possible in different spatial scales ranging from an experimental plot design within one field to a landscape study including several commercial Bt and conventional fields.
- *Outcome:* Effects of the Bt maize on some aspects of life history, abundance, or behavior of non-target animals, their function (e.g. parasitization, predation, pollination, decomposition), or their species richness.

The review will be limited to plot-to-plot and field-to-field comparisons (including field margins). Potential effects on the whole cropping system (crop rotation practices, landscape-level changes, etc.) and off-field effects will not be addressed.

With the systematic review, a broad overview of the available data (and potential data gaps) will become available. In particular, it will identify

- from which countries/continents data are available
- which taxonomic/functional groups of animals have been studied
- which parameters have been measured in field studies
- which Bt proteins were expressed in the plants

This review is one activity of the EU-funded project GRACE (*GMO Risk Assessment and Communication of Evidence*), which aims to provide comprehensive reviews of the evidence on the health, environmental and socio-economic impacts of GM plants – considering both risks and possible benefits (www.grace-fp7.eu). Active stakeholder involvement is a key element of GRACE to guarantee a broader acceptance of the conducted reviews and to ensure their thoroughness and their relevance from a societal perspective. In a workshop, stakeholders were representing 11 academic institutions, 16 competent authorities, 5 civil society organizations, 1 private company, and 1 industry organisation. All stakeholders were invited to comment on the planned review questions and their prioritization as well as on the draft review protocols. According to the received comments, the initial question of this review was reformulated to be open-ended ("change" instead of "decline") and broader ("animals" instead of "arthropods"). In general, stakeholders rated the proposed review question as highly important (4.25 of 5 points). They also indicated that there is medium expert disagreement on the question (3.13 of 5) and that the question is subject to medium public awareness (3.59 of 5). For more details on the stakeholder consultation process see [14,15]. In the further course of the review process, stakeholders will also be consulted for discussion of the review results.

Methods
Data storage
The review process will be documented in the open access database CADIMA, a web-based data portal for evidence synthesis that is under development (www.cadima.info). CADIMA will provide purpose-built online forms for each step of the evidence synthesis process, including both protocols and reviews. CADIMA will allow data exchange with bibliography software (RIS files) and Microsoft Excel for further processing (e.g. for statistical analyses).

Search strategy

Aim of the search

The aim of the literature search is to retrieve as many relevant datasets as possible and in the most unbiased way possible. Because keywords are not standardized in environmental sciences, a high-sensitivity and low-specificity approach will be applied to capture a high proportion of existing datasets. This means that the keywords will be broad. We will search abstracting databases as well as full text databases.

Search strings

Search strings used in the abstracting literature databases (e.g. Web of Science, CAB Abstracts) will be composed of 3 mandatory parts (connected with AND), which are all focused on the "intervention" part of the review question. Terms within one part are linked with "OR". Asterisks will be used to include plural terms when this is required by the database. Quotation marks will be used for multi-word terms (e.g. "*Bacillus thuringiensis*").

Part 1: (maize OR corn OR "Zea mays") Limits the query to the crop for which this review is conducted.

Part 2: (field* OR plot* OR location* OR trial* OR farm-scale OR scouting OR trap* OR sampl* OR monitor*) Limits data to field studies. Because the term "field" is not always present in titles and abstracts, alternative terms, which are often linked to field studies, such as location, plot, or farm-scale, are also used. In addition, the terms scouting or trap refer to methods that are mainly used in field studies.

Part 3: (transgenic OR Bt OR "Bacillus thuringiensis" OR GM OR "genetically modified" OR "genetically engineered" OR "Cry*" OR "Vip*") Returns data on genetically modified plants.

When using more general search engines (e.g. Google Scholar), a fourth part of the search string (referring to the "population" part of the review question) will be added to get more precise hits:

Part 4: ("non-target" or "nontarget" or "natural enemy" or predator or parasitoid or decomposer or pollinator)

Primary focus of the searches will be information available in English (at least on abstract level). Additional searches in other languages (e.g. Spanish, French, German) will also be conducted. Non-English searches will be restricted to search engines that contain non-English information (e.g. Google Scholar). One example in German might be:

Mais AND Feld AND Bt AND Nichtziel* The actual format of the search strings will be adapted to meet the rules of each database individually. The exact search strings and details of how each database was searched will be recorded and provided.

Sources of information

a) Databases to be searched The following abstracting literature databases will be searched:

- ISI Web of Science (WOS) (1900 – 2014) (Thomson Reuters, New York, USA), contains most peer reviewed scientific publications in English language (provided by Web of Knowledge, Thomson Reuters, www.webofknowledge.com).
- BIOSIS (1994–2014) (Thomson Reuters, New York, USA), comprehensive reference database for life sciences (provided by Web of Knowledge).
- Zoological Records (1994–2014) (Thomson Reuters, New York, USA), covers all aspects of animal research (provided by Web of Knowledge).
- CAB Abstracts (1984–2014) (CABI, Wallingford, UK), comprehensive database that also includes more local and non-English publications, such as regional plant protection journals (provided by OvidSP, Wolters Kluwer Health, New York, http://ovidsp.tx.ovid.com/).
- AGRICOLA (1970–2013) (National Agricultural Library, U.S. Department of Agriculture, Beltsville, USA), contains bibliographic records of materials acquired by the National Agricultural Library and cooperating institutions in agricultural and related sciences (provided by OvidSP).
- AGRIS (1975–2013) (Food and Agriculture Organization of the United Nations FAO, Rome, Italy), the International System for the Agricultural Sciences and Technology is an international bibliographic database of national, intergovernmental or international centers (provided by OvidSP).
- ProQuest Dissertations & Thesis A&I, collection of dissertations and theses from around the world (http://search.proquest.com).
- BASE Bielefeld Academic Search Engine, a voluminous search engine especially for academic open access web resources, operated by Bielefeld University Library (www.base-search.net).

General full text search engines:

- GOOGLE SCHOLAR, online search engine to broadly search for scholarly literature (www.scholar.google.com).

- JSTOR, digital library of 1720 academic journals, books, and primary sources (www.jstor.org).

For those search engines the first 200 hits will be checked.

b) Specialist searches Specialist searches will be conducted on the following webpages and web-based databases containing information specifically on environmental effects of GMOs:

- Regulatory agencies (e.g. European Food Safety Authority: www.EFSA.europa.eu; US Environmental Protection Agency www.epa.gov; EU-GMO register: http://gmoinfo.jrc.ec.europa.eu/)
- Project databases (e.g. Cordis: www.CORDIS.europa. eu, Biosaferes: http://biosaferes.icgeb.org/; Gmo-safety: www.Gmo-safety.eu)
- GM-crop databases (e.g. Bibliosafety by ICGEB: http://bibliosafety.icgeb.org/; GM Crop Database by CERA: www.cera-gmc.org; PlantGeneRisk by Testbiotech: www.testbiotech.org/en/database)
- Industry organisations (e.g. Europabio: www.europabio.org)
- Civil society organizations (e.g. GM watch: www.gmwatch.org; Third World Network: www.thirdworldnetwork.net)

c) Bt crops database The Bt crops database generated by Marvier et al. [10] (http://delphi.nceas.ucsb.edu) and Naranjo [12] as well as an updated version provided by Steven E. Naranjo (personal communication) will be used as a basis for the current work.

d) References in previous reviews Review papers and meta-analyses identified during the literature screening process will be checked for potentially useful additional references. A list of review papers that contain relevant references that have not been identified through our literature searches will be provided.

e) Direct contacts In cases where field studies on non-target animals have been identified but data have not been published at all or not in a form usable for the database, corresponding scientists will be contacted and asked to provide data.

f) Non-English searches Full-text databases, such as Google Scholar are also suitable for non-English searches (e.g. German). In addition, potentially useful non-English databases might be used.

Scoping exercise

Systematic reviewing aims to search the literature in a comprehensive and transparent way. While authors of traditional reviews may also search the literature, they usually do not provide records of search terms and searched databases. The most comprehensive traditional review (up to 2009) is the meta-analysis on non-target effects of Bt crops on invertebrates by Naranjo [12], which covers 40 publications on Bt maize. We performed a scoping exercise on 22 February 2013 to check if the search terms suggested for the systematic review are suitable to retrieve the already known literature on this topic. The set of 40 papers was used as a test library against which the new searches were checked. We searched in WOS, CAB Abstracts, Agricola, and Agris with the following search string:

(maize or corn) AND (field* or plot* or location* or trial* or farm-scale or scouting or trap*) AND (transgenic or Bt or "Bacillus thuringiensis" or GM or "genetically modified").

All obtained references were checked for the presence of the references from the test library. The results are listed in Table 1. CAB abstracts delivered the best result. One paper was not found by CAB abstracts, but in both WOS and Agricola. Agris only found 2 papers. All databases combined returned 38 of 40 papers. The 2 papers not obtained were

- Bhatti M et al. Ecological assessment for non-target organisms in the plots of corn rootworm insect-protected corn hybrid containing MON 863 Event: 2000–2001. MRID 457916–01; 2002. p. 1–143 (This is an unpublished industry study that is not indexed in the databases searched).
- Lang A et al. Monitoring of the environmental effects of the Bt gene. Research project sponsored by the Bavarian State Ministry for Environment, Health and Consumer Protection. Institute for Plant Protection, Bavarian State Research Center for Agriculture; Freising-Weihenstephan, Germany; 2005. p 1–111 (This is a published institute report in German that is not indexed in the databases searched).

This scoping exercise demonstrated that the selected search string is suitable to return the vast majority of the literature that was used in the previous meta-analysis and that is indexed in the large literature

Table 1 Literature scoping exercise

Database	WOS	CAB abstracts	Agricola	Agris	All combined
Total hits	1837	1941	1010	242	5030
Papers found	34	37	25	2	38

Number of hits and references from Naranjo [12] (in total 40 references concerning maize only) found with different literature databases (22 February 2013).

databases such as WOS, CAB abstracts, and Agricola. Data that are not found with literature searches (similar to the two studies mentioned previously) might be found when checking the references of previous reviews and the websites and databases on non-target effects of GM crops listed in the section on specialized searches.

Screening of articles

Study inclusion criteria

For the systematic review, articles will be included if fulfilling the following criteria:

General:

- Publication includes original data. Reviews, summaries, abstracts, comment papers, conference proceedings, etc. will only be used if the same data are not available in a peer-reviewed original research paper. Each dataset will be included only once.

Population:

- Non-target animals inhabiting fields (including field margins) are studied (invertebrates, vertebrates), excluding studies exclusively on the target pests *O. nubilalis, S. nonagrioides,* and *H. zea* for maize expressing Cry1, S. *nonagrioides* and *H. zea* for Vip, and *Diabrotica* spp. for Cry3 proteins.

Intervention:

- Study on maize genetically modified to express one or more proteins from *Bacillus thuringiensis* (Cry or Vip proteins).
- Growing of Bt maize under open field conditions (not in laboratory or glasshouse)

Comparator:

- Presence of a non-Bt maize control treatment to which Bt maize is compared. Bt maize not treated with insecticides will be compared to non-Bt maize either treated or untreated and Bt maize treated with insecticides will be compared to treated non-Bt maize.

Outcome:

- Effects of Bt maize on some aspects of life history, abundance, or behaviour of non-target animals, their ecological functions (e.g. parasitization, predation, pollination, decomposition), or their species richness relative to a non-Bt control are measured.

Screening articles: applying inclusion criteria

The hits retrieved by the literature searches described previously will be transferred to Endnote X5 (Thomson Reuters). One Endnote file will be established for each search engine to ensure transparency and repeatability of the search. Subsequently, the hits of all searches will be combined into one database and duplicates will be eliminated automatically. The resulting simplified database will be checked manually. In the first instance, the inclusion criteria will be applied on titles and abstracts to remove spurious citations. At the end of this stage, one database with all remaining potentially useful citations will be created in CADIMA.

For articles remaining in this selection, full texts will be organized and then filtered further according to the inclusion criteria. At this stage, each exclusion of a study for the systematic review will be documented with the reason for exclusion.

Screening articles: quality assurance process

At the beginning of the screening process, a random subset of studies (10%, maximum 200 references) retrieved by the search procedure will be processed independently by a second team member, who applies the same inclusion criteria. The final result will be analysed and documented using Kappa statistics (http://www.vassarstats.net/kappa.html). Studies which were excluded by one, but included by the other reviewer will be documented and the reasons will be discussed in the review team. If the kappa-value is below 0.6 (substantial), a common strategy will be developed and inclusion criteria will be refined and tested to improve reviewer agreement and to minimize discrepancies in the screening process.

Study quality assessment

For each study fulfilling the inclusion criteria of the systematic review, the reliability will be evaluated by judging the risk to systematic bias as low, high, or uncertain. The results of this assessment will be provided.

Possible sources of bias will be addressed for each study including:

- Selection bias (e.g. pre-treatment differences between study groups, differences in baseline characteristics):
  - Field management: High risk if fields or plots selected for the treatments have a different history of management/are owned by different growers, etc.
  - Dispersal of fields/plots: High risk if selected fields are not equally dispersed in a homogenous landscape (e.g. a group of Bt fields close together is compared to a spatially separated group of conventional fields close together; most

Bt fields are near forests, while conventional fields are not or *vice-versa*; Bt fields are isolated from other (maize) crops, while conventional ones are not or *vice-versa*; distance between replicate experimental units is different for the different treatments (Bt crops; control crops); differences for treatments in soil types, neighbouring landscape, shading, altitude, slope, water conditions, etc.).
  ○ Replication: High risk if studies are not replicated or replicates of treatments are not randomized.
  ○ Insecticide applications: High risk if insecticides are applied shortly before the study so that diversity and abundance of non-targets is reduced in the studied fields and differences between treatments may be masked.
- Performance bias (systematic differences between groups e.g. exposure to factors other than the intended intervention):
  ○ Plot size: High risk if plot size is small, in particular for highly mobile animals (e.g. flying species and large mobile species). As a consequence, treatments and controls cannot be separated and the study design does not allow the potential detection of differences between treatments.
  ○ Experimental design: Low risk for (randomized) block designs and replicated split field designs, high risk for unreplicated split-fields and for designs using separated fields (managed by different growers) for the different treatments. Generally, a low number of replicates indicates a high risk of performance bias.
  ○ Cultivars used: Low risk if corresponding near-isolines are used for Bt and conventional treatments, high risk if distantly related cultivars are used because then factors other than Bt may have an effect.
  ○ Expression of Bt protein: High risk if experimental (non-commercial) plants are used and the expression of Bt protein or their effect on the target pest(s) are not measured. Low risk if concentrations are measured or effect on target is demonstrated.
  ○ Application of pesticides: High risk if pesticides (other than insecticides) are potentially applied in one treatment group but not in others or if pesticide applications are not specified (e.g. herbicide or fungicide applications or insecticide applications not accounted for in the treatment comparisons).
- Detection bias (the way outcomes are measured differs between treatment groups):

  ○ Sampling methods: High risk if different methods are used to measure animal populations in the different treatments (unlikely).
  ○ Type and application of methods: High risk if assessments (e.g. counts of individuals per plant) are done by eye and the persons doing the assessments knew how the plots/fields are treated. Lower risk for methods based on traps, such as sticky traps or pitfall traps, which can be evaluated more objectively. However, each trap type has its own limitations and the specimens collected by traps usually do not represent the actual numbers on plants. Furthermore, inadequate use of methods (e.g. inappropriate method for certain taxa, sampling off-peak, etc.) might bias the results.
  ○ Sampling occasions: High risk for low number of sampling dates: especially if population trajectories show a time shift between treatments.
- Attrition bias (imbalance in sample sizes between study groups, missing data):
  ○ Sample size: High risk if the sample sizes differ between treatments (e.g. flooding of traps in some plots, etc.).
  ○ Missing data: High risk if substantially more missing data are present for one treatment than the other.
- Reporting bias (e.g. preferential reporting of positive outcomes):
  ○ Study funding: Higher risk for studies conducted by institutions/organizations/companies with a commercial, political, or ideological interest in a certain outcome of their study (e.g. biotech company might not publish study showing adverse effects of their own product; NGOs with an anti-biotechnology agenda might not publish results showing no adverse effect of GM plant).
  ○ No-effect results may be less frequently published by researchers (file-drawer problem) and thus data may be biased towards studies demonstrating effects. In the area of GMO risk assessment, this is unlikely as studies can be published independent from the outcome. Wolfenbarger et al. [11] did not find an indication for publication bias in their meta-analyses.

For the assessment of study quality, the following characteristics of a study will be extracted and evaluated:

- Selection of field sites/plots (history of management before experiment; location in landscape) → selection bias

- Experimental design → selection bias and performance bias considering layout, randomization, replication, plot size
- Used cultivars, their relationship, and Bt protein concentrations → performance bias
- Pesticide applications in the treatment → performance bias
- Sampling method → detection bias
- Sample sizes → detection bias, attrition bias
- Study funding → reporting bias

Similar to the literature screening process, a random subsample of study quality assessments (10%, maximum 200) will be conducted by 2 members of the review team. The results will be compared with kappa statistics and discussed in the team to reach a consensus decision. Subgroup analyses will be performed that exclude studies with low quality to see if and how they influence the overall results. This will be documented to ensure transparency and reconstructability of the data analyses.

Data extraction strategy

Data extraction will be done in accordance with the previous meta-analysis by Naranjo (2009). All variables that will be extracted are listed in Table 2.

If original data are not available in a form directly usable for the database, estimations or calculations will be done. Those values will also be included in the database together with an explanation of how estimations and calculations were done. Examples are standard deviations calculated from standard errors or peak densities estimated from a seasonal trajectory. Information on the location of the data in the publication and how data were extracted or obtained (including contacts with authors) also will be given. If the presented data of a study are insufficient, authors will be contacted and asked to provide the required data.

The following rules will be applied to extracted data (according to [10-12]):

- Data for comparisons of finest taxonomic resolution will be extracted
- For repeated measurements on the same plots, the seasonal means will be extracted. If they are not available, then values from the time point with the peak of the measured parameter for each treatment will be extracted (peak time may differ for control and Bt).
- All observations will be based on a single season. Thus differences in a parameter represent within-season differences and are not cumulative changes over years. Data from different years will be treated as different observations.

- If data on several life stages of the same species were collected in a study, data on the least mobile, but feeding stage will be used (usually larvae/juveniles).
- If data for one Bt line and multiple control lines are available, data of the near-isoline or closest related line to the Bt line will be used. If data for multiple Bt lines and one control line are available, the values of the control line will be used several times for comparisons with the Bt lines. If data for one Bt treatment and multiple control insecticide treatments are available, the values of the Bt line will be used several times. It will be noted in the database if one treatment is used for different comparisons. The following comparisons will be made: Bt untreated – Conventional untreated; Bt untreated – Conventional insecticide treated; Bt insecticide treated – Conventional insecticide treated.

In a first step, the data collected by Naranjo [12] will be checked to make sure the data were all recorded accurately. A random sample (25%) of newly entered data will be checked by a second member of the review team.

Data analysis

All identified datasets will be presented by means of a narrative summary of the extracted outcomes. All studies will be coded with keywords for certain categories. Tables and figures illustrating all existing evidence will be produced. Examples of how the data will be grouped and presented are: number of years covered, study year, first year and long-term cultivation with Bt-maize, crop and field margin, continent of study, country of study, spatial scale of experiment, experimental design, taxonomic group, functional group, parameter measured, Cry-proteins studied, and studies comparing Bt maize to insecticide treated or untreated conventional maize.

Statistical meta-analyses will be conducted for frequently studied taxonomic groups and measured parameters. Based on the knowledge gained from previous meta-analyses, a sufficient number of datasets is available for arthropods and the parameter abundance. Additional groups and/or parameters will be analysed if the amount of available data allows quantitative assessments (minimum of 5 observations per taxonomic group).

For meta-analyses, the following inclusion criteria will be applied:

- Clearly defined parameter (e.g. abundance of non-target animals) was measured in the field.
- Clearly defined taxonomic group or ecological function was measured.
- Means over plots/fields are available and a measure of variation (SE, SD) as well as sample

Table 2 List of variables to be extracted for the systematic review on non-target effects of Bt maize on non-target animals

Variable name	Definition	Type	Closed terms
article_id	Unique identification number assigned to each publication	Integer	No
author	Author(s) of the listed publication	String	No
publication_year	Year of publication of study	Integer	No
citation	Citation, e.g. journal name, volume and page numbers	String	No
title	Title of the publication	String	No
author_affiliation	Type(s) of institutions that the author(s) are affiliated with (e.g. academic, private sector, government, etc.)	String	Yes
author_institute	Institution of corresponding author (or of first author if no corresponding author was listed)	String	No
was_peer_reviewed	Indicates whether study was published in a peer reviewed journal	String	Yes
Study_funding	Information on funding source of the study (government, private, mix)	String	Yes
country	Country where field study was performed	String	Yes
more_info_from_author	Indicates whether author provided additional details or data	YesNo	Yes
expmt_num	Number of experiment within a study (e.g. different locations, years, etc.)	String	No
data_location	Figure number, table number or page number where means and variation were found	String	No
was_data_scanned	Indicates whether figures were scanned to obtain data values	YesNo	Yes
pip	Insecticidal Bt protein(s) engineered into the transgenic crop	String	Yes
pip_target	Insect order targeted by the Bt protein(s)	String	Yes
event	Transgenic event of the crop tested	String	No
transgenic_hybrid_or_var	Transgenic hybrid or variety name	String	No
nontransgenic_hybrid_or_var	Non-transgenic hybrid or variety name	String	No
nontarget_class	Non-target taxonomic class	String	Yes
nontarget_order	Non-target taxonomic order	String	Yes
nontarget_family	Non-target taxonomic family	String	Yes
nontarget_genus	Non-target taxonomic genus	String	Yes
nontarget_species	Non-target taxonomic species	String	Yes
nontarget_finest grouping	Finest level of taxonomic resolution reported for the nontarget organism(s)	String	Yes
ntarget_f_group	Functional group of nontarget organism(s)	String	Yes
Strata	Specifies if non-target is plant-, ground-, or soil-dwelling	String	Yes
replicate_data_issues	Codes flag for non-independence to be considered for analyses: TGLE = taxonomic group lumped elsewhere; EMUE = experimental means used elsewhere; CMUE = control means used elsewhere	String	Yes
nontarget_final_age_or_stage	Stage or sex of the non-target collected	String	Yes
field_location	Location of field(s) to the level of specificity provided by the author	String	No
Georeference longitude	GPS coordinates of the field (WGS84 decimal format)	Real	No
Georeference latitude	GPS coordinates of the field (WGS84 decimal format)	Real	No
number_of_fields	Number of fields as described by the author	Integer	No
cultivation	Cultivation practices used within the fields (notes on tillage, herbicides, fungicides, etc.)	String	No
site_characterization	Information on site particularities including pressure of target pests, weed infestation, disease infestation, soil characteristics, etc.	String	No
plot_size	Size of replicate plots (in hectares)	Real	No
plot_size_explanation	Explanations for any calculations done to obtain plot size	String	No
is_plot_size_avg	Indicates whether the listed plot size is an average or an estimate	String	Yes
was_study_randomized	Indicates whether the authors indicated that they randomly assigned replicates to treatments	String	Yes
planting_date	Date on which field plots were planted	String	No

Table 2 List of variables to be extracted for the systematic review on non-target effects of Bt maize on non-target animals *(Continued)*

first_sample	Date on which first sample was taken	String	No
last_sample	Date on which last sample was taken	String	No
study_duration_determination	More detailed description on study duration	String	No
years_transgenic	Number of year known for the respective plots known to be grown with the Bt crop	Integer	No
pesticide_name	Brand name of insecticides used	String	Yes
pesticide_active_ingr	Active ingredient for insecticides used	String	Yes
pesticide_spray_rate	Amount of active ingredient per spray	String	No
mechanism_of_pesticide_app	Indicates if insecticide was applied as spray, seed treatment, soil granules	String	Yes
num_of_pesticide_app	Number of insecticide applications	Real	No
is_num_of_pesticide_app_avg	Indicates if the number of insecticide applications is an average	String	Yes
sampling_method_abbrev	Abbreviated description of sampling method	String	Yes
sampling_method_detailed	Detailed description of sampling method	String	No
sampling_frequency	Frequency of repeated samples per replicate field or plot	String	Yes
number_of_sample_days	The number of times that each replicate field or plot was repeatedly sampled over the duration of the experiment	Real	No
num_subsamples	Number of subsamples per true replicate	Real	No
response_variable_abbrev	Major category of response variable	String	Yes
response_variable_detailed	Detailed description of response variable	String	No
true_control_sample_size	True sample size for control treatment	Real	No
true_expmtl_sample_size	True sample size for experimental treatment	Real	No
authors_control_sample_size	Sample size for the control treatment as stated by the author	Real	No
authors_expmtl_sample_size	Sample size for the experimental treatment as stated by the author	Real	No
seasonal_or_peak	Indicates whether values represent seasonal means across multiple sample days or means from peak days	String	Yes
did-we-calc	Indicates whether we calculated the seasonal mean or peak days	YesNo	Yes
calc_method_seas_mean	Explains how we calculated the seasonal mean or peak days	String	Yes
comparison_type	Indicates whether the comparison is 1) untreated Bt vs. untreated control, 2) untreated Bt vs. insecticide treated control, or 3) treated Bt vs. treated control	String	Yes
control_mean	Mean for the control treatment	Real	No
expmtl_mean	Mean for the experimental treatment	Real	No
control_std_err	Standard error for the control treatment	Real	No
expmtl_std_err	Standard error for the experimental treatment	Real	No
control_std_dev	Standard deviation for the control treatment	Real	No
expmtl_std_dev	Standard deviation for the experimental treatment	Real	No
mean_unit	Unit of measurement for the response variable	String	Yes
statistical_test_used	Statistical test used by author	String	Yes
is_effect_significant	Indicates whether a significant effect was detected by the author	String	Yes
warning1	Space for remarks for this record	String	No

Given is the variable name in the database, the definition of the variable, the type, and whether the variable content is restricted to closed (predefined) terms.

size (N) is given for each treatment. Sample size ≥ 2 is required.

- Untreated Bt maize is compared to untreated or insecticide-treated conventional maize or insecticide-treated Bt maize is compared to treated conventional maize. Insecticide treatments include sprays, soil granules or seed coating, depending on the target pest(s).

- Original data are available from the publication (tables, figures) or from the authors upon request.
- Data are presented on an annual basis (not pooled for several years).

Measures of treatment effect

The response variable will be abundance of non-target animals [12]. If enough data are available, other response variables might be used (e.g. diversity indices, measures of ecological function like predation or parasitism). We will report Hedges' d, a weighted mean effect size estimator that is calculated as the difference between an experimental (Bt) and control (non-Bt) mean response divided by a pooled standard deviation and multiplied by a small sample size bias correction term.

Bt and control maize fields will be compared for different ecological functional groups, main taxonomic groups, geographical regions (continents), different Bt proteins (Lepidoptera and Coleoptera specific), commercialized and non-commercialized events.

Separate analyses will be performed for insecticide treated or untreated Bt and conventional maize.

Unit of analysis issues

We will assess if the unit of intervention (Bt maize cultivation, insecticide treatment) is the same as the unit of analysis (plots or fields analysed in a study). For field experiments with Bt maize, clustering issues in the experimental design are not likely to occur.

A data issue that is likely to occur, however, is the multiple use of the same dataset for different comparisons. For data extraction, we defined the rule that if multiple Bt lines are compared to one control line, the values of the control line are used several times and if one Bt treatment and multiple control insecticide treatments are present, the values of the Bt line are used several times. Consequently, for global analyses (comprising all Bt proteins and/or all insecticides) the same datasets might be present in an analysis several times. Furthermore, data on higher taxonomic levels may include data on individual genera or species and might result in a multiple use of the same data. Those datasets will be flagged (see Table 2, variable replicate data issues) in the database during the data extraction process. The proportion of datasets reused several times in one analysis will be given and discussed. Analyses with and without the multiple use of datasets might be conducted within the sensitivity analyses described below.

Dealing with missing data

Studies in which relevant data were recorded but not reported in the publication will be identified and the authors will be contacted and asked if they were able to provide the missing information. If the relevant data (as specified in the inclusion criteria) cannot be obtained or calculated from the given data, the dataset will not be included in the meta-analyses. Any estimates and calculations will be documented in the database.

Quantitative synthesis

The results of the meta-analyses will be presented with the help of figures and tables. The effect size of animal abundance for comparisons of Bt crops with treated or untreated control crops, including 95% confidence intervals, will be illustrated in bar charts. Effect sizes significantly different from 0 as well as the total number of observations will be indicated. Effect sizes will be calculated such that negative effect sizes are associated with lower abundance (or another response parameter) on Bt crops compared with non-Bt controls [12].

A narrative discussion of the findings will be provided. The outcome of the study quality assessment will be discussed.

Assessment of heterogeneity

We will test for heterogeneity across studies using adequate statistical approaches. The influence of variation caused by studying different taxa, environmental conditions (countries, regions, continents), crop managements (tillage, irrigation, pesticide application, rotation), spatial scales, and experimental designs will be explored.

Investigation of heterogeneity

The influence of different parameters causing heterogeneity will be investigated. For parameters causing high heterogeneity, separated meta-analysis of subgroups may be conducted to isolate and help identify the causes of the heterogeneity.

Sensitivity analysis

Sensitivity analyses will be conducted to explore the

- Influence of individual studies: The outcome of the meta-analyses might be influenced by
  - studies that provide many datasets (many years, many comparisons, many collected taxa)
  - studies with high replication (N), which are weighed higher
- Influence of funding sources or affiliation of investigators on the overall results
- Influence of peer-reviewed *vs.* non peer-reviewed studies
- Influence of study quality (high *vs.* low risk studies)
- Influence of studies where peak-days were measured in comparison to studies where seasonal means were measured
- Influence of plot size

- Influence of data on species, genus, family, or order level
- Influence of the number of dates on which samples were collected
- Influence of different transformation events (discussion of differences in Bt protein concentration in different plant parts)

Assessment of publication bias

Effect sizes will be compared for publications with different funding types (see sensitivity analyses described above). Systematic differences would indicate a publication bias depending on funding source of a study.

In addition, Begg's funnel plots might indicate if studies with low precision (high variation) diverge from the pooled mean to a greater extent than studies with high precision. Missing data around effect size 0 might indicate that studies showing any kind of effect (positive or negative) are more likely to get published than no effect studies (file drawer problem).

Abbreviations

Bt: *Bacillus thuringiensis*; CADIMA: *Central Access Database for Impact Analyses of Crop Genetic Improvement Technologies* (Data-portal developed within GRACE, www.cadima.info); EU: European Union; EFSA: European Food Safety Authority; GM: Genetically modified; GMO: Genetically modified organism; GRACE: EU project: *GMO Risk Assessment and Communication of Evidence*, www.grace-fp7.eu; RIS: Research Information System Format (standardized data format for bibliographic data); SD: Standard deviation; SE: Standard error; WOS: Web of Science (Citation index databases provided by Thomson Reuters).

Competing interests

The authors declare that they have no competing interests. The work is funded under the 7th Framework Programme for Research of the European Union (Project number: 311957).

Authors' contributions

MM and JRomeis designed the review question and wrote the review protocol. SEN provided input for the review protocol, provided the template for the data model, and was involved in the adaptation of the data model for the requirements of the current review. CK provided the template for the review protocol and provided input for the current protocol. JRiedel provided input for the current protocol. All authors read and approved the final manuscript.

Acknowledgements

The project has been funded within the GRACE consortium, a EU-FP7 programme project, Grant Agreement KBBE-2011-6-311957. We are grateful to all collaborators within GRACE who provided valuable comments on an earlier version of this review protocol.

Author details

[1]Agroscope, Institute for Sustainability Sciences ISS, Reckenholzstrasse 191, 8046 Zurich, Switzerland. [2]USDA-ARS, Arid-Land Agricultural Research Center, 21881 North Cardon Lane, Maricopa 85138, AZ, USA. [3]Julius Kühn-Institut, Institute for Biosafety in Plant Biotechnology, Erwin-Baur-Str. 27, 06484 Quedlinburg, Germany.

References

1. James C: **Global status of commercialized biotech/GM crops: 2012**. In *ISAAA Brief 44*. Ithaca: International Service for the Acquisition of Agri-Biotech Applications; 2012.

2. Ramankutty N, Evan AT, Monfreda C, Foley JA: **Farming the planet: 1. Geographic distribution of global agricultural lands in the year 2000**. *Glob Biogeochem Cycles* 2008, **22**:GB1003.

3. Meissle M, Romeis J, Bigler F: **Bt maize and integrated pest management – a European perspective**. *Pest Manag Sci* 2011, **67**:1049–1058.

4. Devos Y, Aguilera J, Diveki Z, Gomes A, Liu Y, Paoletti C, du Jardin P, Herman L, Perry JN, Waigmann E: **EFSA's scientific activities and achievements on the risk assessment of genetically modified organisms (GMOs) during its first decade of existence: looking back and ahead**. *Transgenic Res* 2014, **23**:1–25.

5. Nienstedt KM, Brock TCM, van Wensem J, Montforts M, Hart A, Aagaard A, Alix A, Boesten J, Bopp SK, Brown C, Capri E, Forbes V, Köpp H, Liess M, Luttik R, Maltby L, Sousa JP, Streissl F, Hardy AR: **Development of a framework based on an ecosystem services approach for deriving specific protection goals for environmental risk assessment of pesticides**. *Sci Total Environ* 2012, **415**:31–38.

6. Sanvido O, Romeis J, Gathmann A, Gielkens M, Raybould A, Bigler F: **Evaluating environmental risks of genetically modified crops: ecological harm criteria for regulatory decision-making**. *Environ Sci Pol* 2012, **15**:82–91.

7. Garcia-Alonso M, Raybould A: **Protection goals in environmental risk assessment: a practical approach**. *Transgenic Res* 2014, published online doi:10.1007/s11248-013-9760-1.

8. European Food Safety Authority (EFSA): **Guidance on the environmental risk assessment of genetically modified plants. EFSA Panel on Genetically Modified Organisms (GMO)**. *EFSA J* 2010, **8**:1879 [doi:10.2903/j.efsa.2010.1879]

9. Romeis J, Meissle M, Sanvido O: **Ecological Impact of genetically modified organisms**. *IOBC/WPRS Bull* 2008, **33**:158.

10. Marvier M, McCreedy C, Regetz J, Kareiva P: **A meta-analysis of effects of Bt cotton and maize on nontarget invertebrates**. *Science* 2007, **316**:1475–1477.

11. Wolfenbarger LL, Naranjo SE, Lundgren JG, Bitzer RJ, Watrud LS: **Bt crop effects on functional guilds of non-target arthropods: A meta-analysis**. *PLoS One* 2008, **3**:e2118.

12. Naranjo SE: **Impacts of Bt crops on non-target invertebrates and insecticide use patterns**. *CAB Rev Pers Agric Vet Sci Nutr Nat Resour* 2009, **4**:11.

13. Duan JJ, Lundgren JG, Naranjo S, Marvier M: **Extrapolating non-target risk of Bt crops from laboratory to field**. *Biol Lett* 2010, **6**:74–77.

14. Kohl C, Craig W, Frampton G, Garcia-Yi J, van Herck K, Kleter GA, Krogh PH, Meissle M, Romeis J, Spök A, Sweet J, Wilhelm R, Schiemann J: **Developing a good practice for the review of evidence relevant to GMO risk assessment**. *IOBC/WPRS Bull* 2013, **97**:49–56.

15. Smets G, Spök A, GRACE team members: **Assessing the evidence of health, environmental and socio-economic Impacts of GMOs - GRACE stakeholder consultation on good review practice in GMO impact assessment. Part 1: overall process and review questions**. Available online: http://www.grace-fp7.eu/content/reports-study-plans-consultation-documents [Accessed 11 April 2014].

12

Socioeconomic and environmental effects of China's Conversion of Cropland to Forest Program after 15 years: a systematic review protocol

Lucas Gutiérrez Rodríguez[1*], Nick Hogarth[1], Wen Zhou[1], Louis Putzel[1], Chen Xie[2] and Kun Zhang[2]

Abstract

Background: Agricultural activities on sloping lands have historically led to forest loss and degradation in China which, coupled with industrial pressures on the environment, were deemed responsible for catastrophic flooding events in the late 1990s. After these events, China's forest policy underwent a significant reorientation towards ecological conservation and rural development, a process epitomized by the Conversion of Cropland to Forest Program (CCFP). Launched in 1999, the CCFP integrates both socioeconomic and environmental objectives with the aim of reforesting smallholder cropland on sloping lands, while compensating farmers with payments for their lost income. Following 15 years of implementation, it is timely to conduct a comprehensive assessment of the state of knowledge about the CCFP's impacts on human populations and the environment.

Methods/design: The primary research question asks "What socioeconomic and environmental effects has the Conversion of Cropland to Forest Program had on human populations and land resources during its first 15 years in China?" We use a theory of change and a Population-Intervention-Comparator-Outcome (PICO) framework to structure our systematic review, where populations of interest consist of both human populations and land resources targeted by the program, while the intervention of interest is the CCFP as defined by its component activities, including compensatory subsidies, skill-training, and enforcement with field checks. Outcomes are defined as both the socioeconomic and environmental impacts of the program. We will conduct a search for relevant English and Chinese language literature on Scopus, Web of Science, CAB Abstracts, AGRIS (FAO), and the China National Knowledge Infrastructure. Search results will be screened for relevance in a two stage process (titles and abstracts, followed by full texts) based on predefined eligibility criteria, and then further assessed for potential sources of bias. Extraction of data from those studies that have passed full-text screening will follow a coding protocol based on the PICO framework, and quantitative and qualitative analyses of the extracted data will be conducted and synthesized. Finally, a narrative report will present the findings of the review, alongside a geographic map illustrating the coverage of included studies compared with the actual implementation area of the CCFP.

Keywords: Conversion of Cropland to Forest Program, Sloping Land Conversion Program, Grain for Green, Payment for ecosystem services, Land use change, Afforestation, Soil erosion, Flooding, Poverty alleviation and Social equity

* Correspondence: l.grodriguez@cgiar.org
[1]Center for International Forestry Research (CIFOR), Jalan CIFOR, Situ Gede, Sindang Barang, Bogor, Barat 16115, Indonesia
Full list of author information is available at the end of the article

Background

The Conversion of Cropland to Forest Program (CCFP), also known as the Sloping Land Conversion Program (SLCP) or 'Grain for Green', was initiated in a context of ecological crisis and rising environmental awareness in China [1]. In 1997 there was a severe 267 day drought in the Yellow River basin [2], followed in 1998 by massive floods that devastated both the Yangtze and Songhua River basins, resulting in 3,600 deaths, 13.2 million people left homeless, and widespread economic impacts [3,4].

In addition to the extraordinary weather conditions occurring between 1997–98 caused by ENSO (El Niño Southern Oscillation) [3], these flooding events were associated with growing pressures from human activities [5,6], particularly the over-logging of natural forests and conversion of forests on steep slopes into farmland [7,8]. The authorities mainly attributed this disaster to unsustainable logging practices in State Forest Farms [7] and the conversion of forestland into cropland on steep slopes by smallholders throughout the catchments [4]. In response, the central government radically reoriented its forest policy by moving from a focus on timber production to a strategy involving conservation, restoration and livelihoods. A range of new programs related to forest conservation and environmental restoration followed, which together are known as the Priority Forestry Programs (or the Six Key National Forestry Programs). The first two of these programs to be introduced, and the most far-reaching, are the Natural Forest Protection Program[a] (NFPP) and the CCFP.

The NFPP was launched in 1998 to ban logging in the upper reaches of the Yangtze River and upper-middle reaches of the Yellow River, and the launch of the CCFP soon followed, with pilot sites introduced in the Yangtze and Yellow River basins in 1999. While the former program had the objective of reducing timber harvests, the latter aimed to restore vegetation on sloping croplands and lands classified as "wasteland" or "barren land" used by smallholder farmers [9].

The original intention of the CCFP was to reduce flooding and soil erosion; however the program was revised after a few years of operation to emphasize the improvement of rural livelihoods and poverty alleviation, in line with the emerging focus of the national poverty reduction strategy [10-12]. The CCFP can thus be conceptualized as an afforestation program or a large-scale forest Payment for Ecosystem Services (PES) scheme with a compensatory approach towards upstream areas inhabited by economically less-advantaged populations, who play a key role in providing downstream users with forest ecosystem services. The scheme represents an important monetary compensation from both central and local governments to these upstream smallholders.

Through a large-scale conversion of land use (from sloping cropland into forestland) and economic reorientation (from on-farm towards off-farm sectors) in upstream areas, the CCFP is designed to provide ecosystem goods and services, initially to upstream populations and in the long run to downstream populations.

Since the CCFP's inception, compensating smallholders for the opportunity cost of converting their sloping cropland into forest has been the core operational mechanism of the program. At the beginning of the program, compensations included a one-time payment for the purchase of saplings or seeds, an annual living allowance paid per unit area of cropland enrolled, and an annual grain/cash subsidy (with different amounts for households in the Yangtze River watershed and the Yellow River watershed regions[b]) [9,13]. The payment period of this three-tiered compensation system also depends on the type of land-use to be established, with two years of payments provided for converting cropland into grasslands[c], five years for converting cropland into forests of 'economic trees' (trees with direct economic returns) and eight years for converting cropland into forests of 'ecological trees' (trees with higher use restrictions). Program participants are paid conditionally upon maintaining a tree-survival rate higher than a minimum set at between 70% and 85%, depending on local criteria, which is verified by annual site inspections [13].

The nature of the CCFP's interventions has since evolved from this three-tiered subsidy system to its current simplified form, with a single cash payment now integrating the former grain compensation and livelihood-allowance subsidies, whereas seedling subsidies have been removed from the CCFP intervention. Apart from the compensation delivered to farmers, half of CCFP investment has been used on complementary activities such as cropland improvement, replanting on CCFP land, rural energy, etc. With regard to policy enforcement, the central government sets national-level compensation standards, while provincial governments may make further contributions for higher farmer compensations. The CCFP is implemented by county-level Forestry Bureaus, which were responsible for determining the sloping lands eligible for conversion and later allocating funds to those households willing to engage in the CCFP.

The CCFP is currently being implemented in 25 provinces (1,897 counties), has already afforested more than 25 million hectares (comprised of 9.27 million hectares of cropland and 15.8 million hectares of barren land classified as 'wasteland'), and provides direct subsidies to 32 million households (around 124 million people) [14,15]. In terms of its scale and magnitude, with 298 billion CNY (~42.82 billion USD) already invested between 1999–2013 [16], the CCFP is one of the most significant forest policies implemented in the developing world [9].

Objectives of the systematic review

After 15 years of implementation (1999–2014), it is timely to conduct a systematic evaluation of the program's impacts on both human populations and land resources. The objective of the systematic review is to provide evidence from the literature that could be used to actively inform the CCFP's design and future implementation, while identifying research gaps and new testable hypotheses so as to strengthen its positive impacts and minimize negative ones on both human populations and land resources. The systematic review will contribute by reviewing and analyzing not only the English-language CCFP literature, but also the data available within Chinese bibliographic databases. This systematic evaluation is also an important part of the Center for International Forestry Research (CIFOR)'s emerging Sloping Lands in Transition project (SLANT), which examines smallholder re/afforestation and sustainable forest management across several countries in the Asia-Pacific region.

Participants from CIFOR, China National Forestry Economics and Development Research Center (FEDRC) of the State Forestry Administration, Beijing Forestry University (BFU) and Forest Trends held together a stakeholder meeting in Kunming, China, in April 2014 to discuss research objectives and methods of this systematic review protocol (please refer to Additional file 1, list of participants). Special attention was given to defining the *populations* that might have been affected by the CCFP during its implementation period, and the actual *interventions* along with the potential *comparators* and *outcomes* of interest, a process that helped us to define our primary and secondary research questions. At the stakeholder meeting, we further compiled a list of contextual factors that might affect the implementation and outcomes of the CCFP, and discussed recurrent themes as found across a sample of articles.

Research questions

The primary research question of the systematic review is:

What socioeconomic and environmental effects has the Conversion of Cropland to Forest Program had on human populations and land resources during its first 15 years in China?

The secondary questions that the systematic review intends to find evidence for are as follows:

- How effective has the CCFP been in achieving its own stated objectives of soil erosion control, flood prevention and poverty reduction?

- Under which circumstances would/have farmers revert(ed) forestland back to cropland?
- Are there any unintended socioeconomic/ environmental outcomes?

Methods

Theory of change

A conceptual understanding of the CCFP using the methods of the theory of change can help to explain the cause-and-effect interactions between CCFP interventions and its expected socioeconomic and environmental outcomes (see Figure 1). The diagram below was designed after scoping the CCFP literature and following discussions with CCFP researchers and monitoring and evaluation specialists from the State Forestry Administration at the stakeholder meeting in Kunming, held in April 2014.

The first step in CCFP implementation began with the selection of households and sloping lands for participation in the program, which was undertaken by county-level forestry bureaus. Under the CCFP, smallholders are expected to become forest stewards on former agricultural and barren sloping lands. The institutional regulators are the central government which transfers economic resources to provincial governments which then, in turn, transfer the necessary funds to county-level forestry bureaus; these bureaus are responsible for CCFP implementation on the ground to provide smallholders with compensation for converting their cropland and barren sloping lands. On these agricultural and barren sloping lands, smallholders planted 'economic' or 'ecological' trees, a conversion that was actively facilitated through the delivery of a livelihood allowance, subsidies for purchasing tree saplings, and skill-training to plant the selected species.

After the conversion, it was expected that smallholders may experience an increase in available time (freed-up labor), which they could use to either intensify agricultural production on their lands or pursue off-farm work in urban areas; options that will be mediated by the degree of social equity among households and individuals in the CCFP implementation area (related to both intra-household and inter-household power relations across age, education level, gender, income and ethnicity heterogeneity factors). These options are expected to increase household incomes, which are further supplemented by compensation for any lost agricultural income from land conversion (paid following field checks by the county-level forestry bureau). The CCFP thus aims for livelihood change through reduced dependence on sloping agricultural lands, which will ultimately lead to a generalized poverty reduction provided that social equity and ecosystem functions are actively promoted [17]. In the medium term, the delivery of

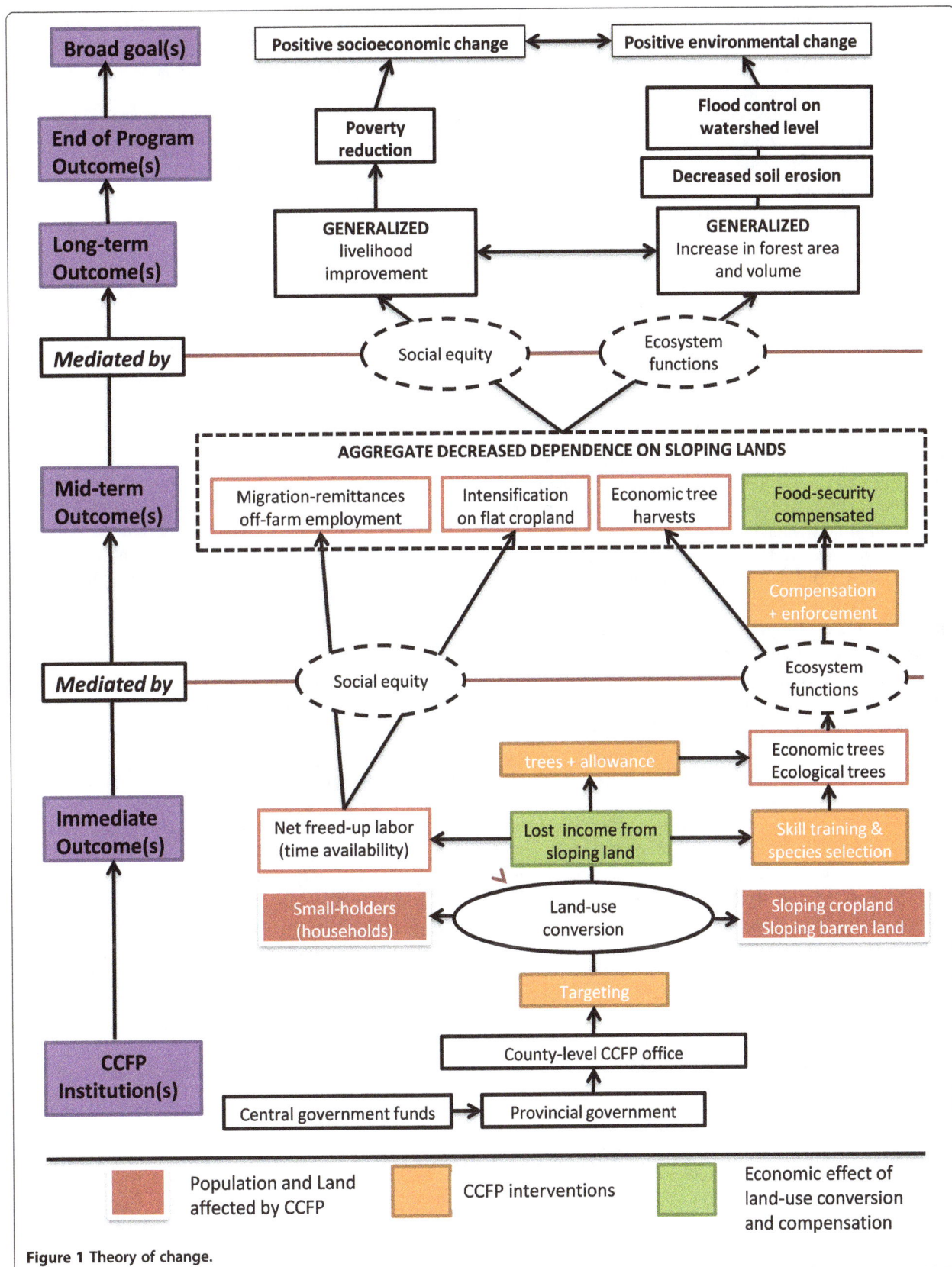

Figure 1 Theory of change.

these annual subsidies will compensate smallholders for the opportunity cost of such livelihood changes.

In terms of environmental outcomes, the core assumption of the CCFP is that increased forest area and timber volume on sloping lands will lead to a decrease in erosion and thereby a decrease in flood risk at the watershed level. Thus the higher subsidies paid for planting 'ecological' trees rather than 'economic' trees should lead to a greater incentive for reforestation on longer timescales. Nonetheless, both types of planted forest will contribute to the reduction of soil erosion on sloping lands. Moreover, skills training and monitoring provided by county forestry bureaus are expected to lead to higher tree survival rates, as will the selection of suitable tree species for individual sites. The targeting of suitable households and the degree of farmer voluntarism in participating in the CCFP will also affect the longevity of land conversion and thus the achievement of its broader environmental goals. It is expected that farmers who have sufficient livelihood alternatives to agriculture (ie. availability of non-targeted farmland or sources of off-farm income) and willingly choose to participate will be less likely to reconvert lands back to agriculture after subsidies end. On the other hand, if disadvantaged farmers and groups are not effectively targeted [17], this could also be a deterrent for achieving both the environmental and socioeconomic goals of the program, i.e. to produce a generalized socio-ecological readjustment towards soil conservation, flood prevention and poverty reduction. Finally, the targeting of suitable lands is critical for the success of the CCFP, as sloping lands that have already experienced considerable degradation may be difficult to rehabilitate through tree planting alone, and suitable sloping lands may not always be targeted if they are difficult to reach (and thus monitor).

On the basis of the CCFP's theory of change we are going to evaluate the typology, methods, geographical coverage (systematic map), and the extent of the socio-economic and environmental effects brought about by the program during its first 15 years of implementation (systematic review). Subsequently, within the systematic review, we will evaluate the effectiveness of the CCFP in achieving both its socioeconomic and environmental objectives, as defined by soil erosion control, flood prevention and poverty reduction. Moreover, we will also assess

the range of both intended and unintended outcomes, including studies that find forest reconversion to cropland and account for its explanatory factors.

PICO framework

To operationalize our research questions, theoretical hypotheses and database searches, we have further defined a Population-Intervention-Comparator-Outcome (PICO) model (see Table 1).

Populations

Our target human population consists of CCFP participant households and their individual members. Our target land resources population consists of CCFP enrolled lands (cropland, wasteland, ecological trees, economic trees).

Interventions

As the CCFP is enacted through multiple activities, interventions of interest consist of CCFP subsidies paid to smallholders for land converted, skill-training for local farmers, and enforcement of CCFP implementation (field-based checks on compensation delivery and household compliance with tree-survival rates).

Comparators

These are defined as both human populations and land resources that have not been exposed to the CCFP intervention (i.e. non-participant households and non-enrolled lands), with whom/which human populations/land resources exposed to the CCFP intervention might be potentially compared. Both households and lands prior to receiving CCFP interventions can also be compared to human populations/land resources post-CCFP intervention. Other types of comparators might also include macro-level comparisons between upstream interventions and upstream non-interventions or upstream socio-ecosystems prior to and following the CCFP intervention, and also comparisons between upstream and downstream socio-ecosystems. All these comparators will be used for analysis whenever there are available studies that can provide these primary data (i.e. on the actual socioeconomic and environmental effects of the CCFP).

Table 1 PICO elements of the systematic review

Population(s)	Intervention(s)	Comparator(s)	Outcome(s)
CCFP households and their individual members	CCFP (subsidies , skill-training, and enforcement with field checks	Non-participant households, households prior to CCFP implementation	Socioeconomic outcomes (changes in households' income structure, migration, etc.)
CCFP enrolled lands (cropland/ wasteland/ecological trees/ economic trees)	CCFP (subsidies , skill-training, and enforcement with field checks	Non-enrolled sloping lands, lands prior to CCFP implementation	Environmental outcomes (changes in water discharge, soil erosion, flood risk, local biodiversity, etc.)

Outcomes

Socioeconomic outcomes of the CCFP include impacts on household production (income, labor allocation, employment), household consumption, land tenure, food security and nutrition, social equity, farmers' autonomy in decision-making, power relations (including between income groups, ethnic groups, gender), and rural out-migration and remittances.

Environmental outcomes include impacts on watersheds (floods, discharge rates, filtration), soil (erosion, nutrients), changes in forest cover and standing volume, tree-survival rate, changes in tree biomass and carbon storage, changes in biodiversity, changes in energy sources (biomass, coal, hydro, solar), and other land-use and cover changes (LUCC).

Other socioeconomic and environmental outcomes reported in the literature will be noted. These potential interactions and hypotheses have been explained in detail within the 'theory of change' section and, whenever there are available studies, socioeconomic-environmental interactions will also be reported.

Search strategy

Our search strategy has been structured according to the Collaboration for Environmental Evidence's guidelines [18] and a PICO framework to consider the CCFP's impacts on both human populations and land resources.

Searches

During our initial literature scoping, we assessed the breadth of the CCFP's current bibliography and determined that Chinese research databases contain an enormous body of potentially relevant literature. Around 500,000 hits were initially identified within the Chinese China National Knowledge Infrastructure (CNKI) databases when employing the phrase 退耕还林 (CCFP) as a search term. This huge number would later be reduced to about 3,500 hits by making use of *population, intervention* plus *comparator* and *outcome* search terms within the frame of our research strategy.

Languages: Searches will be conducted in English and Chinese. Spanish was also used in our scoping search searches, but as there were no meaningful results (no published literature), Spanish has been removed from the final search strategy.

Time frame: Searching will be limited to studies published or produced in and after 1999, the first year of CCFP implementation

Search terms: See Table 2 below for a comprehensive list of search terms as organized by their Population, Intervention, Comparator, and Outcome categories.

Search strings and/or combinations of searches Search terms from each of the population, intervention,

comparator and outcome categories were combined using the Boolean command OR, then combined in a comprehensive search string using the Boolean command AND. Our searches will be adjusted for the specific requirements/features of each database and their specific truncation and/or wildcard symbols. For instance, as Google Scholar does not allow the use of complex search strings, the following intervention terms will be used: Conversion of Cropland to Forest Program (CCFP), Sloping Land Conversion Program (SLCP), Grain for Green, Upland Conversion Program, China. The Chinese search strategy was reviewed by a subject specialist librarian at the University of Michigan (see Additional file 2, for a detailed account of the search strings that have been employed in this protocol).

Estimating the comprehensiveness of the searches

Our scoping searches confirmed our previous expectations that Chinese databases would retrieve a far higher number of results than English databases, and that this difference is striking in quantitative terms. Initial search results showed Chinese results to be over a thousand orders of magnitude greater than English results. After refining our search strings, around 900 results were identified from English databases (486 hits in Web of Science, 253 hits in Scopus, 144 hits in CAB abstracts and 21 hits in AGRIS, or 879 unique hits after duplicate removal) compared to around 3,500 hits (titles) from Chinese databases.

Publication databases We aim to identify CCFP peer-reviewed articles, CCFP doctoral theses and CCFP master theses, and other study types through the following sources:

- Web of Science, Scopus, CAB Abstracts, AGRIS (FAO).
- CNKI or 中国知网, which includes China Academic Journals Full-text Database, China Doctoral Dissertations Full-text Database, China Masters' Theses Full-text Database, China Core Newspapers Full-text Database, China Proceedings of Conference Full-text Database.

Internet searches We will use Google Scholar to conduct internet searches. In developing these methods, the first 200 studies listed were retrieved for screening as a test of the search strategy. The comprehensiveness of the databases versus the internet as a source of articles determined to be relevant through screening will be reported in the review.

Specialist searching for grey literature, contacts and organizations: During the searching process we will

Table 2 PICO search terms

Population search terms	English	Chinese
Human population	household, farmer, family peasant	农户，农民
Land resources	sloping land, cropland, wasteland, economic forest/tree, ecological forest/tree, land use, soil water, basin	坡地，耕地，荒地，经济林，生态林，土地使用，土壤，水，流域
Intervention search terms	**English**	**Chinese**
	Conversion of Cropland to Forest, Sloping Land Conversion Grain for Green Upland Conversion	退耕还林
Comparator search terms	**English**	**Chinese**
Human comparators	Participant, non-participant intra-household, upstream user downstream user, uphill resident, lowland resident cross-sectional, comparison comparative, longitudinal space, time, panel data	参与者，非参与者，家庭成员/农户成员，上游使用者，下游使用者，山地居民，低地(平原)居民 横截面(数据)，比较，纵向(数据)，空间，时间，面板数据
Land resources comparators	Enrolled, non-enrolled upstream, downstream uphill, lowland cross-sectional, comparison comparative, longitudinal time series, space, time panel data.	退耕地，未退耕地，上游，下游，山地，低地(平原)，横截面(数据)，比较，纵向(数据)，空间，时间，面板数据
Outcome search terms	**English**	**Chinese**
Socioeconomic outcomes	household production household consumption food security, nutrition livelihood equity, power relations, equality, Gini, gender, intra-household, ethnic decision making, governance voluntary migration, remittances.	农户生产/家庭生产，农户消费/家庭消费，粮食安全/粮食保障，营养民生/生计，公平，权力关系，平等，基尼系数，性别平等，家庭内关系，少数民族，政府决策，治理，自愿程度/志愿程度，迁移，汇款.
Environmental outcomes	watershed, floods, discharge rate, soil, filtration, erosion, soil nutrient deforestation, forest degradation, afforestation, reforestation forest cover, survival rate biodiversity, biomass, carbon energy, land use and land use change	流域，洪灾/水灾/洪水灾难，流量，土壤，土壤渗滤，水土流失/土壤流失/土壤侵蚀，土壤养分，森林砍伐/毁林，滥砍滥伐，森林退化，造林，再造林，森林覆盖率，林木成活率/林木保存率，生物多样性，生物量，碳汇量，能源，土地利用和，土地利用变化

identify key institutions/organizations that could potentially be involved in conducting research studies linked to the CCFP. Afterwards, we will search for additional reports delivered by these institutions/organizations' websites. Moreover, we have also issued a call for grey literature, with both English and Chinese language brochures being circulated online and hard copies distributed at relevant meetings and conferences [19]. In the meantime, an advisory group formed by the people who took part in the stakeholder meeting in Kunming will also provide key inputs in recommending relevant reports/datasets that may be unpublished or not found in our searching.

Inclusion criteria

Relevant subjects: Both human populations and land resources are to be included as relevant populations, including: CCFP participant households, their individual members and their CCFP enrolled lands (cropland, wasteland, ecological trees, economic trees). Grasslands are excluded from our analysis since they no longer form part of the CCFP, they are under the administration of the Ministry of Agriculture, and because they contribute to significantly different environmental outcomes as compared with forests.

Relevant interventions: These include CCFP compensation subsidies, skill training for local farmers, and enforcement work with field checks. When possible, we will retrieve all information on other types of subsidies that might have an impact on household livelihoods and the environment. Broadly speaking, the Natural Forest Protection Program (NFPP) does not overlap with the CCFP, as the former is related to state forestland whereas the latter mainly occurs over collective forestland. Therefore, the NFPP is not included in our analysis (although it is taken into account as a contextual factor).

Relevant comparators: We are interested in assessing the existing evidence comparing the effects of the CCFP between *participating and non-participating* CCFP households. This systematic review will simultaneously consider the available evidence about CCFP land resources' comparators such as both *enrolled and non-enrolled* lands (under the management by both types of households dwelling upstream). This systematic review will also use the available empirical data to track those 'before-and-after' comparators in both human populations (i.e. the socioeconomic status of both participant and non-participant households *before and after* the CCFP interventions) and land resources (i.e. the environmental status of both enrolled and non-enrolled lands before and after the CCFP intervention). At a more aggregated level, we will synthetize the available evidence on the effects of the CCFP on watershed with-intervention upstream and without-intervention upstream socio-ecosystems (i.e. both between intervention and non-intervention watershed regions, and between pre-CCFP and post-CCFP watershed regions), and also between upstream and downstream socio-ecosystems.

Relevant outcomes: From the stakeholder workshop and initial literature scoping, we have identified a number of relevant socioeconomic outcomes; including CCFP impacts on: household production and consumption; changes in household land tenure; changes in food security and nutrition; social equity (between and within households), farmers' autonomy in decision-making, power relations (income groups, intra-household and gender levels, ethnic groups); migration and remittances. With regards to environmental outcomes of interest, we have identified the following: floods and watershed discharge rates; soil filtration, erosion and nutrient cycling; deforestation and forest degradation on slopes (should farmers revert converted land back to cropland); forest cover, afforestation and reforestation; tree-survival rates, biodiversity, biomass and carbon storage; changes in household energy sources (biomass, coal, hydro, solar, etc.); and land-use cover change (LUCC) dynamics. Studies assessing potential or future outcomes of CCFP, including model projections or other predictions of program impact, will not be included as this review only seeks to assess the actual impacts of CCFP implementation (those which have already taken place). Socioeconomic-environmental interactions will be reported, whenever there are available studies on this issue.

Relevant types of study design: Primary studies using quantitative and qualitative methods will be considered; these can include experimental and quasi-experimental designs, case–control experiments and broad sample-size surveys of participant and non-participant populations (cross-sectional analyses), surveys of populations prior to and following CCFP implementation (longitudinal analyses), and individual case studies of populations that have been targeted for CCFP interventions. Studies must use primary data to present actual impacts that have already happened, and are causally linked or correlated to the CCFP interventions. Primary studies concerning farmers' perceptions of CCFP impacts will also be included, provided that a robust and reliable methodology was used, as these perceptions can be used as a proxy for measuring certain socioeconomic impacts. Modeling exercises that use primary data to calculate actual impacts shall be included for further analysis, whereas models that project potential or future impacts will not be included (although they will be collected in a separate folder for future analysis).

With regards to qualitative evidence, we will consider the following design/methods: participant and non-participant observations, structured, semi-structured, and unstructured interviews, focus group discussions, and qualitative data from surveys and questionnaires. On the other hand, with regard to quantitative evidence, we will consider the following design/methods: direct measurements of observed phenomena, including use of geo-spatial technologies (GIS and remote sensing) as well as the use of polls, questionnaires, and surveys where answers are restricted to given choices. Finally, studies that will not be considered for data extraction include reviews, meta-analyses, summary studies, theoretical and methodological framework studies, and editorials and commentaries, although these will be considered in our background and discussion.

Study screening Given the big volume of references expected from the Chinese literature database, we will first perform title screening, then abstract screening, and finally full-text screening of retrieved search results according to the inclusion criteria stated above. For the English literature database, we will first conduct title-and-abstract screening and later full-text screening. At the beginning of each stage of screening, four reviewers for English (NH, WZ, LP and LGR) and three reviewers for Chinese (CX, KZ and LGR) will review a sample of 50 studies to conduct kappa analysis on their screening decisions. Should the kappa statistic fall below 0.6, the reviewers will discuss points of disagreement and conduct a second round of screening. This will take place for both the Chinese and English language literature, with LGR coordinating pilot screening in both languages. Once an acceptable level of agreement is reached, the remainder of studies will be screened in each stage. Members of the advisory group will also randomly review the screening, and for a 25% selection of the scoped files, tests will be conducted first on the abstracts and then on the full-texts to ensure screening decisions remain consistent. Then full-text reading and extraction of qualitative and quantitative information will proceed into several categories (see next subsection, study quality assessment).

Potential effect modifiers and reasons for heterogeneity
During the stakeholder workshop in Kunming (April 2014), a set of independent socioeconomic variables with potential influence on CCFP outcomes was defined, including: household members' age, gender, education level, income group and ethnicity. These variables will be useful to assess cross-household and intra-household heterogeneity of CCFP impacts. For instance, we can assess if impacts of the CCFP equally affect households with different income levels, or whether different members of the same household experience different socioeconomic impacts. At the same time, a set of independent environmental

variables with potential effects on CCFP outcomes at the targeted land plots were also defined, including: orientation, slope, size, distance to home, weather/climate, altitude and latitude. Socioeconomic-environmental interactions among these factors can be especially relevant for the CCFP implementation process.

Study quality assessment (critical appraisal)

Those studies that meet our inclusion criteria through the full-text screening will be assessed based on the following five quality criteria:

1. Data collection methods are thoroughly explained and clear and replicable.
2. Qualitative or quantitative analysis methods are thoroughly explained and clear and replicable; key terms and variables are well defined.
3. Sample size is well explained and representative of the population.
4. Results/conclusions are logically derived and supported by presented evidence.
5. Confounding factors are considered and well explained.

We will document individual study quality based on each of these five criteria, and report on the overall quality of the evidence base in our systematic review. For each study, we will also record yes/no answers for each criteria, where "yes" is equal to a score of one and "no" equal to a score of zero. Each study will thus have a quality assessment score of 0 to 5, where scores of 3 to 5 will be considered acceptable while studies with scores of 0 to 2 will be considered low quality. For our systematic review, we will consider and compare the outcomes from both sets of studies to determine whether the low quality studies demonstrate significantly different results from those of acceptable quality studies, and whether their inclusion in our final analysis leads to any change in our overall assessment. Assessment of the studies against these criteria is a strong indicator of the presence or absence of most types of potential bias. In addition, we will check for patterns of correspondence between authors' affiliations and specific findings so as to identify additional sources of potential bias. We will also determine whether there are discernable biases in results that correspond to the study designs used (i.e. control/counterfactual, longitudinal study, or case study).

Data extraction strategy

For all studies that have met our critical appraisal criteria after full-text screening, we will proceed with extracting both quantitative and qualitative data for both socioeconomic and environmental outcomes, following the general structure of our PICO framework

and using the set of factors of interest that were raised in discussions at the stakeholder meeting in Kunming. The data extraction categories are as follows:

Study metadata and methodology:

- Bibliographic information: author, year, title, institution of the lead author.
- Type of study: quantitative/qualitative study, or both (mixed methods).
- Comparative methods: cross-sectional, longitudinal, or both.
- Geographic location (county and GPS coordinates whenever available)
- Time-span covered by the study.

Population:

- Type of population: human population, land resources population, or both.
- Unit of comparative analysis (scale): household/individual, village/community, county, provincial or national levels.
- Sample size and land area: number of households covered by study or land area covered by the study.

Intervention:

- Type and duration of intervention: compensation subsidies plus tree-sapling provision, skill-training, enforcement with field checks (one or multiple intervention types can be present).

Outcomes:

- Socioeconomic outcome categories: changes in upstream household production and production structure (as measured by income, labor, employment); changes in household consumption and household income structure; changes in household land tenure; changes in social equality (Gini coefficient) and intra-household equality, both of them across income levels, gender, age groups and educational levels; changes in household migration and remittances; enforcement (voluntary/compulsory and degree of tree species selection).
- Environmental outcome categories: changes in upstream forest cover and standing volume, tree survival rates, changes in measures of biodiversity (species richness, composition, and abundance), tree biomass and carbon storage; changes in upstream soil erosion and soil nutrient content; changes in upstream household energy use and energy structure; changes in upstream land-use and

land-cover change (NDVI and leaf-area indices); changes in downstream discharge rates and floods; frequency of natural disasters; trade-offs among ecosystem services.

- Socioeconomic-environmental interactions (among the aforementioned outcomes), whenever there are available studies on this issue.

Potential effect modifiers and reasons for heterogeneity:

Socioeconomic factors: household members' age, gender, education, income group and ethnicity:

- Age: average and percentage distribution across several ranges (over 20, between 20–40, between 40–60, and over 60).
- Gender: percentage of women/men
- Education: average and percentage distribution across several ranges (primary school or less, middle school, high school or above).
- Income group: locally-defined low, middle and high income groups.
- Ethnicity: Han/non-Han
- Environmental factors: land orientation, slope, size, distance to home, and elevation of land plots
- Other: Voluntarism of participation in CCFP

Data synthesis and presentation

One narrative synthesis report, i.e. a systematic review, will be produced relying on both qualitative and descriptive statistics so as to assess the available evidence on the CCFP's socioeconomic and environmental outcomes. Using descriptive statistics, we will first present the results of each screening stage (title, abstract, and full text screening) and will also use statistics to show the results of the quality assessment (high quality vs. low quality studies). Secondly, we will organize and synthesize the data according to the types of research conducted on the Conversion of Cropland to Forest Program, through: 1) categorizing the empirical evidence as socioeconomic and/or environmental; 2) classifying its research methods in typologies; and 3) presenting the geographical distribution of studies throughout the country. Qualitative data on both socioeconomic and environmental outcomes will be grouped by the individual measures they address (i.e. income change, forest cover change, etc.) and synthesized narratively alongside quantitative data on the same measures. In so far as we have sufficient data to perform meta-analysis of quantitative data of outcome measures (and particularly their correlation with human and environmental population characteristics), we will do so.

With regard to the main research question, the socioeconomic and environmental outcomes of the CCFP will

be also presented in a searchable database, provided along with a geographic map that identifies and locates the available evidence from studies across China. GIS and Remote Sensing techniques will be used to document the locations of included studies within China to compare the coverage of CCFP evaluations with the actual area of its implementation.

As for the secondary research questions, the existing results on the program's effectiveness will be first synthesized and accordingly presented in terms of its achievements in poverty reduction, soil erosion control and flood prevention. Second, conclusions will be made in terms of conditions that may have led to the reconversion of sloping lands to agriculture. Third, CCFP's unintended socioeconomic and environmental effects will be synthesized and presented in order to guide future program implementation and research, and also to uncover possible knowledge gaps and new hypotheses on the program's impacts.

Endnotes

[a]The NFPP was approved in 1998 so as to stop natural forest loss and degradation [15]. The introduction of this 'logging ban' policy meant the re-structuring of state-owned forestry enterprises, into which government subsidies have been channeled to compensate laid-off workers and alleviate the economic crisis faced by these companies in the late 1990s.

[b]Since 2004, grain transfers were completely replaced by cash.

[c]Although the CCFP initially included the conversion of cropland into grassland, this land-use transformation no longer forms part of the program, and so has become a different program which is currently under the Ministry of Agriculture.

Additional files

Additional file 1: Participants in stakeholder meeting, Kunming, April 2015.

Additional file 2: Search String Combinations.

Abbreviations
BFU: Beijing Forestry University; CCFP: Conversion of Cropland to Forest Program; CIFOR: Center for International Forestry Research; CNKI: China National Knowledge Infrastructure; ENSO: El Niño Southern Oscillation; FEDRC: Forestry Economics and Development Research Center (China State Forestry Administration); NFPP: Natural Forest Protection Program; PES: Payments for Ecosystem Services; SLCP: Sloping Land Conversion Program.

Competing interests
The authors declare that they have no competing interests.

Authors' contributions
All authors took part in the stakeholder meeting held in Kunming in April 2014. NH carried out initial database searches of English studies, and provided key research inputs to this protocol. LP and CX provided key research guidance for the definition of primary and secondary questions. LP,

NH, WZ, and KZ participated in pilot Kappa tests for study screening. LGR carried out literature scoping in English, Chinese and Spanish databases, and led the writing of this draft protocol. All authors reviewed the final version of this protocol before submission. All authors read and approved the final manuscript.

Acknowledgments
Special thanks to Peng Daoli (Beijing Forestry University), Michael Bennett (Forest Trends), Wang Jianan (FEDRC), Wang Jiang (BFU), Christine Padoch (CIFOR), Kiran Asher (CIFOR) and Yustina Artati (CIFOR), who participated and contributed to the stakeholder meeting held in Kunming (April 12th 2014). We also want to express our gratitude to Liangyu Fu (University of Michigan), who reviewed our Chinese search strategy. The review team would like to thank CIFOR's Evidence Based Forestry Initiative and the UK Department for International Development (DfID) for financing this research through its KNOWFOR program grant.

Author details
[1]Center for International Forestry Research (CIFOR), Jalan CIFOR, Situ Gede, Sindang Barang, Bogor, Barat 16115, Indonesia. [2]China National Forestry Economics and Development Research Center, State Forestry Administration, Hepingli Dongjie No. 18, Beijing 100714, China.

References
1. Xi XL, Fan LH, Deng XM. The situation of environmental awareness in China. Analysis of results from a survey of Chinese citizens (In Chinese). Zhongguo gongzhong huanjing yishi zhuangkuang - gongzhong diaocha jieguo pouxi Zhongguo ruan kexue-gongzhong juece-huanjing baohu ji qi chanye zhuanti yantao. 1998;9:24–30.
2. Gao YC. Analysis on reasons for the Yellow River's dry-up and its eco-environmental impacts. J Environ Sci. 1998;10(3):357–64.
3. Qian Y, Glantz M. The 1998 Yangtze Floods: the use of short-term forecasts in the context of seasonal to interannual water resource management. Mitig Adapt Strat Glob Chang. 2005;10:159–82.
4. Yin RS, Xu JT, Li Z, Liu C. China's Ecological Rehabilitation: The Unprecedented Efforts and Dramatic Impacts of Reforestation and Slope Protection in Western China. China Environment Series. 2005;7:17–32.
5. Li WH. Yangtze's floods and ecological construction (In Chinese). Changjiang hongshui yu shengtai jianshe Ziran ziyuan xuebao. 1999;14(1):1–8.
6. Shi DM. Analysis on the relationship between soil erosion and flood disasters in the Yangtze basin (In Chinese). Changjiang liuyu shuitu liushi yu honglao zaihai guanxi pouxi Turang qinshi yu shuitu baochi xuebao. 1999;5(1):1–7.
7. Liu C. An economic and environmental evaluation of the Natural Forest Protection Program. China National Forest Economics and Development Research Center (FEDRC), State Forestry Administration (SFA) 2002 (working paper).
8. Xu ZG, Xu JT, Deng XZ, Huang JK, Uchida E, Rozelle S. Grain for Green versus Grain: Conflict between Food Security and Conservation Set-Aside in China. World Dev. 2006;34(1):130–48.
9. Uchida E, Xu JT, Rozelle S. Grain for Green: Cost-Effectiveness and Sustainability of China's Conservation Set-Aside Program. Land Econ. 2005;81(2):247–64.
10. Gauvin C, Uchida E, Rozelle S, Xu JT, Zhan JY. Cost-Effectiveness of Payments for Ecosystem Services with Dual Goals of Environment and Poverty Alleviation. Environ Manage. 2010;45:488–501.
11. China Green Times. 10 years of monitoring, sending our country 'a postcard of data'. Looking back 10 years of monitoring the socioeconomic benefits of the key forestry programs (In Chinese). Shi nian jiance, jigei zuguo de shuju mingxinpian——linye zhongdian gongcheng shehui jingji xiaoyi jiance shinian huigu. *Zhongguo lvse shibao* 2014-January-2nd. Retrieved from: http://www.greentimes.com/green/news/zhuanti/lyzdgc/content/2014-01/02/content_242505.htm.
12. CCICED. Implementing the Natural Forest Protection Program and the Sloping Land Conversion Program: Lessons and policy recommendations. In: Xu J, Katsigris E, White TA, editors. China Council for International Cooperation on Environment and Development; Western China Forests and Grassland Task Force. 2002.
13. Bennett M, Xie C, Hogarth N, Peng DL, Putzel L. China's Conversion of Cropland to Forest Program for Household Delivery of Ecosystem Services: How Important is a Local Implementation Regime to Survival Rate Outcomes? Forests. 2014;5(9):2345–76.
14. Song C, Zhang YL, Mei Y, Liu H, Zhang ZQ, Zhang QF, et al. Sustainability of forests created by China's sloping land conversion program: a comparison among three sites in Anhui, Hubei and Shanxi. Forest Policy Economics. 2014;38:161–7.
15. Delang CO, Wang W. Chinese Forest Policy Reforms After 1998: The Case of the Natural Forest Protection Program and the Slope Land Conversion Program. Int For Rev. 2013;15(3):290–304.
16. Zhai DL, Xu JC, Dai ZC, Cannon CH, Grumbine RE. Increasing tree cover while losing diverse natural forests in tropical Hainan, China. Reg Environ Change. 2014;14:611–21.
17. Pascual U, Phelps J, Garmendia E, Brown K, Corbera E, Martin A, et al. Social equity matters in payments for ecosystem services. Bioscience. 2014;64 (11):1027–36.
18. Collaboration for Environmental Evidence. Guidelines for Systematic Review and Evidence Synthesis in Environmental Management. Version 4.2. In: Environmental Evidence. 2013. http://environmentalevidence.org/wp-content/uploads/2014/06/Review-guidelines-version-4.2-final.pdf.
19. Evidence-Based Forestry. http://www1.cifor.org/ebf/reviews/current-reviews/systematic-reviews/chinas-conversion-of-cropland-to-forest-program-ccfp.html.

13

What are the impacts of reindeer/caribou (*Rangifer tarandus* L.) on arctic and alpine vegetation? A systematic review

Claes Bernes[1*], Kari Anne Bråthen[2], Bruce C Forbes[3], James DM Speed[4] and Jon Moen[5]

Abstract

Background: The reindeer (or caribou, *Rangifer tarandus* L.) has a natural range extending over much of Eurasia's and North America's arctic, alpine and boreal zones, yet its impact on vegetation is still unclear. This lack of a common understanding hampers both the management of wild and semi-domesticated reindeer populations and the preservation of biodiversity. To achieve a common platform, we have undertaken a systematic review of published studies that compare vegetation at sites with different reindeer densities. Besides biodiversity, we focused on effects on major plant growth forms.

Methods: Searches for literature were made using online publication databases, search engines, specialist websites and bibliographies of literature reviews. Search terms were developed in English, Finnish, Norwegian, Russian and Swedish. Identified articles were screened for relevance based on titles, abstracts and full text using inclusion criteria set out in an *a priori* protocol. Relevant articles were then subject to critical appraisal of susceptibility to bias. Data on outcomes such as abundance, biomass, cover and species richness of vegetation were extracted together with metadata on site properties and other potential effect modifiers.

Results: Our searches identified more than 6,000 articles. After screening for relevance, 100 of them remained. Critical appraisal excluded 60 articles, leaving 40 articles with 41 independent studies. Almost two thirds of these studies had been conducted in Fennoscandia. Meta-analysis could be made of data from 31 of the studies. Overall, effects of reindeer on species richness of vascular plants depended on temperature, ranging from negative at low temperature to positive at high temperature. Effects on forbs, graminoids, woody species, and bryophytes were weak or non-significant, whereas the effect on lichens was negative. However, many individual studies showed clear positive or negative effects, but the available information was insufficient to explain this context dependence.

Conclusions: We see two pressing matters emerging from our study. First, there is a lack of research with which to build a circumpolar understanding of grazing effects, which calls for more studies using a common protocol to quantify reindeer impacts. Secondly, the highly context-dependent outcomes suggest that research and management have to consider local conditions. For instance, predictions of what a management decision would mean for the effects of reindeer on vegetation will have to take the variation of vegetation types and dominant growth forms, productivity, and grazing history into account. Policy and management have to go hand-in-hand with research in individual cases if the dynamics between plants, animals, and humans are to be sufficiently understood.

Keywords: Reindeer, Caribou, *Rangifer tarandus*, Forbs, Grasses, Graminoids, Woody species, Lichens, Bryophytes, Species diversity, Herbivory, Grazing, Browsing, Tundra

* Correspondence: claes.bernes@eviem.se
[1]Mistra Council for Evidence-Based Environmental Management, Royal Swedish Academy of Sciences, P.O. Box 50005, SE-104 05 Stockholm, Sweden
Full list of author information is available at the end of the article

Background

Reindeer and reindeer husbandry

The reindeer (*Rangifer tarandus* L.) has a natural range extending over much of Eurasia's and North America's arctic, alpine and boreal zones. In considerable parts of this region, reindeer are the only large herbivores. In the 20th century, the species was also introduced into several areas where it never occurred naturally, including a number of islands in the Arctic and South Atlantic.

Rangifer tarandus is the only species of the genus *Rangifer*, but it includes several subspecies. The Eurasian subspecies are referred to as reindeer, while those native to North America generally are known as caribou. We will normally use the term caribou only when specifically referring to studies from North America.

Wild reindeer are still numerous in parts of the world, notably in Canada and Alaska. In northern Europe and Siberia, however, the majority of reindeer populations have been domesticated or semi-domesticated for several centuries. Here, they are herded by indigenous and local peoples. Some large populations of wild reindeer are still present in Russia, and a few small ones remain in southern Norway and southeastern Finland, but in Sweden all reindeer are semi-domesticated.

Over the seasons, many reindeer herds migrate over large distances between summer and winter pastures, and between pastures with different vegetation within the seasonal ranges. Reindeer in Sweden normally spend the snow-free season foraging on alpine tundra, in the forest-tundra ecotone, or in subalpine birch forests, whereas they spend the winter in boreal coniferous forests.

The quality of the summer ranges are very important for the growth and condition of the reindeer, including how well they can survive subsequent difficult winters. However, the quality of the winter pastures is usually a strong determining factor for the population size of reindeer [1]. During some winters, foraging is made difficult by ice or deep snow, and herd sizes can therefore vary considerably from one decade to another (although in some areas, domesticated reindeer may be given supplementary food through parts of the winter [2]). In Sweden, the number of reindeer has oscillated repeatedly between c. 150,000 and c. 300,000 over the last 125 years, with a long-term average of about 225,000 (see Figure 1). Similar numbers are currently found in Norway and Finland. These statistics refer to sizes of post-slaughter winter herds. In summer, the numbers are considerably higher due to calving during spring.

Impacts of reindeer on arctic and alpine vegetation

Being the most numerous large herbivores in circumpolar areas, reindeer play a pivotal role in their ecosystems. Through their effects on vegetation and carnivore populations, reindeer affect several ecosystem processes, while also providing essential ecosystem services to indigenous peoples [6].

The arctic and alpine tundra, where reindeer find much of their food, constitutes one of the most marginal habitats on earth [7]. For primary producers, conditions here are nutrient-limited and climatically extreme. The vegetation is dominated by growth forms that require only small amounts of nutrients (e.g. lichens and

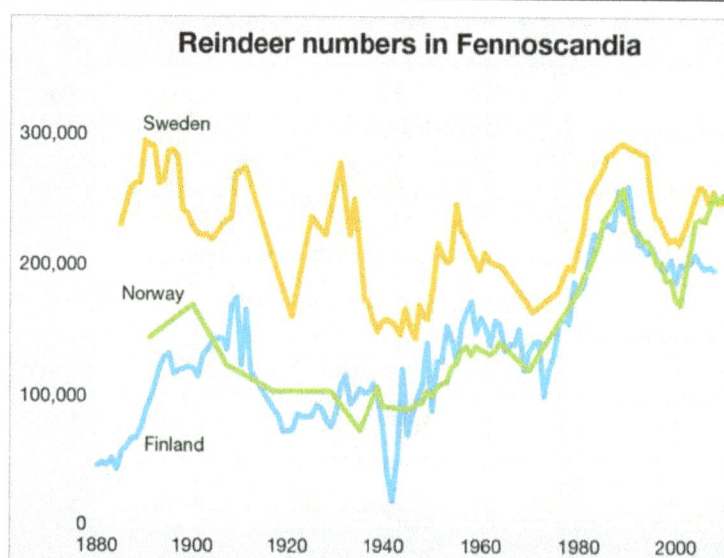

Figure 1 Reindeer numbers in Fennoscandia. The diagram shows total populations of semi-domesticated reindeer in Sweden, Norway and Finland following the autumn slaughter. After calving in spring, herds are significantly larger. Data from Statistics Sweden, Norwegian Directorate for Reindeer Husbandry, Finnish Reindeer Herder's Association, Finnish Game and Fisheries Research Institute (A. Ermala) and [3-5].

bryophytes) or are able to conserve nutrients (e.g. dwarf shrubs [8]).

Since reindeer are among the few herbivores able to digest lichens [9] and bryophytes [10], they are highly adapted to the vegetation of the tundra ecosystem [11]. Along with species of *Cetraria*, reindeer lichens (*Cladonia* and *Cladina* spp.) form a prominent part of the ground vegetation in many polar areas [12]. However, while lichens usually make up a substantial part of the reindeer diet, especially in winter, the animals may well survive without them if other food is available [11,13]. In summer, reindeer prefer vascular plants such as graminoids (grasses, sedges and rushes), forbs, and leaves of shrubs and deciduous trees. It is on summer pastures dominated by such plants that reindeer gain enough weight to survive the long winter, a season when icing events sometimes block their access to food resources almost entirely.

Reindeer can therefore be seen as seasonally adaptable ruminants and as intermediate feeders, and their migratory nature enables them to locate and utilise pulses of nutrients in space and time [14,15]. Nutrient pulses mainly occur after snowmelt, facilitating the growth of nutritious plants such as forbs and graminoids [16,17]. Reindeer are able to feed under spring-like conditions during large parts of the growing season, either by following snowmelt as it advances to higher altitudes or latitudes [18] or by changing preference during the summer season between growth forms with different phenological timing [19].

The impacts of reindeer herbivory can apparently be strong enough to cause transitions between vegetation states in tundra ecosystems [11], such as changes from lichen- to bryophyte- to graminoid-dominated vegetation. Shifts from lichen- to bryophyte-dominated stages following intensive grazing and trampling have been documented in a vast number of reindeer ranges [6,11]. Evidence for transitions to a graminoid-dominated state has been found in experimental studies of the effects of reindeer activity [20], or where reindeer behaviour has been manipulated, e.g. along fences regulating reindeer migration [21,22]. Yet, this evidence has not been corroborated by studies assessing rangelands of freely roaming semi-domesticated reindeer [23,24]. Some studies indicate that reindeer grazing (particularly by semi-domesticated herds in Eurasia) may counteract climatically induced encroachment of trees and shrubs in tundra [6,25-27], even to the extent of limiting populations of shrub-dependent bird species [28]. However, others have found the impacts on vegetation by caribou in North America to be minor [14].

These seemingly inconsistent results may reflect the fact that *Rangifer* grazing systems are particularly variable, spanning vast areas with both domestic and wild herds, with introduced reindeer populations as well as native ones, with many different management systems, and with large climatic and biotic gradients. The response of vegetation to herbivory depends on factors such as productivity [29] and the long-term history of grazing [30], and these factors are known to vary considerably in areas where reindeer occur.

Changes in the impact of reindeer on vegetation currently cause concern in several regions, both where populations have been reduced and where they have reached historic highs [31,32]. For instance, increased population densities have reduced the abundance of palatable forage plants in some summer ranges, with consequences for reindeer calf weights [23]. Lack of forage has even been claimed to contribute to reindeer losses to predators [33] in areas where no assessment of reindeer impact on vegetation has been made. On the other hand, a reduction of the grazing pressure could have negative effects on biodiversity if it means that reindeer are no longer able to control shrub encroachment [34].

Effects of climate change on vegetation and reindeer grazing

Climate change will likely have strong effects on arctic ecosystems. Temperature changes occur at a much faster pace in the Arctic than in the world as a whole [35], causing a rapid increase in terrestrial biomass. This so-called 'greening of the Arctic' [6,36-38] involves both range expansions and increased *in situ* growth of tall shrubs, treeline trees and graminoids. The changes are not evenly distributed over the Arctic, however. Recent estimates [35,39] indicate substantial greening over about a third of the area (the North American High Arctic and the east European Arctic), browning within a few percent of the area, and no significant change within about half of the Arctic. The reasons for these differences vary, but they include differential warming, moisture changes, herbivory, industrial development and legacies from past land use [35].

The net effects of these changes on reindeer populations are not easy to predict [40]. Since plant species respond in different ways to climate change, novel ecosystems may arise [38]. The expansion of tall shrubs may have negative effects on field-layer plants due to competition, forage quality may change as plant phenology and nutrient cycling are altered [17], and reindeer migrations between summer and winter ranges may be affected. The effects of reindeer grazing on shrub expansion [27] might cause climatic feedbacks through albedo changes [41]. These could add to the highly dynamic nature of the tundra system.

Shifting perspectives on the impacts of reindeer grazing

In Sweden, public opinion on how reindeer grazing affects mountain vegetation has shifted during the last few decades. In the 1990s, several well-publicised records of grazing-related vegetation degradation helped to form a widespread perception that some mountain areas were overutilised, and a concern that Swedish reindeer husbandry was not sustainable [42]. This was, for instance, reflected in a Swedish government bill stating that some areas had become overgrazed over a long time because of 'an imbalance between reindeer numbers and available forage' [43].

In other parts of Fennoscandia, severe overexploitation of reindeer ranges was noted, particularly on lichen heaths in Finnmark in northernmost Norway and in Finnish Lapland [44,45], and on summer ranges dominated by vascular plants in Finnmark [23]. The damage on the lichen heaths was caused by a change in seasonal grazing from winter to summer, with lichens being worst affected due to their sensitivity to trampling during the snow-free season, whereas the reduction of palatable vascular plants was an effect of increased reindeer numbers.

More recently, however, the impact of reindeer grazing on mountain vegetation was subject to re-evaluation in Sweden. Analyses of available data on reindeer numbers and grazing effects indicated that the fears of overgrazing were based on local damage around a few enclosures and fences. Some of the effects were due to trampling on lichen-dominated vegetation, while others involved vegetation dominated by vascular plants, but no evidence of large-scale overutilisation of reindeer ranges in the Swedish mountains could be found [42]. The present-day consensus is that overgrazing of Swedish reindeer ranges has been temporary and local, and that it rarely has caused permanent damage. Recent evidence from Finnmark's winter rangelands points to the same conclusion [46]. Drawing on a literature review, Linkowski & Lennartsson [47] concluded that even heavy grazing during a limited period can promote the diversity of alpine vegetation in the long run.

Moreover, the Swedish Parliament has adopted an environmental quality objective for the mountains. One of the specifications of this objective declares that it is essential to preserve 'a mountain landscape characterised by grazing' [48], referring to the conservation of key ecological functions in the landscape. However, no details have been given on how this specification is to be interpreted in ecological terms. For instance, one study suggests that grazing impacts on species richness are small, while effects on rare species and species composition (i.e. changes of relative species abundances) are stronger [49]. It is not clear how this translates into a 'landscape characterised by grazing'.

Rationale for a systematic review

The variation in the impacts of reindeer on vegetation between studies and regions demonstrates that it is challenging to predict the ecological consequences of various forms of management of both domesticated and wild reindeer populations. The lack of a comprehensive assessment of how vegetation is affected by reindeer suggests that there is a need to evaluate the ecological significance of reindeer grazing through a systematic review.

In Sweden especially, the recent re-evaluation of what reindeer grazing means for arctic and alpine vegetation is another reason why it is imperative to examine the scientific support for today's prevailing opinions on this issue. The need to interpret and clarify the environmental quality objective for the Swedish mountains also contributed to the decision to launch the present review.

In this review, we use a systematic approach to synthesise available evidence on the impacts of reindeer herbivory. Systematic reviews are designed to avoid bias and permit quantitative conclusions by means of meta-analysis. The ultimate aim of this review is to facilitate evidence-based management of reindeer grazing systems, with a particular focus on Fennoscandian conditions.

To the best of our knowledge, no systematic review of how reindeer grazing affects vegetation in treeless areas has been performed earlier. Our review was designed to include studies from any arctic or alpine region where reindeer are present, either as native or as introduced populations, provided that the data are informative for Fennoscandian conditions (e.g. by referring to vegetation types similar to those found in Fennoscandia). The review design was established in detail in an *a priori* protocol [50]. It follows the guidelines for systematic reviews issued by the Collaboration for Environmental Evidence [51].

Stakeholder involvement

This review was proposed by the Swedish Environmental Protection Agency. Prior to completion of the review protocol, a meeting was arranged with stakeholders with an interest in reindeer husbandry and environmental aspects of reindeer herbivory in Sweden [50]. Several suggestions made by the stakeholders were adopted by the review team, e.g. that the review should not be restricted to impacts on biodiversity but should consider other aspects of vegetation too, and that it should include vegetation in subalpine birch forests as well as treeless mountain areas. We have thus covered studies on treelines and on the forest-tundra ecotone, including subalpine birch forests but not coniferous forests at lower elevations. Moreover, it was pointed out that overgrazing of reindeer pastures is a questionable concept. Being perspective-driven, its definition tends to vary between stakeholders [52], and no attempt to define or apply the concept of overgrazing has been made in this review.

Before submission, peer review, revision and final publication of the protocol, a draft version was open for public review at the EviEM website in November 2012. Comments were received from about ten stakeholders, most of them Swedish scientists or environmental managers.

Objective of the review

The primary aim of this review is to clarify how grazing, browsing and trampling by reindeer (or caribou) affect the vegetation of arctic, subarctic, alpine and subalpine areas, including the forest-tundra ecotone. We would like to point out that an understanding of the reasons behind variations in reindeer grazing pressure on vegetation is outside the scope of this review. Such an understanding would, for instance, require analyses of the entire annual range of the grazing system, including the use of winter pastures in the boreal forest, variations in reindeer management, historical land use, external pressures from other land users, and political, legal, and societal drivers. This was not possible to achieve within the time, resource, and data constraints of our review.

Primary question

What are the impacts of reindeer/caribou (Rangifer tarandus L.) on arctic and alpine vegetation?

Components of the primary question:

- *Subject (population):* Vegetation (as a whole, or divided into major groups such as graminoids, forbs, dwarf-shrubs, lichens, mosses etc.) in alpine/subalpine areas or arctic/subarctic tundra, including the forest-tundra ecotone.
- *Exposure:* Herbivory (including grazing, browsing and trampling) by reindeer (or caribou). Reindeer density (number of reindeer per unit area) is used as a quantification of the intensity of herbivory.
- *Comparator:* Lower (or no) herbivory by reindeer (or caribou).
- *Outcome:* Change of vegetation. Relevant aspects of vegetation include cover (abundance), biomass, diversity (e.g. species richness), structure, composition (at both species and functional group levels) and productivity.

Methods
Searches for literature

Searches for relevant literature have been made using online publication databases, search engines, specialist websites and bibliographies of literature reviews. As far

as possible, the search strings specified below were applied throughout the searches using online databases, search engines and specialist websites. In several cases, though, they had to be simplified as some sites can handle only a very limited number of search terms or do not allow the use of 'wildcards' or Boolean operators.

Full details of the search strings used and the number of articles found at each stage of the search are provided in Additional file 1.

Search terms

A scoping exercise identified the following search terms as being most closely related to the primary question:

Exposure: herbivory, graz*, brows*, trampl*
Agent: reindeer, caribou, *Rangifer*

The terms within each category ('exposure' and 'agent') were combined using the Boolean operator 'OR'. The two categories were then combined using the Boolean operator 'AND'. An asterisk (*) indicates wildcard truncation.

Searches were also made for Swedish, Norwegian, Finnish and Russian counterparts of the above terms. The following search strings were used:

- *English:* (herbivory OR graz* OR brows* OR trampl*) AND (reindeer OR caribou OR *Rangifer*)
- *Swedish:* renbet* OR ((herbivori OR bet* OR tramp*) AND (renar OR caribou OR *Rangifer*))
- *Norwegian:* reinbeit* OR renbeit* OR ((beit* OR gressing OR tramp*) AND (*rein OR *ren OR reinsdyr OR rensdyr OR karibu OR caribou OR *Rangifer*))
- *Finnish:* (laidun* OR tallata OR talloa OR polkea) AND (poro OR *Rangifer*)
- *Russian:* (пастбище OR пастись OR выпасать OR выбирать OR высматривать OR вытаптывать) AND (олень OR оленеводство)

No time, language or document type restrictions were applied.

In addition to the exposure and agent terms mentioned above, the following terms for 'subject' had been tested during the scoping exercise:

vegetation, vascular, plant*, herb*, forb*, gramin*, lichen*, moss*, bryophyte*, flora, shrub*, tree*, forage, tundra, alpine, subalpine, arctic, subarctic, heath*, pasture*, rangeland*

However, it was found that searches using the exposure and agent terms alone were specific enough to return a fully manageable amount of articles. Including the above subject terms would have restricted the search and reduced the number of hits by a factor of about two. The subject terms were therefore excluded – the loss of specificity was judged to be less important than the increase of sensitivity.

Publication databases

The search included the following online publication databases:

1) Academic Search Premier
2) Agricola
3) Arctic & Antarctic Regions (EBSCOhost)
4) Arto (reference database of Finnish articles)
5) Biological Abstracts
6) BioOne
7) COPAC
8) Directory of Open-Access Journals
9) GEOBASE and GeoRef (Engineering Village)
10) IngentaConnect
11) JSTOR
12) Melinda (union catalogue of Finnish libraries)
13) Scopus
14) SpringerLink
15) SwePub (academic publications at Swedish universities)
16) Web of Science
17) Wiley Online Library

To identify relevant literature in bibliographic databases, systematic reviews normally use searches in titles, abstracts and keywords of the indexed publications [51]. To an increasing extent, however, such databases now also allow searches in the full text of available articles. For the purpose of checking whether full-text searching identifies relevant articles more efficiently and/or completely than conventional searching, we made both kinds of searches in three of the databases (Academic Search Premier, JSTOR and Scopus). The other fourteen databases were searched at title/abstract/keyword level only.

Search engines

An Internet search was also performed using the following search engines:

Google (www.google.com)
Google Scholar (scholar.google.com)
Dogpile (www.dogpile.com)
Scirus (www.scirus.com)

In each case, the first 100 hits (based on relevance) were examined for appropriate data. Potentially useful documents that had not already been found in publication databases were recorded.

Specialist websites

Websites of the specialist organisations listed below were searched for links or references to relevant publications and data, including grey literature. Potentially useful documents that had not already been found using publication databases or search engines were recorded.

Alaska Department of Natural Resources (dnr.alaska.gov)
Alberta Conservation Association (www.ab-conservation.com)
Alberta Reindeer Association (www.albertareindeer.com)
Arctic Centre (University of Lapland) (www.arcticcentre.org)
Arctic Council (www.arctic-council.org)
Bioforsk (www.bioforsk.no)
Bureau of Land Management, US Dept. of the Interior (www.blm.gov)
Conservation of Arctic Flora and Fauna (CAFF) (www.caff.is)
Environment Canada (www.ec.gc.ca)
European Commission Joint Research Centre (ec.europa.eu/dgs/jrc)
European Environment Agency (www.eea.europa.eu)
Finland's environmental administration (www.environment.fi)
Finnish Environment Institute (SYKE) (www.environment.fi)
Finnish Game and Fisheries Research Institute (www.rktl.fi)
Food and Agriculture Organization of the United Nations (www.fao.org)
Greenland Institute of Natural Resources (www.natur.gl)
GRID Arendal (www.grida.no)
International Centre for Reindeer Husbandry (icr.arcticportal.org)
International Union for Conservation of Nature (www.iucn.org)
Ministry of Natural Resources of the Russian Federation (www.mnr.gov.ru)
Natural Resources Canada (www.nrcan.gc.ca)
Nordic Council for Reindeer Husbandry Research (*Rangifer* journal) (site.uit.no/rangifer)
Nordic Council of Ministers (www.norden.org)
Northern Research Institute (NORUT) (www.norut.no)
Norwegian Directorate for Nature Management (www.dirnat.no)
Norwegian Institute for Nature Research (NINA) (www.nina.no)
Norwegian Polar Institute (www.npolar.no)
Norwegian Wild Reindeer Centre (www.villrein.no)
Reindeer Herders' Association (www.paliskunnat.fi)

Reindeer Research Program, University of Alaska (reindeer.salrm.uaf.edu)

Reindriftsforvaltningen (www.reindrift.no)

Reinportalen (www.reinportalen.no)

Russian Guild of Ecologists (www.ecoguild.ru)

Russian Regional Environmental Centre (www.rusrec.ru)

Sámediggi (Finnish Sami Parliament) (www.samediggi.fi)

Sámediggi (Norwegian Sami Parliament) (www.sametinget.no)

Sámi Reindeer Herders' Association of Finland (www.beboedu.fi)

Sápmi (Sami Parliament in Sweden) (www.eng.samer.se)

Swedish Environmental Protection Agency (www.naturvardsverket.se)

Swedish University of Agricultural Sciences (SLU) (www.slu.se)

United Nations Environment Programme (www.unep.org)

United States Environmental Protection Agency (www.epa.gov)

United States Fish and Wildlife Service (www.fws.gov)

University of Alaska Anchorage (www.uaa.alaska.edu)

Other literature searches

Relevant literature was also searched for in bibliographies of literature reviews by Forbes & Kumpula [31], Moen & Danell [42], Linkowski & Lennartsson [47] and Suominen & Olofsson [53]. Potentially useful documents that had not already been found in online sources were recorded. A few more articles were brought to our attention by stakeholders.

Search update

An update to the literature searches was made one year after the main searches. The update involved searches for articles in English using a subset of the publication databases and search engines listed above (see Additional file 1).

Screening

Screening process

Articles found by searches in publication databases were evaluated for inclusion at two or three successive levels. The literature identified by full-text searches in three databases was first assessed by title by a single reviewer (CB). In cases of uncertainty, the reviewer chose inclusion rather than exclusion.

The articles found to be relevant based on title were then combined with those identified by title/abstract/keyword searches in the fourteen other publication databases. After removal of duplicates, these articles were assessed by abstract, again by a single reviewer (CB) who in cases of uncertainty tended towards inclusion.

A second reviewer (JS) assessed a subset consisting of 20% of the abstracts, and the agreement between the two reviewers' assessments was checked with a kappa test. The outcome, $\kappa = 0.565$, indicated a 'moderate' agreement [54], but since the inconsistency had almost entirely been caused by the main reviewer being more inclusive than the second one, it seemed safe to proceed with the screening without modification or further specification of the inclusion/exclusion criteria.

Next, each article found to be relevant on the basis of abstract was judged for inclusion by a reviewer studying the full text. This task was shared by all members of the review team. The articles were randomly distributed within the team, but some redistribution was then made to avoid having reviewers assess studies authored by themselves or written in an unfamiliar language. Studies found by other means than database searches were also entered at this stage in the screening process. Doubtful cases – articles that the reviewer could not include or exclude with certainty even after having read the full text – were discussed and decided on by the entire team.

A list of all studies rejected on the basis of full-text assessment is provided in Additional file 2 together with the reasons for exclusion. This file also contains a list of articles that we failed to find in full text.

Study inclusion criteria

Each study had to pass each of the following criteria in order to be included at any of the screening stages:

- *Relevant subject(s):* Vegetation in alpine/subalpine areas or arctic/subarctic tundra, including the forest-tundra ecotone. Reindeer may also occur in boreal coniferous forests, but studies of vegetation in such regions were not included; nor studies of reindeer herbivory on meadows formerly used for cattle or sheep grazing.
- *Relevant types of exposure:* Grazing, browsing or trampling by reindeer. Modern reindeer husbandry may also affect vegetation through disturbances caused by reindeer herders' all-terrain vehicles, but such impacts are not considered by this review.
- *Relevant types of comparator:* Lower or no grazing, browsing or trampling.
- *Relevant types of outcome:* Change in cover, abundance, biomass, diversity (including species richness), structure, composition or productivity of vegetation. Studies of single plant species and of the soil seed bank were also included.
- *Relevant types of study:* Any primary field study comparing vegetation in areas and/or time periods with different degrees of reindeer herbivory.

Remote-sensing studies have been included, but not simulation-modelling studies or field studies of simulated herbivory, since these do not represent direct impacts of reindeer.

The review protocol indicated that manipulative studies as well as purely observational ones were to be considered. However, while we have included experiments where fences were used to keep reindeer out from certain areas, we have chosen to exclude studies involving artificial removal or transplantation of vegetation, again in order to focus on effects of reindeer herbivory under natural conditions. We have also excluded studies where differences in grazing pressure have been inferred from vegetation properties, since such conclusions would introduce circular reasoning if used in this review.

At screening on full text, the following inclusion criterion was also applied:

- *Language:* Full text written in English, Swedish, Norwegian, Danish, Finnish, German or Russian.

Potential effect modifiers and reasons for heterogeneity
The following potential effect modifiers were considered and recorded:

Latitude and longitude
Elevation
Annual mean temperature
Annual mean precipitation
Soil moisture (dry/mesic/wet)
Soil/bedrock type
Vegetation type
Reindeer subspecies involved
Seasonality of reindeer grazing
Domestication status of the reindeer
Presence and species identity of other herbivores
Control for small herbivores (using small-mesh exclosures)
History of herd (e.g. whether native or introduced)
Study design and experimental treatment
Study and intervention timescale and seasonality

Study quality assessment
Articles that remained included after full-text screening were subject to critical appraisal as described below. This appraisal was made by the four ecologists in the review team (KAB, BF, JM, JS) and double-checked by the fifth member of the team (CB). Uncertain cases were discussed and decided on by the entire team.

Before critical appraisal, the articles had been redistributed among the reviewers based on where the studies had been carried out. All studies from a specific region

were assessed by the same reviewer, which made it easier to detect any redundancies between them.

Articles sorted under the categories listed below were considered to have high susceptibility to bias and were therefore excluded from the review.

- *Methodology inadequately described.*
- *Inappropriate comparator* (comparison between different seasons, use of small-mesh exclosures that prevented grazing by small mammals as well as reindeer, or comparator difficult to interpret for the purposes of this review). Studies that compare grazing in different seasons have usually been made in areas where summer and winter ranges are separated by a fence. Since the effects of grazing on bare and snow-covered ground are entirely different, such a study design makes it difficult to judge which of the two ranges that is more heavily grazed, even if reindeer densities are known in both of them.
- *No replication at lowest level* (no replication of exclosures or site comparators). Studies based on comparison of so-called reindeer-herding districts or regions have not been excluded, however, even though study units representing such a district may be seen as pseudoreplicates in a strict statistical sense. Since the study units represent different geographical contexts within the district and thus also differ in terms of biology and reindeer impact, the interpretation of them as pseudoreplicates is not justified in an ecological sense.
- *No data on variability.*
- *Vegetation data difficult to interpret* (such as when methods used to assess vegetation have differed between sites or sampling seasons).
- *Reindeer data difficult to interpret* (such as when conclusions on the presence of reindeer have been based on weak and circumstantial evidence).

Since checks for redundancy were made during critical appraisal, this was added as a cause for exclusion, although it is not strictly a quality criterion:

- *Primary data redundant* (data also published elsewhere).

In accordance with the review protocol, notes were also made on certain other quality aspects (such as whether plot locations were randomised and well-matched or not), but since these aspects were considered less important than those listed above, they were not chosen as exclusion criteria. The duration of exposure differences (i.e. how long differences of grazing pressure had been maintained) was recorded too, but it was handled as a potential effect modifier rather than a measure of study quality.

According to the review protocol, studies were to be categorised as having high, medium, or low susceptibility to bias based on the critical appraisal. However, since we adopted fairly strict exclusion criteria, there seemed to be no need for a further quality grading of studies that fulfilled quality standards well enough to remain included.

A list of the studies that were rejected due to high susceptibility to bias is provided in Additional file 2 together with the reasons for exclusion.

Data extraction strategy

Outcome means and measures of variation (standard deviation, standard error, confidence intervals) have been extracted from tables and graphs in the included articles, using image analysis software when necessary (software used included Graphclick for Mac, http://www.arizona-software.ch/graphclick/, and WebPlotDigitizer, http://arohatgi.info/WebPlotDigitizer/). Data were extracted not only on major functional or taxonomic groups of vegetation and on vegetation as a whole, but also on individual species and genera whenever such information was available.

Most studies in this field compare vegetation in areas that for a long time have been subject to different levels of reindeer herbivory, or vegetation inside and outside areas that have been fenced to exclude reindeer. Thus, they are usually 'CI' (Comparator/Intervention) studies describing effects of reindeer herbivory relative to a control site. Other studies have a 'BA' (Before/After) design – they present data on vegetation before and after reindeer exclusion or over a period when herbivory has changed. A few studies combine these two approaches in 'BACI' (Before/After/Comparator/Intervention) designs, where site control and intervention comparisons are made both before and after herbivory has changed.

Where time-series of data were available, we only extracted the most recent results (plus pre-intervention data from BA and BACI studies). Similarly, we only extracted data from sites under high and low grazing pressure even if data from sites under intermediate pressure were also available.

In a few cases, study authors have been asked to supply vegetation data in digital format. This was done where useful data had been published in graphs from which they were difficult to extract accurately enough, or when it was known or assumed that considerable amounts of relevant but unpublished data could be available in addition to the published results. In cases where raw data were received, summary statistics have been calculated by us. This has e.g. enabled us to consider data on the diversity of vegetation even from studies where the published outcomes only include biomass or abundance.

Data on potential effect modifiers and other metadata were extracted from the included articles whenever available, but climatic data were downloaded from the WorldClim database [55].

Data on reindeer densities are often incomplete or entirely absent in studies of the effects of reindeer grazing. Many of these studies simply describe sampling sites as being subject to 'grazing' or 'no (or lighter) grazing'. Some authors have actually estimated local reindeer densities, using counts of animals, trampling indicators or counts of reindeer droppings, but the two latter types of data cannot readily be transformed to reindeer numbers per unit area.

In the absence of reliable local information, therefore, we have used data on average reindeer densities at a regional level (e.g. mean densities in entire reindeer-herding districts). Where such data were not provided by study authors, we have calculated them ourselves for the appropriate time periods using herd sizes and range sizes retrieved from external sources, such as Reindriftsforvaltningen (the Norwegian reindeer herding administration) and Svensk-norska renbeteskommissionen (the Swedish-Norwegian commission on reindeer pastures). We are well aware that these data must be used and interpreted with care, since herding districts may cover thousands of square kilometres and a study site may be far from representative of average conditions in the district where it is located.

The extraction of data was carried out by the four ecologists in the review team and then double-checked by the fifth team member. Each pair of BA or CI outcomes (and each quadruple of BACI outcomes) was recorded in a separate row of an Excel spreadsheet together with data on reindeer densities and all appropriate metadata, including data on effect modifiers.

Data synthesis and presentation

The impacts of reindeer exposure on vegetation were analysed using meta-analytical approaches. Meta-analyses were carried out using the metafor package [56] within the R environment v. 3.0.2 [57]. Standardised mean difference (SMD) effect sizes were derived for all responses using Hedges' g statistic. The effect sizes were calculated as the difference between the mean response at high exposure to reindeer and the mean response at low exposure to reindeer divided by the pooled standard deviation. Positive effect sizes thus indicate that the response parameter was higher at high reindeer exposure than at low exposure.

We calculated summary effect sizes by using random effects models. Models were developed for the main groups of vegetation (lichens, bryophytes, forbs/herbs, graminoids and woody plants) and the most common aspects of vegetation assessed (cover, abundance, biomass, height and productivity) as well as for bare-ground cover and species richness of vascular and non-vascular plants. Where applicable, subgroup models were developed for further vegetation groups within each of the main groups (e.g. for deciduous and ever-green shrubs). Heterogeneity was estimated by the Hedges' method, and data are presented in forest plots showing mean effect sizes and 95% confidence intervals. Models were weighted by the inverse of the variance.

We used univariate mixed effects models in order to test whether the impact of reindeer exposure on vegetation varied with reindeer density, vegetation cover or mean annual temperature. The independent variables were fitted as modifiers in the meta-analytical models. Four separate measures of reindeer density were used: (1) the high-exposure reindeer density, (2) the absolute difference in reindeer density between high and low exposures, (3) the relative difference in reindeer density (100 × [high density − low density] / high density) and (4) the accumulated exposure difference estimated as the absolute difference in reindeer density multiplied by the duration of the exposure difference. We investigated whether the impact of reindeer exposure varied with the cover of lichens and bare ground only, as there was not enough data to do this for other vegetation groups. Lichen and bare-ground cover effect sizes were fitted against the average cover of lichens and bare ground (meaning that the standardised mean difference in lichen cover was used as the dependent variable and the average lichen cover across both exposures as the independent variable).

Results
Review descriptive statistics
Searches, screening and quality assessment
The main searches for literature were conducted between 19 October and 8 December 2012, and an update was made on 2 November 2013.

Full-text searches with English search terms in three publication databases (Academic Search Premier, JSTOR and Scopus) returned a total of 8,039 articles (6,638 after removal of duplicates). After title screening of these articles, 618 of them remained. Searches based on title/abstract/keywords in fourteen other databases returned a total of 1,323 articles (772 after removal of duplicates). Removal of the overlap between the outcomes of the two different search approaches left a total of 1,197 unique publications. After screening based on the abstracts of these articles, 376 of them remained. About two thirds of the exclusions were due to absence of relevant vegetation data.

Searches using search engines returned 9 potentially relevant articles (8 found with English search terms, 1 with Norwegian ones) in addition to those that already

had been identified. Similarly, searches on specialist websites located another 9 potentially useful publications (7 were found using English search terms and 2 using Norwegian ones). An additional 15 potentially relevant articles were found in bibliographies of literature reviews, while 6 more were added following contacts with stakeholders.

This resulted in a total of 415 articles to be screened on a full-text basis. After screening, 96 of them were still included. At this stage, the most common reasons for exclusion were that studies dealt with other aspects of reindeer than their herbivory, that no relevant vegetation data were reported, or that no primary observational data were presented at all. In 28 cases, publications had to be excluded because they were not found in full text.

When the search was updated in 2013, 6 new potentially relevant articles were found, 4 of which were included after screening on full text.

Finally, quality assessment was made of the 100 articles that had passed the screening process, and 60 of them were then excluded. Common reasons for exclusion were inadequate methodological descriptions, and vegetation or reindeer data that were difficult to interpret. In 12 cases, articles were excluded since they reported data that could also be found elsewhere.

The 40 finally included articles are listed in Table 1. Since one of them (van der Wal & Brooker [20]) reports on three different studies, two of which were found to be useful in this review, the total number of included studies is 41. See Figure 2 and Additional file 3 for further details on search results and outcomes of the screening process and quality assessment.

Sources of included articles
Nearly all of the 40 articles included in this review were found in publication databases. Of the 39 included publications that had been identified during the main searches in late 2012, 37 were returned by at least one of the databases searched (see Additional file 4).

The three publication databases where full-text searches had been made were also searched based on title/abstract/keywords. Of the 37 included articles found in 2012, 34 were retrieved by at least one of the full-text searches, whereas 33 were found by at least one of the searches on title/abstract/keywords in the same three databases. The single included study caught by the former searches but not by the latter ones was a remote-sensing analysis of vegetation in reindeer-herding districts [58] where 'grazing' was mentioned in the full text but not in the abstract. In general, the articles found by full-text searches but not by title/abstract/keywords searches had little or no relevance to the topic of this review. Since the full-text searches required about a week of extra work, mainly spent screening more than 6,000 titles and

Table 1 Articles included in the systematic review

Authors	Year	Title	Study area	Ref.
Bråthen & Oksanen	2001	Reindeer reduce biomass of preferred plant species	N Norway	[73]
Bråthen et al.	2007	Induced shift in ecosystem productivity? Extensive scale effects of abundant large herbivores	N Norway	[23]
Cahoon et al.	2012	Large herbivores limit CO_2 uptake and suppress carbon cycle responses to warming in West Greenland	Greenland	[68]
den Herder et al.	2004	Effects of reindeer browsing on tundra willow and its associated insect herbivores	Finland	[74]
Dormann & Skarpe	2002	Flowering, growth and defence in the two sexes: Consequences of herbivore exclusion for Salix polaris	Svalbard	[75]
Eriksson et al.	2007	Use and abuse of reindeer range	Sweden	[62]
Eskelinen & Oksanen	2006	Changes in the abundance, composition and species richness of mountain vegetation in relation to summer grazing by reindeer	Finland	[76]
Gaare et al.	2006	Overvåking av vinterbeiter i Vest-Finnmark og Karasjok: Ny beskrivelse av fastrutene	N Norway	[59]
Gonzalez et al.	2010	Large-scale grazing history effects on Arctic-alpine germinable seed banks	N Norway	[69]
Gough et al.	2008	Long-term mammalian herbivory and nutrient addition alter lichen community structure in Alaskan dry heath tundra	Alaska, USA	[77]
Grellmann	2002	Plant responses to fertilization and exclusion of grazers on an arctic tundra heath	N Norway	[78]
Hansen et al.	2007	Ungulate impact on vegetation in a two-level trophic system	Svalbard	[61]
Jandt et al.	2003	Western Arctic Caribou Herd winter habitat monitoring and utilization, 1995-1996	Alaska, USA	[60]
Johansen & Karlsen	2005	Monitoring vegetation changes on Finnmarksvidda, Northern Norway, using Landsat MSS and Landsat TM / ETM+ satellite images	N Norway	[58]
Kitti et al.	2009	Long- and short-term effects of reindeer grazing on tundra wetland vegetation	N Norway/ Finland	[66]
Lehtonen & Heikkinen	1995	On the recovery of mountain birch after Epirrita damage in Finnish Lapland, with a particular emphasis on reindeer grazing	Finland	[79]
Manseau et al.	1996	Effects of summer grazing by caribou on composition and productivity of vegetation: Community and landscape level	Canada	[80]
Moen et al.	2009	Variations in mountain vegetation use by reindeer (Rangifer tarandus) affects dry heath but not grass heath	Sweden	[63]
Nellemann et al.	2000	Cumulative impacts of tourist resorts on wild reindeer (Rangifer tarandus tarandus) during winter	S Norway	[81]
Nellemann et al.	2001	Winter distribution of wild reindeer in relation to power lines, roads and resorts	S Norway	[82]
Olofsson & Strengbom	2000	Response of galling invertebrates on Salix lanata to reindeer herbivory	N Norway	[83]
Olofsson et al.	2001	Effects of summer grazing by reindeer on composition of vegetation, productivity and nitrogen cycling	N Norway/ Finland	[21]
Olofsson et al.	2004	Importance of large and small mammalian herbivores for the plant community structure in the forest tundra ecotone	N Norway/ Sweden	[84]
Olofsson & Oksanen	2005	Effects of reindeer density on vascular plant diversity on North Scandinavian mountains	N Norway/ Finland/Sweden	[49]
Olofsson	2006	Short- and long-term effects of changes in reindeer grazing pressure on tundra heath vegetation	N Norway	[67]
Olofsson et al.	2009	Herbivores inhibit climate-driven shrub expansion on the tundra	N Norway/ Sweden	[26]
Olofsson et al.	2013	Complex biotic interactions drive longterm vegetation dynamics in a subarctic ecosystem	Sweden	[85]
Pajunen	2009	Environmental and biotic determinants of growth and height of arctic willow shrubs along a latitudinal gradient	Yamal, Russia	[86]
Pedersen & Post	2008	Interactions between herbivory and warming in aboveground biomass production of arctic vegetation	Greenland	[87]
Post & Pedersen	2008	Opposing plant community responses to warming with and without herbivores	Greenland	[70]
Post	2013	Erosion of community diversity and stability by herbivore removal under warming	Greenland	[88]
Ravolainen et al.	2010	Additive partitioning of diversity reveals no scale-dependent impacts of large ungulates on the structure of tundra plant communities	N Norway	[24]

Table 1 Articles included in the systematic review *(Continued)*

Ravolainen *et al.*	2011	Rapid, landscape scale responses in riparian tundra vegetation to exclusion of small and large mammalian herbivores	N Norway	[64]
Tømmervik *et al.*	2009	Above ground biomass changes in the mountain birch forests and mountain heaths of Finnmarksvidda, northern Norway, in the period 1957-2006	N Norway	[89]
Tømmervik *et al.*	2012	Rapid recovery of recently overexploited winter grazing pastures for reindeer in northern Norway	N Norway	[46]
van der Wal *et al.*	2001	Differential effects of reindeer on high Arctic lichens	Svalbard	[90]
van der Wal & Brooker	2004	Mosses mediate grazer impacts on grass abundance in arctic ecosystems	Svalbard	[20]
Vistnes *et al.*	2004	Effects of infrastructure on migration and range use of wild reindeer	S Norway	[91]
Zamin & Grogan	2012	Birch shrub growth in the low Arctic: The relative importance of experimental warming, enhanced nutrient availability, snow depth and caribou exclusion	Canada	[92]
Zamin & Grogan	2013	Caribou exclusion during a population low increases deciduous and evergreen shrub species biomass and nitrogen pools in low Arctic tundra	Canada	[65]

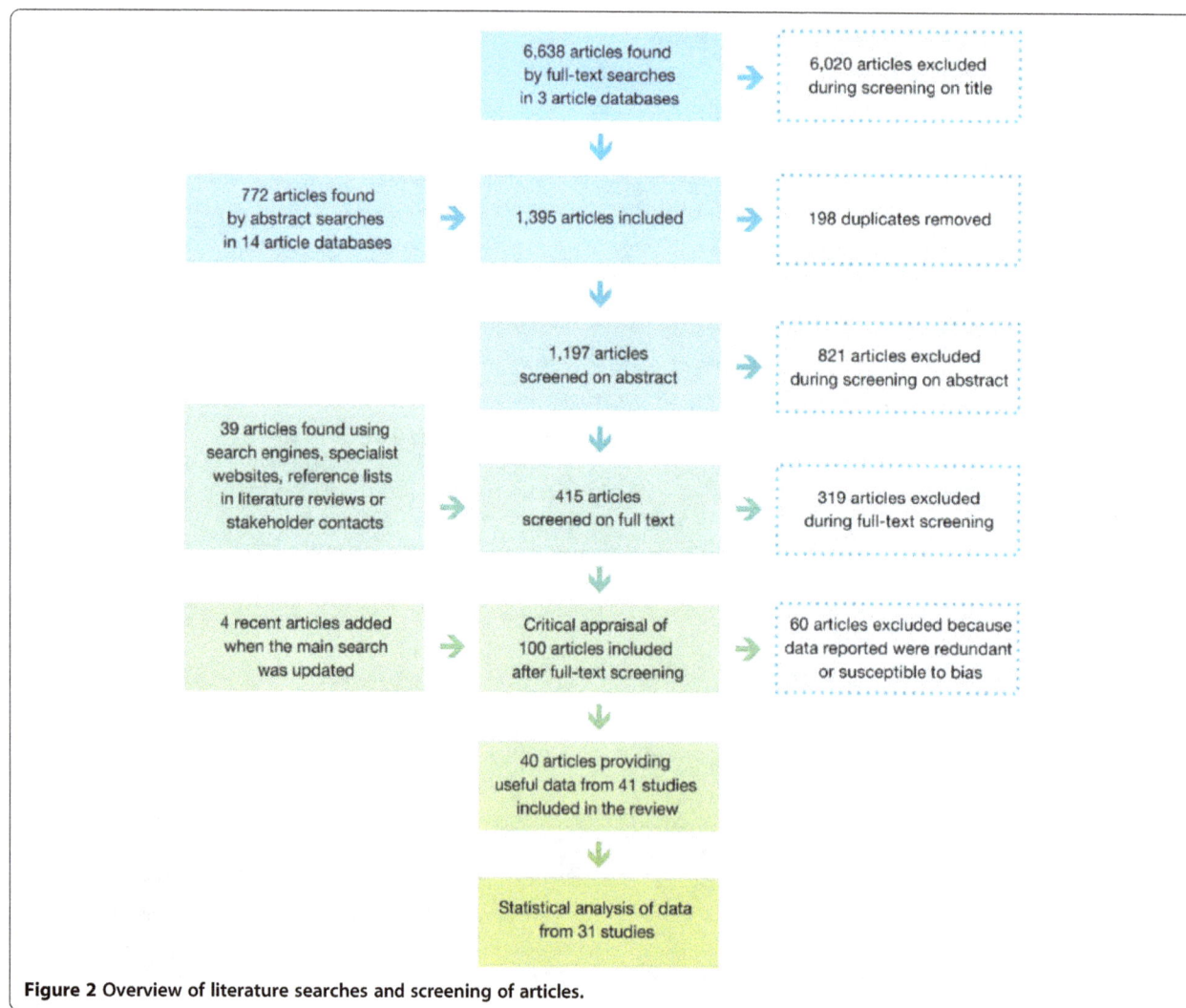

Figure 2 Overview of literature searches and screening of articles.

some 400 abstracts, we conclude in retrospect that this effort did not pay off well enough to be justified.

The two included publications that had not been retrieved from any of the databases, Gaare *et al.* [59] and Jandt *et al.* [60], were found at specialist websites belonging to the Norwegian Institute for Nature Research and the Bureau of Land Management of the US Dept. of the Interior, respectively. The report by Gaare *et al.* [59] is the only included one that is written in a non-English language (Norwegian), and even this one was found using English search terms.

The searches that used non-English search strings returned very few potentially relevant publications that had not already been identified by other means. A total of about 30 articles in Swedish, Danish, Norwegian or Finnish were considered during the initial stages of this review, but a majority of them had been found in review bibliographies or through searches with English search terms, and none of them made it through both full-text screening and critical appraisal, with the one exception mentioned above.

While the articles screened on full text included several publications dating from the 1980s or earlier, a large share of the older articles were excluded during this stage of screening or during critical appraisal (see Figure 3). All but two of the 40 articles finally included were published in 2000 or later.

Overall characteristics of included studies

Although the searching and screening processes involved no geographical limitations, 25 of the 41 included studies were conducted in Fennoscandia (see Figures 4, 5 and 6). The other ones were carried out in Svalbard, Greenland, Canada, Alaska or Russia. Most of the studies (31 of them) were conducted in treeless terrain such as tundra or alpine areas, but 2 studies were carried out in subalpine birch forests, and 8 studies reported data from both treeless areas and birch forests.

As a consequence of the uneven geographic distribution of the studies, the majority (26 of them) dealt with herbivory by native Eurasian reindeer (*Rangifer tarandus tarandus*), either semi-domesticated or wild. The North American studies were all concerned with wild caribou (*R. t. caribou* or *R. t. groenlandicus*). The reindeer on Svalbard (*R. t. platyrhynchus*) are wild and native to the archipelago, but one of the studies conducted there (Hansen *et al.* [61]) was made on the Brøggerhalvøya peninsula, where Svalbard reindeer were reintroduced in 1978 after a century of absence.

About half of the studies (21 of them) were made in areas where reindeer herbivory mainly took place during summer, whereas winter grazing was the subject of 5 studies. In 13 cases, reindeer were present during several seasons or throughout the whole year. Note that the winter grazing considered in this review was confined to treeless areas and birch forests, as studies in coniferous forests were outside our scope.

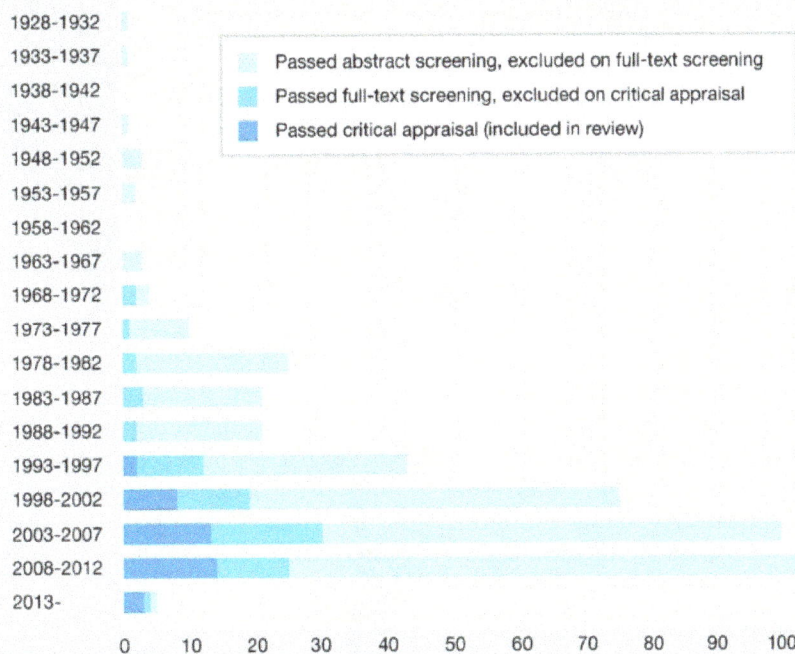

Figure 3 Year of publication of articles that passed abstract screening.

Figure 4 Sites where studies included in this review were carried out.

Of the 41 included studies, 30 had a CI design, based on sampling inside and outside reindeer exclosures, in districts with different reindeer densities, or at various distances from a border fence. BA design had been applied in 6 studies, 2 of which were based on remote sensing. The remaining 5 studies had a BACI design based on sampling inside and outside exclosures. See Figure 6 and Additional file 5 for further details on the characteristics of included studies.

Quantitative data have been extracted from 35 of the 41 included studies. These data consist of a total of 2,143 pairs of BA or CI outcomes (or quadruples of BACI outcomes). Almost three quarters of these outcomes (1,595 of them) originate from four of the studies [23,24,62,63] and have been supplied as raw data.

Most of the extracted outcomes are comparisons of the cover (762 cases), abundance (668 cases), biomass (271 cases), Shannon or Simpson diversity (242 cases), or species richness (137 cases) of a group or species of vegetation that had been exposed to different levels of reindeer herbivory. See Table 2 for an overview of the most frequently covered groups and species.

Figure 5 Fennoscandian sites, districts or ranges where studies included in this review were carried out. Numbers refer to Norwegian reindeer-herding districts (see Table B in Additional file 5).

Nearly 80% of the extracted data refer to single species rather than functional or taxonomic groups of vegetation, but an individual species was rarely covered by more than 1–3 of the included studies. For that reason, we eventually decided not to analyse single-species data in this review. Instead, we have focused on the 455 comparisons of vegetational groups that were available.

Narrative synthesis

We begin with a narrative synthesis in order to provide context and background for the quantitative meta-analyses that follow. An overview of the included studies can be found in Table A in Additional file 5, with Table B in the same file providing data on the sites or regions where the studies were carried out. The tables are subdivided based on the geographic distribution of the studies.

Study locations

Norway — 15
Finland — 4
Sweden — 3
More than one Fennoscandian country — 4
Svalbard — 5
Greenland — 4
Canada — 3
USA (Alaska) — 2
Russia — 1

Reindeer populations studied

Native semi-domesticated Eurasian reindeer — 24
Native wild Eurasian reindeer — 3
Wild Svalbard reindeer — 5
Native wild caribou — 9

Seasonality of herbivory

Summer — 21
Spring/autumn — 2
Winter — 5
Year round — 6
Different seasons — 7

Study design

BACI (exclosures) — 5
CI (exclosures) — 15
CI (studies along fences) — 3
CI (regional comparisons) — 12
BA (field studies) — 4
BA (remote sensing) — 2

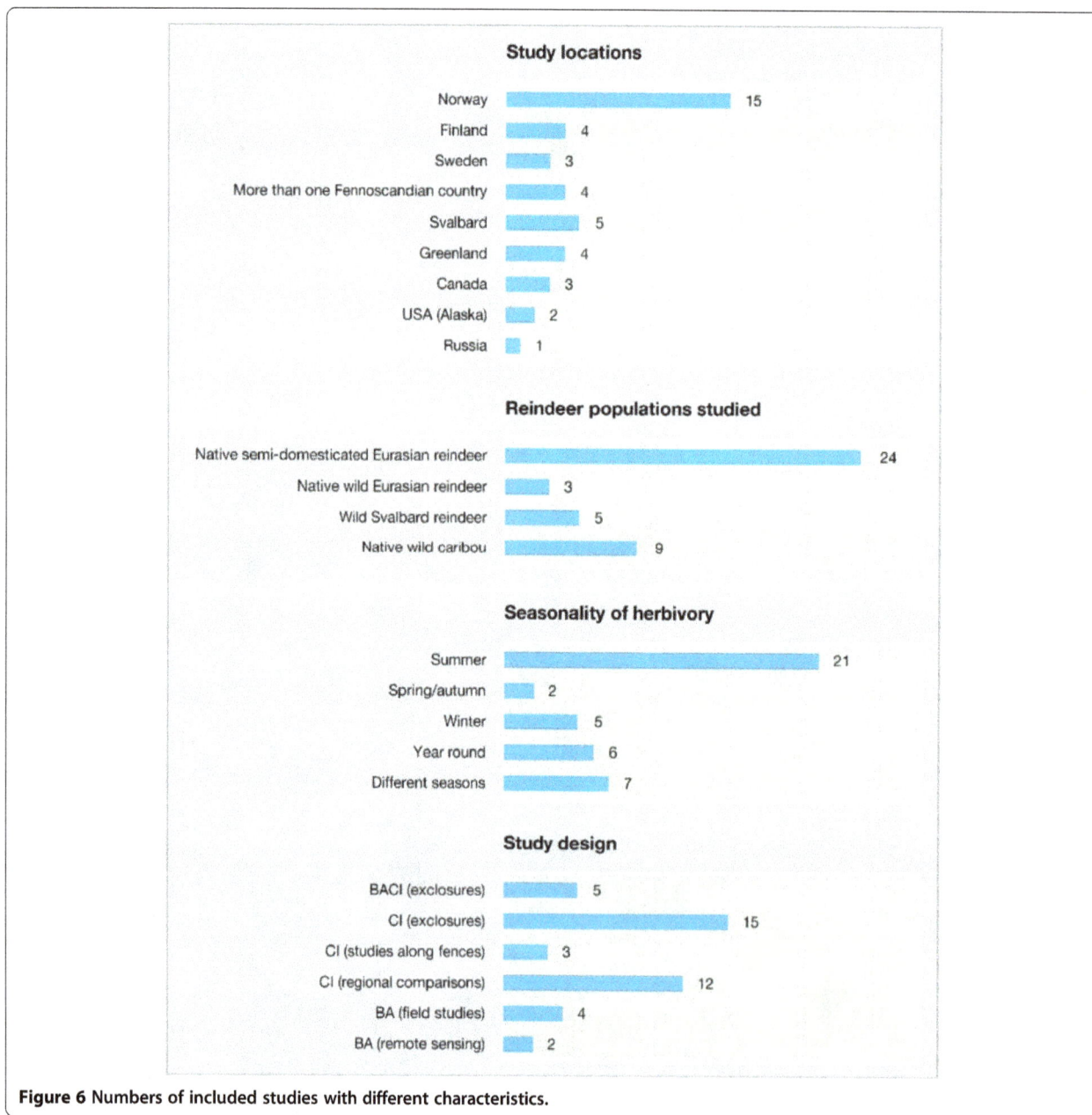

Figure 6 Numbers of included studies with different characteristics.

One of the columns in Table A summarises the effects of reindeer herbivory as reported by the respective authors of the studies. The results often show divergent responses of the vegetation. For instance, Olofsson *et al.* [21] found increased graminoid cover with increased herbivory at two sites but a non-significant response at two other sites. Similarly, Ravolainen *et al.* [64] found a negative response of forb biomass at one site and a non-significant response at another site. Responses also varied between studies. For instance, Zamin & Grogan [65] showed positive effects on species richness of vascular plants, while Olofsson & Oksanen [49], Kitti *et al.* [66] and Olofsson [67] all showed non-significant effects.

In Finnmark, northern Norway, an interesting remote-sensing study was done on the recovery of lichen-dominated vegetation after a decline in reindeer densities [46]. In the 1980s, the reindeer populations had more than doubled in the area [58]. This caused a strong decline in lichen cover over a large area (>15,000 km^2); average lichen cover in five herding districts in the area changed from 25% in 1973 to 1.6% in 2000 as estimated through remote sensing [58]. Beginning around 1990, the reindeer populations were reduced [42], and Tømmervik *et al.* [46] showed that the recovery of the lichen cover was very rapid. Cover increased 8.6-fold from 1998 to 2005, and the increase

Table 2 Numbers of extracted comparisons

	Abundance	Biomass	Cover	Species richness	Shannon diversity	Simpson diversity
Total vegetation		12 (4)		18 (4)	15 (2)	17 (3)
Bare ground			34 (6)			
Groups of vegetation						
Vascular plants		5 (2)	14 (2)	21 (5)	19 (3)	15 (2)
Shrubs, deciduous	13 (3)	5 (2)				
Herbs/forbs	14 (4)	8 (4)	20 (5)			
Graminoids	2 (2)	4 (3)	11 (7)			
Grasses	13 (3)	5 (2)				
Sedges		5 (2)				
Cryptogams, non-vascular				15 (2)	15 (2)	15 (2)
Lichens	3 (2)	7 (5)	50 (14)	5 (2)		
Bryophytes	3 (2)		21 (6)	5 (2)		
Mosses			26 (6)			
Liverworts			20 (2)			
Single species						
Agrostis capillaris	11 (2)					
Alectoria ochroleuca		4 (2)	3 (2)			
Betula nana	15 (5)	12 (4)	21 (5)			
Bistorta vivipara	12 (2)		3 (2)			
Calluna vulgaris	7 (2)					
Carex bigelowii	11 (2)		7 (2)			
Carex spp.	13 (3)	5 (2)	8 (3)			
Cetraria islandica			17 (4)			
Cetraria nivalis		4 (2)	12 (3)			
Cladina alpestris			3 (2)			
Cladina mitis			3 (2)			
Cladina rangiferina			17 (4)			
Deschampsia flexuosa	13 (3)	10 (2)	14 (2)			
Dicranum spp.	3 (2)					
Empetrum	15 (5)	10 (2)	16 (3)			
Festuca ovina	13 (3)		9 (2)			
Hylocomium splendens			9 (2)			
Juncus trifidus	13 (3)		9 (2)			
Lycopodium annotinum	5 (2)					
Nardus stricta	12 (3)					
Phegopteris connectilis	5 (2)					
Phyllodoce caerulea	14 (4)					
Polytrichum spp.	3 (2)					
Rubus chamaemorus	11 (2)	4 (2)	2 (2)			
Salix herbacea	12 (3)	4 (2)	4 (2)			
Salix spp.			7 (3)			
Solidago virgaurea	12 (3)					
Sphagnum spp.			11 (2)			

Table 2 Numbers of extracted comparisons *(Continued)*

Trientalis europaea	12 (3)		
Vaccinium myrtillus	14 (4)	10 (2)	
Vaccinium uliginosum		12 (3)	15 (4)
Vaccinium vitis-idaea	15 (5)	10 (2)	16 (3)
Viola biflora			3 (2)

Figures in brackets indicate the number of studies from which data have been extracted. Data are shown for groups (or species) and aspects of vegetation covered by at least 2 studies.

was faster on leeward ridges than on exposed ones. The increase rate was inversely related to changes in reindeer densities, and positively related to mean summer precipitation. The authors conclude that the rapid transition from barren ground to a flourishing lichen-dominated vegetation suggests that vegetation degradation by grazing and trampling is reversible [46].

Quantitative synthesis/Meta-analysis

We have performed quantitative syntheses (meta-analyses) of data extracted from 31 of the 41 included studies. Some of the 10 studies that appear in the narrative synthesis (Additional file 5) but not in the quantitative one have only reported on responses of single species, or on species aggregations that were not, or were poorly, replicated in other studies, such as the data on leaf area index in Cahoon *et al.* [68] or seed bank density in Gonzales *et al.* [69]. There were also a few studies that could not be used in meta-analysis since lack of information on outcome deviances or sample sizes made it impossible to calculate effect sizes.

Like the narrative synthesis, the meta-analyses unraveled a great divergence among responses to reindeer exposure (data on a total of eight vegetation categories, such as lichens, graminoids, etc., are presented in Figures 7 and 8 and Additional files 6 and 7). Both significantly positive and significantly negative average responses could be found in all vegetation categories that were considered. However, most studies had large confidence intervals that included zero effect size. Overall responses (average standardised mean differences of all cover, abundance and biomass data combined) were small. Despite the divergence between studies, the overall response to reindeer exposure was significantly negative for herbs/forbs and lichens. Vascular species richness also responded negatively to an increased grazing pressure.

Funnel plots were created to visually check for systematic heterogeneity and publication bias in the data set. No publication bias was detected.

Effects on growth forms and bare ground

For herbs/forbs, the overall response to reindeer exposure, including all vegetation categories and aspects

(cover, biomass and abundance), was significantly negative with a standardised mean difference (SMD) of −0.28 (CI: −0.48, −0.09). However, responses of individual plant aspects were not significantly different from zero. Some studies show a high variation of effects. For instance, the study by Bråthen *et al.* [23] showed both significant decreases and increases of forb abundances in different pairwise comparisons between herding districts (mean effect size ranging from −1.01 to 0.43).

For graminoids, the overall response was close to zero, although grass abundance showed a significant negative overall effect with a standardised mean difference of −0.25 (CI: −0.46, −0.03). For this group too, responses varied within and between studies, vegetation categories and aspects. For instance, Bråthen *et al.* [23] and Tømmervik *et al.* [46] found significant negative effects in some districts (mean effect size ranging from −1.73 to −0.51), while Olofsson *et al.* [21], Jandt *et al.* [60] and Post & Pedersen [70] found significant positive effects elsewhere (mean effect size ranging from 1.54 to 3.85).

Woody plants showed a non-significant overall response to reindeer exposure. Again, however, responses varied within and between studies, with some individual comparisons showing significant negative responses and others significant positive responses.

Lichens showed a significant negative overall response (SMD: −1.14, CI: −2.03, −0.25)). However, even for this group, which is well known to respond negatively to grazing and trampling, abundance showed a non-significant but positive response to reindeer exposure (SMD: 0.34, CI: −0.45, 1.12). Bryophytes showed no overall response to reindeer herbivory, while the cover of bare ground showed an overall (non-significant) tendency to increase (SMD: 0.27, CI: −0.06, 0.59).

Effects in relation to reindeer densities

We used weighted meta-regressions of effect sizes against four different ways of measuring intervention strength (grazing pressure) to see if they explained some of the divergence in the results. As measures of grazing pressure, we used the reindeer density in the high-exposure treatment, the absolute and relative differences between high and low densities, and the product of absolute density difference and duration of exposure

Summary of vegetation responses to reindeer herbivory

Mean and 95% C.I.

Herbs/forbs, cover	−0.10 [−0.41 , 0.22]
Herbs/forbs, biomass	−0.29 [−0.73 , 0.16]
Herbs/forbs, abundance	−0.22 [−0.50 , 0.07]
Herbs/forbs, overall random effects model	**−0.28 [−0.48 , −0.09]**
Sedges/rushes, abundance	0.05 [−0.17 , 0.27]
Sedges, cover	0.14 [−0.42 , 0.70]
Sedges, biomass	0.04 [−0.40 , 0.48]
Sedges, abundance	0.00 [−0.93 , 0.94]
Grasses, silica-rich, biomass	0.55 [−0.62 , 1.72]
Grasses, cover	0.03 [−0.42 , 0.48]
Grasses, biomass	0.12 [−0.32 , 0.57]
Grasses, abundance	−0.25 [−0.46 , −0.03]
Graminoids, cover	0.25 [−0.31 , 0.81]
Graminoids, biomass	2.10 [−1.00 , 5.20]
Graminoids, abundance	0.83 [−0.60 , 2.26]
Graminoids, overall random effects model	**0.02 [−0.12 , 0.15]**
Trees, cover	0.38 [−0.34 , 1.10]
Shrubs, evergreen, biomass	0.37 [−0.17 , 0.91]
Shrubs, evergreen, abundance	0.01 [−0.31 , 0.33]
Shrubs, deciduous, cover	−0.12 [−0.57 , 0.34]
Shrubs, deciduous, biomass	−0.20 [−0.98 , 0.58]
Shrubs, deciduous, abundance	−0.25 [−0.63 , 0.13]
Shrubs (non-forage), cover	−0.82 [−3.59 , 1.94]
Ericoids, deciduous, abundance	−0.18 [−0.40 , 0.04]
Dwarf shrubs, evergreen, abundance	−0.35 [−1.29 , 0.59]
Dwarf shrubs, cover	−0.40 [−1.05 , 0.25]
Woody plants, overall random effects model	**−0.14 [−0.32 , 0.03]**
Lichens, cover	−0.58 [−1.08 , −0.07]
Lichens, biomass	−5.87 [−11.42 , −0.31]
Lichens, abundance	0.34 [−0.45 , 1.12]
Lichens, overall random effects model	**−1.14 [−2.03 , −0.25]**
Liverworts, cover	0.07 [−0.32 , 0.46]
Liverworts, abundance	−0.69 [−2.75 , 1.36]
Bryophytes, cover	0.02 [−0.18 , 0.22]
Bryophytes, abundance	0.36 [−2.12 , 2.84]
Bryophytes, overall random effects model	**−0.02 [−0.23 , 0.18]**
Bare ground, random effects model	**0.27 [−0.06 , 0.59]**

−11 −10 −9 −8 −7 −6 −5 −4 −3 −2 −1 0 1 2 3 4 5

Standardised mean difference between high- and low-exposure data

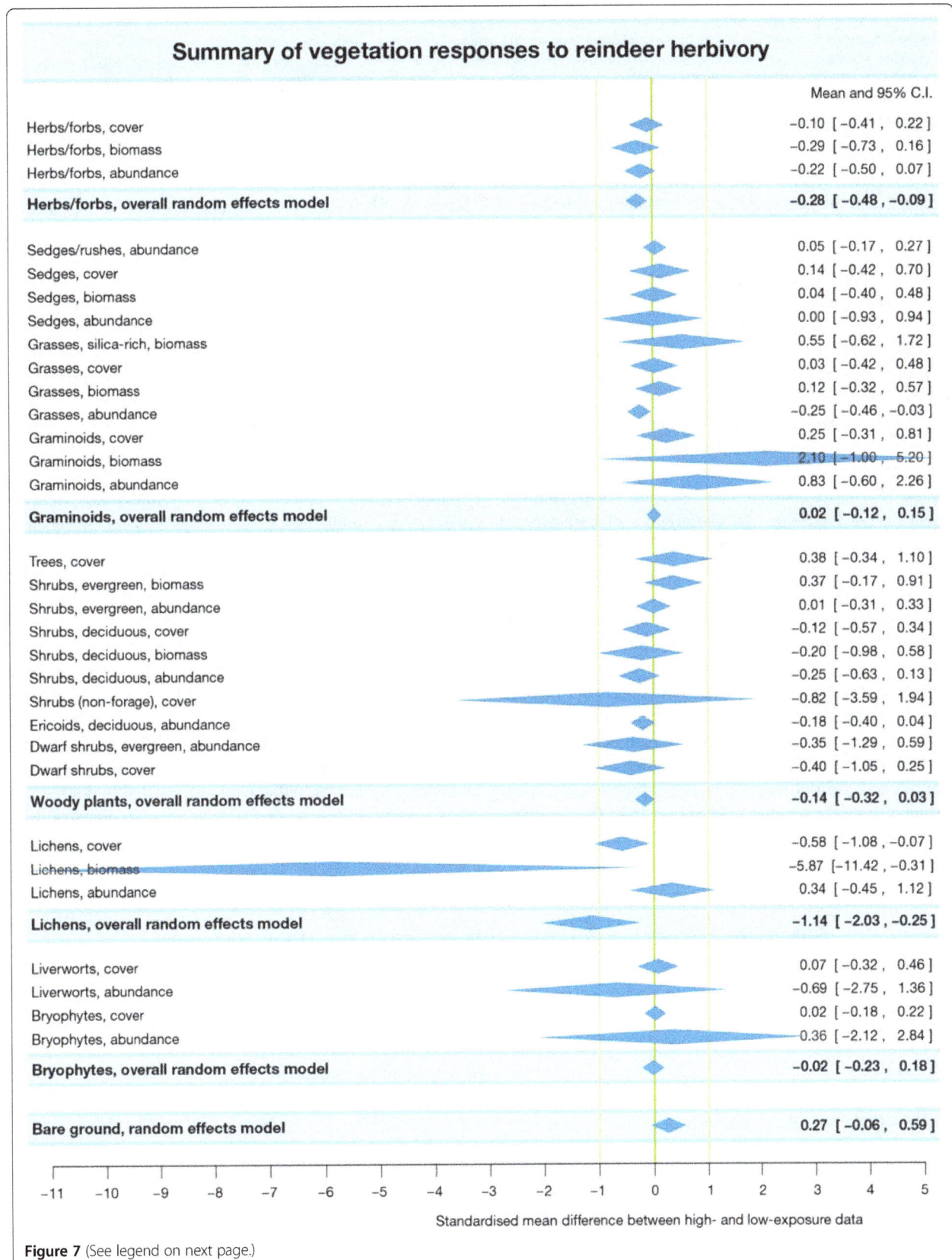

Figure 7 (See legend on next page.)

Figure 7 Summary of meta-analyses of vegetation responses to reindeer herbivory. Subdivisions are based on groups and aspects of vegetation. Positions and widths of the diamond-shaped symbols indicate means and 95% confidence intervals of effect sizes. A position to the left of the green zero line indicates that the cover, biomass or abundance of vegetation is lower at high exposure to reindeer herbivory than at low exposure, and vice versa. Symbols not intersected by the zero line represent statistically significant effects (p < 0.05). Overall models combine data on cover, biomass and abundance. Individual data are presented in forest plots in Additional file 6

difference (i.e. duration of the experiment). The latter measure can be seen as the difference in number of 'reindeer years'. None of the measures of grazing pressure showed any significant relationship with any of the effect sizes that we had calculated, suggesting either that the reindeer densities were too coarsely estimated, or that reindeer density is context-dependent. We illustrate the lack of pattern with meta-regressions for lichens (all aspects combined) and cover of bare ground (see Additional file 8).

However, when differences of lichen cover are plotted against reindeer densities, it becomes clear that the largest differences occurred at sites where lichen cover was highest (these are also the sites with the lowest reindeer densities; see Figure 9, left). Differences of bare-ground cover in response to reindeer exposure also only occurred at sites where reindeer densities were low (see Figure 9, right).

Indeed, the effect of reindeer on lichen cover was negatively related to average lichen cover, whilst the effect of reindeer on bare-ground cover was positively related to average bare-ground cover (Figure 10). Sites with high reindeer densities already had very low lichen cover (perhaps as a result of grazing or trampling before the studies began) as well as a higher cover of bare ground, and did not respond to the exposure differences during the study period. This shows that the composition of the vegetation is very important in determining the effects of herbivory.

Effects on diversity

Vascular plant species richness showed a significant negative overall response to reindeer exposure (Figure 8 and Additional file 7; SMD: −0.15, CI: −0.25, −0.06), but there were no relationships with any of the measures of reindeer densities (see Additional file 8). However, when exploring the diverging responses (ranging from significantly positive to significantly negative), we found them to be significantly related to mean annual temperature (Figure 11; est = 0.138, p = 0.001), suggesting that reindeer exposure tends to increase richness at warmer (more productive) sites and decrease richness at colder (less productive) sites.

Non-vascular cryptogam species richness also showed a negative overall response, but this was not significant (Figure 8 and Additional file 7; SMD: −0.35, CI: −1.02, 0.31), and we found no relationship to reindeer exposure or temperature.

Stratification by other variables (effect modifiers)

We tested the effects of the most commonly reported effect modifiers on cover, since that was the aspect of vegetation most frequently represented in our data set. There was not enough data on other combinations of effect modifiers, aspects, and vegetation types to analyse them in a meaningful manner. The clearest differences were related to domestication – wild reindeer tended to have stronger impacts than semi-domesticated ones. For instance, wild reindeer had a negative effect on lichen cover (SMD: −3.85, CI: −7.17,

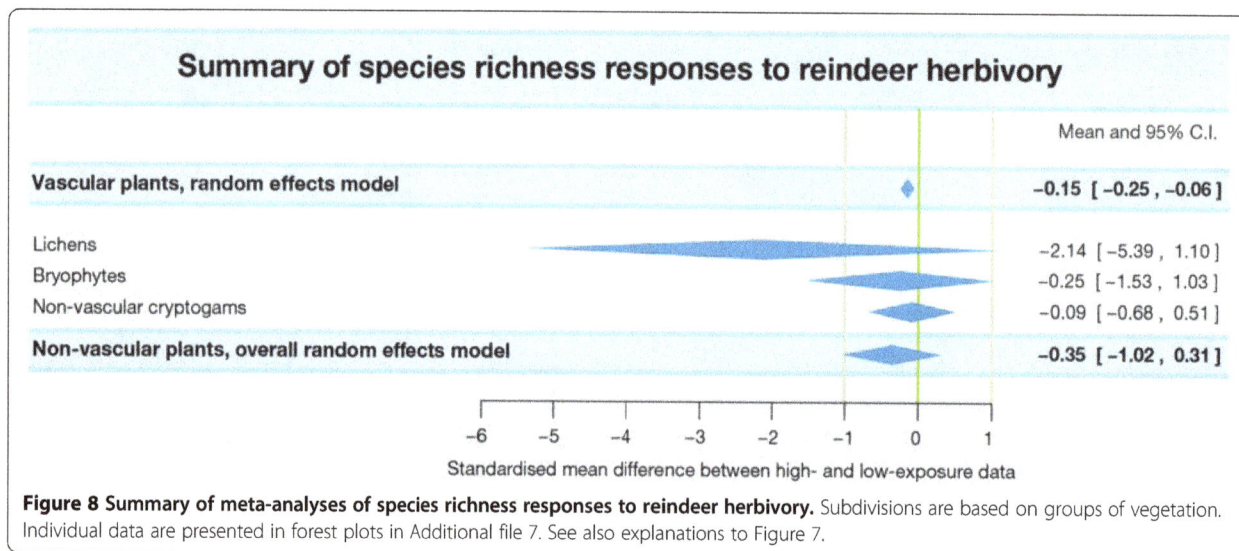

Figure 8 Summary of meta-analyses of species richness responses to reindeer herbivory. Subdivisions are based on groups of vegetation. Individual data are presented in forest plots in Additional file 7. See also explanations to Figure 7.

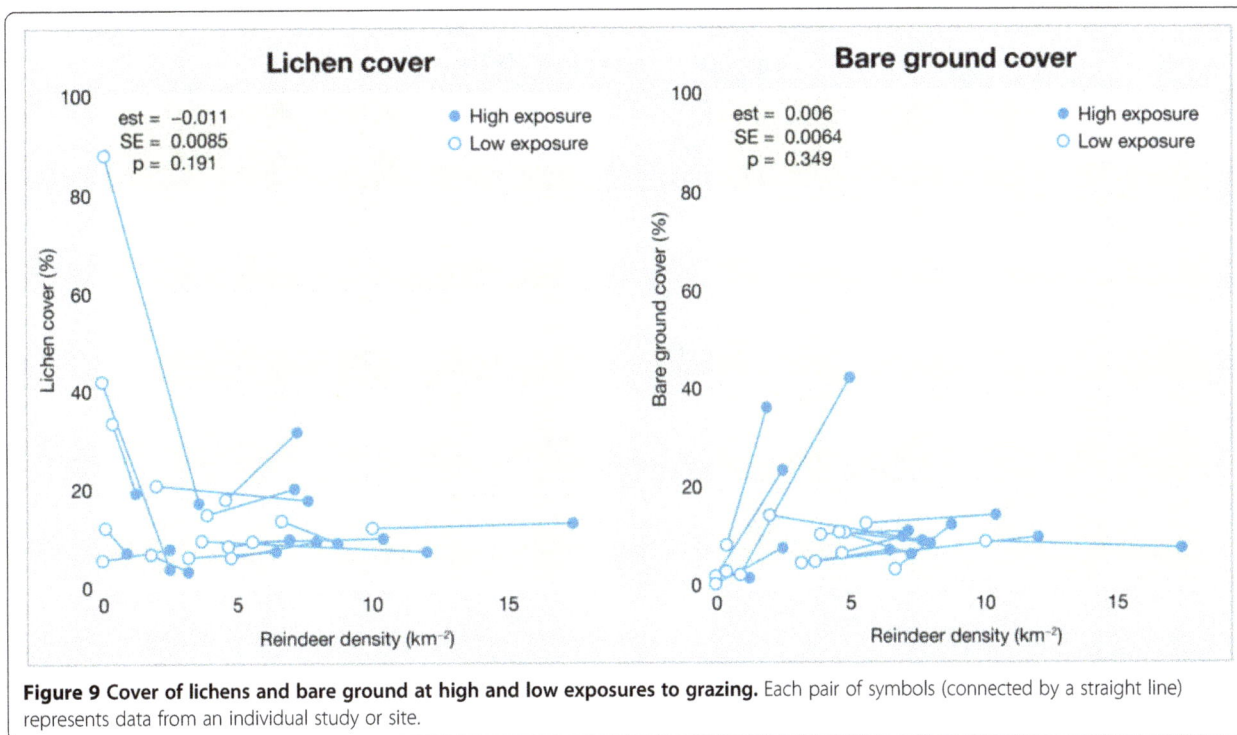

Figure 9 Cover of lichens and bare ground at high and low exposures to grazing. Each pair of symbols (connected by a straight line) represents data from an individual study or site.

–0.53; 5 studies), while semi-domesticated reindeer had strongly varying effects ranging from negative to positive, giving a mean effect size of zero (CI: –0.10, 0.11). Wild reindeer had a positive effect on the cover of graminoids (SMD: 0.74, CI: 0.30, 1.18), while semi-domesticed reindeer had a negative effect (SMD: –0.28, CI: –0.51, –0.05).

We also tested effects of soil moisture (dry, mesic, wet), habitat type (tundra, forest-tundra ecotone, birch forest), seasonality of grazing (summer-autumn-winter), and exposure type (fencing, exclosures, area comparisons, etc.) on the cover of various groups of vegetation, but no patterns emerged.

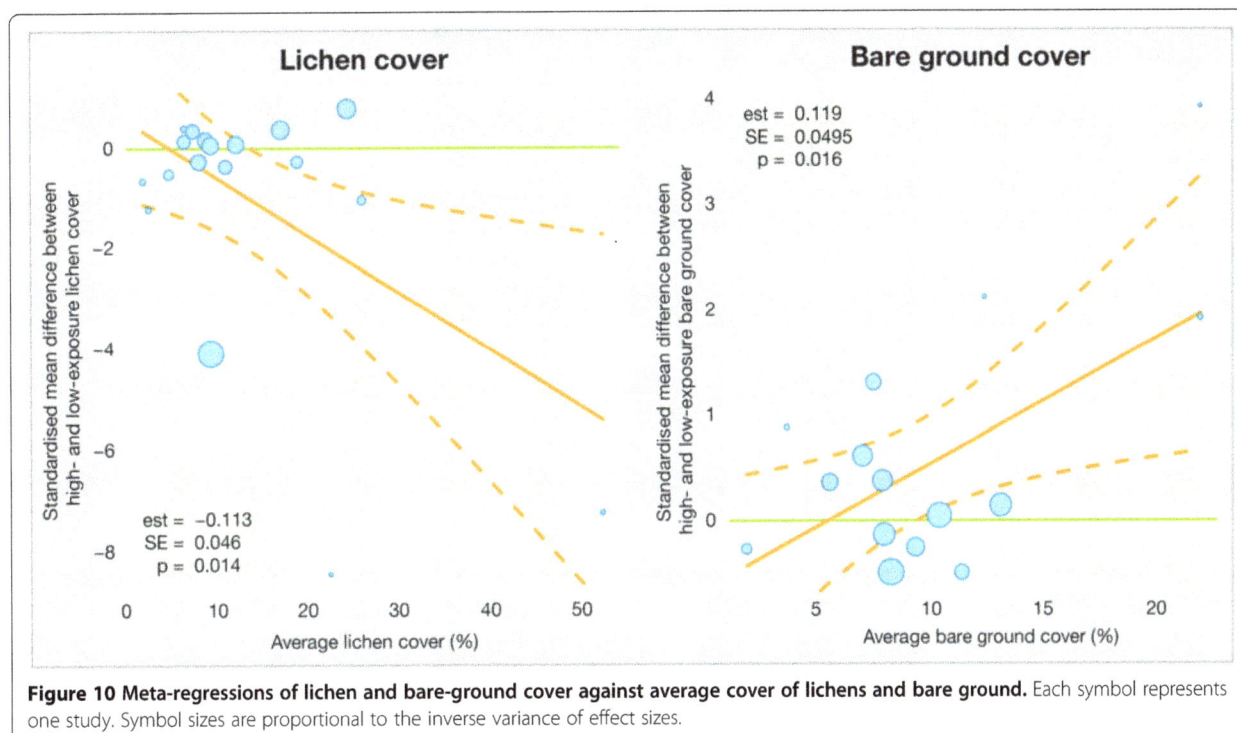

Figure 10 Meta-regressions of lichen and bare-ground cover against average cover of lichens and bare ground. Each symbol represents one study. Symbol sizes are proportional to the inverse variance of effect sizes.

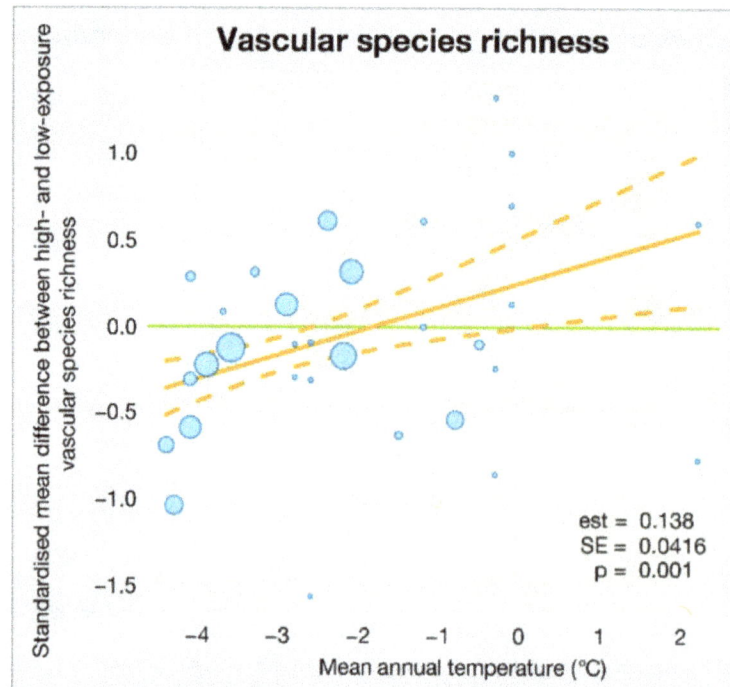

Figure 11 Meta-regression of vascular species richness against mean annual temperature. Each symbol represents one study. Symbol sizes are proportional to the inverse variance of effect sizes.

Discussion

Overall we found large variations in the effects of reindeer on vegetation, and outcomes seem difficult to predict. However, we did identify a few patterns. 1) We did not find strong evidence that reduction of shrubs by reindeer is a general phenomenon. 2) The overall effect of reindeer on graminoids was neutral, indicating that this supposedly grazing-tolerant growth form is not always promoted by reindeer grazing. 3) We found a reduction of forbs, despite the typically low abundance of this growth form. This indicates that forbs are both highly selected by reindeer and vulnerable. They may be an important indicator group of vascular plants. 4) We were able to corroborate the many studies showing that lichens are vulnerable to reindeer activities. 5) Bryophytes were not vulnerable to reindeer exposure.

In conclusion, the effect of reindeer grazing on arctic-alpine vegetation appears to be context-dependent. What, then, could cause this context dependence, i.e. what can explain the heterogeneity of our results?

Reasons for heterogeneity

The distribution of reindeer encompasses large environmental gradients, ranging from low-productive to highly productive sites, from dry to wet environments and from forests to high alpine tundra, with an entire vegetation mosaic being present within these gradients. All different vegetation types that occur here cannot be expected to

respond in the same way to grazing and trampling. For instance, Proulx & Mazumder [71] showed that species richness tended to decline under grazing at low-productive sites, but increase at more productive sites. This is mirrored in our study, where vascular-plant species richness was found to be related to mean temperature. Further, the studies included in our analysis ranged from dry ridge vegetation to riparian herb meadows and mesic birch forests, with many different plant communities occurring within those vegetation types. Unfortunately, there was not enough replication at this scale, nor enough consistency in the presentation of plant communities, to analyse grazing effects for each community type.

Seasonality in grazing may also affect the results, especially if grazing in snow-covered terrain is included. However, only about 4% of the data that we extracted concerned winter grazing, and most of these data were not used in the meta-analyses since they referred to responses of single species. It is thus unlikely that differences in seasonal grazing can explain the heterogeneity we found.

Variations in management systems may be another reason for heterogeneity. Our dataset encompasses impacts of both wild and semi-domesticated reindeer, but with a strong bias towards semi-domesticated herds (>90% of the extracted data). However, management systems were confounded with the initial vegetation state;

for instance, wild reindeer tended to occur in areas with high lichen cover, which precluded us from testing their influence on the outcomes. Even among semi-domesticated herds, management systems vary considerably. Reindeer husbandry operates under many different drivers, forms of legislation, market situations and historical legacies within and between different countries. Grazing pressure may thus vary strongly between different herding districts and systems.

It has been suggested that reindeer have relatively small and subtle overall effects on vegetation in summer grounds. Estimates of biomass removal vary from 0.6% to 2% of net primary productivity, and such levels are unlikely to cause large changes in the vegetation [e.g. 1]. The reasons for the low overall effects could include a decoupling of population dynamics and food resources in the more forage-abundant season, [1,72]. Reindeer numbers are generally determined by winter resources, and the summer resources constitute a pulse of plant biomass growth that the reindeer populations cannot fully respond to. However, reindeer do not use the landscape randomly, and grazing pressure may be much higher than average in preferred sites and vegetation types. For instance, Bråthen & Oksanen [73] estimated a forage removal of preferred plants of the order of 60%.

Heterogeneity is also likely due to the fact that woody plants, often comprising the dominant species in dwarf-shrub tundra [8] and being a characteristic component of reindeer ranges, have highly differential qualities. Ranging from evergreen, allelopathic species such as *Empetrum nigrum* to deciduous ericoids, *Betula nana* and species of *Salix*, the woody plants can be sorted in categories of palatability to reindeer, and hence they are likely to be differentially affected by reindeer herbivory. Exceptions may occur close to fences or in corrals, where reindeer densities are locally increased and where trampling may even have an impact on *Empetrum* [21,49].

Review limitations

Most of the meta-analyses that we conducted are based on a limited number of studies, and they are heavily influenced by a few large-scale investigations (especially Olofsson *et al.* [21,26], Bråthen *et al.* [23], Eriksson *et al.* [62] and Ravolainen *et al.* [24,64]), mostly conducted in northern Norway and Sweden. This, together with the lack of a common research protocol (researchers have measured several aspects of vegetation in dissimilar ways), makes it difficult to summarise results across all studies included in our systematic review. For instance, we found data on 16 different types of species groupings among woody plants (including e.g. non-forage shrubs and deciduous ericoids). When combined with different methods of measuring vegetation (abundance, cover, and

biomass), this resulted in 23 different subgroups in the meta-analysis. These kinds of study-specific subgroupings and methodologies limit our ability to disentangle under what circumstances reindeer have positive, neutral or negative effects on various aspects of vegetation. This is not a trivial problem, since there are also large environmental and social gradients that need to be addressed.

The studies included in our review are fairly recent (mostly dating from 2000 or later). We have very little data from extended experiments based on true replication that could help us to examine the long-term effects of reindeer herbivory. The included studies are also limited both geographically and in terms of vegetation types, and thus not representative of the distribution of reindeer. Most of the data come from tundra sites, and the database is weak on effects on treelines and subalpine forests.

Most of the studies have merely categorised reindeer densities or grazing pressures at treatment and control sites as high or low, without attempting to quantify them further. However, as Bråthen *et al.* [23] show, what is regarded as high in one area can be regarded as low in another, making comparisons difficult. In cases where authors have not reported any data on reindeer densities, we have tried to recalculate them from official population records for entire herding districts. Such data are well correlated with indices of grazing pressure in herding districts in Finnmark [23,24], but much less so in the herding districts in Sweden. In fact, this lack of correlation in Sweden was the basis for the Moen *et al.* study [63], where detailed pellet counts within one herding district were used to distinguish areas with different grazing pressures. As shown in that study, vegetation responses to grazing may vary even within the same district and vegetation type, depending on the actual landscape use of the reindeer. Our estimates of reindeer densities are thus very coarse and possibly misleading.

Conclusions
Implication for policy/management

While our review has gathered a large body of research into vegetation responses to reindeer grazing, we still have to conclude that the evidence base is too weak and scattered to inform policy or management in a detailed way. Some of the reasons for this are small sample sizes, short duration of experiments, limited geographic distribution of studies, difficulties in determining reindeer population densities, and a lack of common research protocols. The included studies are not standardised and representative enough that they can be used as a basis for specific recommendations regarding reindeer ranges in Sweden, nor in the circumpolar region as a whole.

However, an important point that we can make is that vegetation responses to reindeer grazing are context-dependent. For instance, the differential effects of reindeer on vegetation found in this review are most likely due to the variation of vegetation types and dominant growth forms, productivity, and grazing history, just to mention a few important factors. This suggests that there is no panacea that can cover all combinations of factors. Policy and management have to go hand-in-hand with research in individual cases if the dynamics between plants, animals, and humans are to be sufficiently understood.

Implication for research and monitoring

It is unlikely that further research will be able to improve the evidence base much unless a more rigorous research protocol, specifically aimed at the questions raised in this review, becomes generally accepted and implemented. We provide some suggestions for achieving this:

- Adopt a standardised way of aggregating plant species into growth forms and/or functional groups.
- Agree on the aspects of vegetation (biomass, abundance or cover) that should have priority.
- Adopt common measures of vegetation.
- Develop common measures of reindeer densities and grazing/browsing and trampling pressures.
- Implement the protocol in monitoring programs as well as in specific research programs.
- Be very clear in the methods section of an article on how the experiment has been done, and include more information in table and figure captions (such as the number of replicates that the means are based on, and the type of variation that is shown).
- Deposit raw data in public data repositories.
- Include some measure of plant productivity.
- Include some measure of plant palatability and plant nutrition status.
- If possible, adopt a gradient approach to investigate variations in responses.
- External factors, such as the history of land use and reindeer husbandry, should be carefully documented for each study site.

Additional files

Additional file 1: Literature searches.

Additional file 2: Excluded articles.

Additional file 3: Reasons for article exclusions.

Additional file 4: Sources of included articles.

Additional file 5: Narrative tables summarising included studies.

Additional file 6: Forest plots, cover/biomass/abundance.

Additional file 7: Forest plots, species richness.

Additional file 8: Meta-regressions against reindeer density.

Competing interests

The authors declare that they have no competing interests.

Authors' contributions

All authors participated in the drafting, revision and approval of the manuscript.

Acknowledgements

We wish to thank Robert Björk, Johan Olofsson, Virve Ravolainen and Christina Wegener for helpful responses to our requests for raw data. We also thank Annika Hofgaard for her contributions as chair of the review project during its initial phases, and Neal Haddaway for his advice concerning review methodology during the development of the review protocol. Moreover, we are grateful for suggestions and comments received from stakeholders representing the Swedish Environmental Protection Agency, the Ministry for Rural Affairs, the Sami Parliament, the county administrative boards of Jämtland and Norrbotten, Stockholm University, the Swedish Biodiversity Centre, the Swedish Species Information Centre, the Swedish Polar Research Secretariat, and Ájtte (Swedish Mountain and Sami Museum). This systematic review has been financed by the Mistra Council for Evidence-Based Environmental Management (EviEM). EviEM is funded by the Swedish Foundation for Strategic Environmental Research (Mistra) and hosted by the Royal Swedish Academy of Sciences. The review process has been approved by the EviEM Executive Committee, but the authors are solely responsible for the contents and conclusions of the review.

Author details

[1]Mistra Council for Evidence-Based Environmental Management, Royal Swedish Academy of Sciences, P.O. Box 50005, SE-104 05 Stockholm, Sweden. [2]Department of Arctic and Marine Biology, University of Tromsø, NO-9037 Tromsø, Norway. [3]Arctic Centre, University of Lapland, P.O. Box 122, FIN-96101 Rovaniemi, Finland. [4]University Museum, Norwegian University of Science and Technology, NO-7491 Trondheim, Norway. [5]Department of Ecology and Environmental Science, Umeå University, SE-901 87 Umeå, Sweden.

References

1. Moen J, Andersen R, Illius AW. Living in a seasonal environment. In: Danell K, Bergström R, Duncan P, Pastor J, editors. Large Herbivore Ecology, Ecosystem Dynamics and Conservation. Cambridge: Cambridge University Press; 2006. p. 50–70.
2. Helle TP, Jaakkola LM. Transitions in herd management of semi-domesticated reindeer in northern Finland. Ann Zool Fennici. 2008;45:81–101.
3. Renbetesmarksutredningen. Renbetesmarkerna. Stockholm: SOU 12; 1966.
4. Tømmervik H, Riseth JÅ. Historiske tamreintall i Norge fra 1800-tallet fram til i dag. Tromsø: NINA; 2011. Rapport 672.
5. Kortesalmi JJ. Poronhoidon synty ja kehitys Suomessa. Tampere: Tammer-Paino Oy; 2007.
6. Ims RA, Ehrich D, Forbes BC, Huntley B, Walker DA, Wookey PA, et al. Terrestrial Ecosystems. In: Meltofte H, editor. Arctic Biodiversity Assessment. Status and Trends in Arctic Biodiversity. Akureyri: Conservation of Arctic Flora and Fauna; 2013. p. 385–440.
7. McNaughton SJ, Oesterheld M, Frank DA, Williams KJ. Ecosystem-level patterns of primary productivity and herbivory in terrestrial habitats. Nature. 1989;341:142–4.
8. Walker DA, Raynolds MK, Daniels FJA, Einarsson E, Elvebakk A, Gould WA, et al. The circumpolar arctic vegetation map. J Vegetation Sci. 2005;16:267–82.
9. Storeheier PV, Mathiesen SD, Tyler NJC, Olsen MA. Nutritive value of terricolous lichens for reindeer in winter. Lichenologist. 2002;34:247–57.
10. Sørmo W, Haga ØE, Gaare E, Langvatn R, Mathiesen SD. Forage chemistry and fermentation chambers in Svalbard reindeer (*Rangifer tarandus platyrhynchus*). J Zool. 1999;247:247–56.
11. van der Wal R. Do herbivores cause habitat degradation or vegetation state transition? Evidence from the tundra. Oikos. 2006;114:177–86.
12. Dahlberg A, Bültmann H. Fungi. In: Meltofte H, editor. Arctic Biodiversity Assessment. Status and Trends in Arctic Biodiversity. Akureyri: Conservation of Arctic Flora and Fauna; 2013. p. 354–71.

13. Nieminen M, Heiskari U. Diets of freely grazing and captive reindeer during summer and winter. Rangifer. 1989;9:17–34.

14. Jefferies RL, Klein DR, Shaver GR. Vertebrate herbivores and northern plant communities. Reciprocal influences and responses. Oikos. 1994;71:193–206.

15. Iversen M, Fauchald P, Langeland K, Ims RA, Yoccoz NG, Bråthen KA. Phenology and cover of plant growth forms predict herbivore habitat selection in a high latitude ecosystem. PLoS One. 2014;9(6):e100780.

16. Chapin FS, BretHarte MS, Hobbie SE, Zhong HL. Plant functional types as predictors of transient responses of arctic vegetation to global change. J Vegetation Sci. 1996;7:347–58.

17. Wookey PA, Aerts R, Bardgett RD, Baptist F, Bråthen KA, Cornelissen JHC, et al. Ecosystem feedbacks and cascade processes: understanding their role in the responses of Arctic and alpine ecosystems to environmental change. Glob Chang Biol. 2009;15:1153–72.

18. Mårell A, Hofgaard A, Danell K. Nutrient dynamics of reindeer forage species along snowmelt gradients at different ecological scales. Basic and Appl Ecol. 2006;7:13–30.

19. Iversen M, Bråthen KA, Yoccoz NG, Ims RA. Predictors of plant phenology in a diverse high-latitude alpine landscape: growth forms and topography. J Vegetation Sci. 2009;20:903–15.

20. van der Wal R, Brooker RW. Mosses mediate grazer impacts on grass abundance in arctic ecosystems. Function Ecol. 2004;18:77–86.

21. Olofsson J, Kitti H, Rautiainen P, Stark S, Oksanen L. Effects of summer grazing by reindeer on composition of vegetation, productivity and nitrogen cycling. Ecography. 2001;24:13–24.

22. Olofsson J, Stark S, Oksanen L. Reindeer influence on ecosystem processes in the tundra. Oikos. 2004;105:386–96.

23. Bråthen KA, Ims RA, Yoccoz NG, Fauchald P, Tveraa T, Hausner V. Induced shift in ecosystem productivity? Extensive scale effects of abundant large herbivores. Ecosystems. 2007;10:773–89.

24. Ravolainen V, Yoccoz N, Bråthen KA, Ims RA, Iversen M, Gonzalez V. Additive partitioning of diversity reveals no scale-dependent impacts of large ungulates on the structure of tundra plant communities. Ecosystems. 2010;13:157–70.

25. Cairns DM, Moen J. Herbivory influences tree lines. J Ecol. 2004;92:1019–24.

26. Olofsson J, Oksanen L, Callaghan T, Hulme PE, Oksanen T, Suominen O. Herbivores inhibit climate-driven shrub expansion on the tundra. Global Change Biol. 2009;15:2681–93.

27. Ravolainen VT, Bråthen KA, Yoccoz NG, Nguyen JK, Ims RA. Complementary impacts of small rodents and semi-domesticated ungulates limit tall shrub expansion in the tundra. J Appl Ecol. 2014;51:234–41.

28. Ims RA, Henden JA. Collapse of an arctic bird community resulting from ungulate-induced loss of erect shrubs. Biol Conserv. 2012;149:2–5.

29. Olff H, Ritchie ME. Effects of herbivores on grassland plant diversity. Trends Ecol Evol. 1998;13:261–5.

30. Milchunas D, Sala O, Lauenroth WK. A generalized model of the effects of grazing by large herbivores on grassland community structure. Am Nat. 1988;132:87–106.

31. Forbes BC, Kumpula T. The ecological role and geography of reindeer (Rangifer tarandus) in northern Eurasia. Geography Compass. 2009;3/4:1356–80.

32. Forbes BC, Stammler F, Kumpula T, Meschtyb N, Pajunen A, Kaarlejärvi E. High resilience in the Yamal-Nenets social-ecological system, West Siberian Arctic, Russia. PNAS. 2009;106:22041–8.

33. Tveraa T, Ballesteros M, Bårdsen BJ, Fauchald P, Lagergren M, Langeland K, et al. Rovvilt og reindrift – Kunnskapsstatus i Finnmark. Tromsø: NINA; 2012. Rapport 821.

34. Swedish EPA. Förslag till en strategi för miljökvalitetsmålet Storslagen fjällmiljö. Naturvårdsverket, Stockholm: Naturvårdsverket; 2014.

35. Larsen JN, Anisimov OA, Constable A, Hollowed A, Maynard N, Prestrud P, et al. Polar regions. In: Climate Change 2014: Impacts, Adaptation, and Vulnerability. Working Group II contribution to Intergovernmental Panel on Climate Change – 5th Assessment Report. Geneva: WMO/UNEP; 2014. Chapter 28.

36. ACIA. Arctic Climate Impact Assessment. Cambridge: Cambridge University Press; 2005.

37. Verbyla D. The greening and browning of Alaska based on 1982–2003 satellite data. Global Ecol Biogeogr. 2008;17:547–55.

38. Macias-Fauria M, Forbes BC, Zetterberg P, Kumpula T. Eurasian Arctic greening reveals teleconnections and the potential for structurally novel ecosystems. Nat Clim Chang. 2012;2:613–8.

39. Xu L, Myneni RB, Chapin III FS, Callaghan TV, Pinzon JE, Tucker CJ, et al. Temperature and vegetation seasonality diminishment over northern lands. Nat Clim Chang. 2013;3:581–6.

40. Moen J. Climate change: effects on the ecological basis for reindeer husbandry in Sweden. Ambio. 2008;37:304–11.

41. Cohen J, Pullianen J, Menard CB, Johansen B, Oksanen L, Luojos K, et al. Effect of reindeer grazing on snowmelt, albedo and energy balance based on satellite data analyses. Remote Sens Environ. 2013;135:107–17.

42. Moen J, Danell Ö. Reindeer in the Swedish mountains: An assessment of grazing impacts. Ambio. 2003;32:397–402.

43. Swedish Ministry of the Environment. Hållbar utveckling i landet fjällområden. Stockholm: Government Bill 1995/96:226; 1996.

44. Johansen B, Tømmervik H. Finnmarksvidda – vegetasjonskartlegging. Tromsø: FORUT; 1992.

45. Käyhkö J, Pellikka P. Remote sensing of the impact of reindeer grazing on vegetation in northern Fennoscandia using SPOT XS data. Polar Res. 1994;13:115–24.

46. Tømmervik H, Bjerke JW, Gaare E, Johansen B, Thannheiser D. Rapid recovery of recently overexploited winter grazing pastures for reindeer in northern Norway. Fungal Ecol. 2012;5:3–15.

47. Linkowski WI, Lennartsson T: Renbete och biologisk mångfald – kunskapssammanställning. Luleå: Länsstyrelsen i Norrbottens län, Rapport 18/2006; 2006.

48. Swedish Ministry of the Environment. Svenska miljömål. Miljöpolitik för ett hållbart Sverige. Stockholm: Government Bill 1997/98:145; 1998.

49. Olofsson J, Oksanen L. Effects of reindeer density on vascular plant diversity on North Scandinavian mountains. Rangifer. 2005;25:5–18.

50. Bernes C, Bråthen KA, Forbes BC, Hofgaard A, Moen J, Speed JDM. What are the impacts of reindeer/caribou (Rangifer tarandus L.) on arctic and alpine vegetation? A systematic review protocol. Environ Evidence. 2013;2:6.

51. Collaboration for Environmental Evidence: Guidelines for systematic review and evidence synthesis in environmental management. Version 4.2, p. 37. Environmental Evidence: www.environmentalevidence.org/Documents/Guidelines/Guidelines4.2.pdf; 2013.

52. Mysterud A. The concept of overgrazing and its role in management of large herbivores. Wildl Biol. 2006;12:129–41.

53. Suominen O, Olofsson J. Impacts of semi-domesticated reindeer on structure of tundra and forest communities in Fennoscandia: A review. Ann Zool Fennici. 2000;37:233–49.

54. Landis JR, Koch GG. The measurement of observer agreement for categorical data. Biometrics. 1977;33:159–74.

55. Hijmans RJ, Cameron SE, Parra JL, Jones PG, Jarvis A. Very high resolution interpolated climate surfaces for global land areas. Int J Climatol. 2005;25:1965–78.

56. Viechtbauer W. Conducting meta-analyses in R with the metafor package. J Stat Softw. 2010;36:1–48.

57. R Core Team: R: A language and environment for statistical computing. Vienna: R Foundation for Statistical Computing; 2013. http://www.R-project.org/

58. Johansen B, Karlsen SR. Monitoring vegetation changes on Finnmarksvidda, Northern Norway, using Landsat MSS and Landsat TM/ETM+ satellite images. Phytocoenologia. 2005;35:969–84.

59. Gaare E, Tømmervik H, Bjerke JW, Thannheiser D. Overvåking av vinterbeiter i Vest-Finnmark og Karasjok: Ny beskrivelse av fastrutene. Trondheim and Tromsø: Norsk institutt for naturforskning; 2006. Rapport 204.

60. Jandt RR, Meyers CR, Cole M J: Western Arctic Caribou Herd winter habitat monitoring and utilization, 1995–1996. Anchorage: Bureau of Land Management, US Dept. of the Interior, BLM-Alaska Open File Report 88; 2003.

61. Hansen BB, Henriksen S, Aanes R, Sæther B-E. Ungulate impact on vegetation in a two-level trophic system. Polar Biol. 2007;30:549–58.

62. Eriksson O, Niva M, Caruso A. Use and abuse of reindeer range. Acta Phytogeogr Suecica. 2007;87:1–88.

63. Moen J, Boogerd C, Skarin A. Variations in mountain vegetation use by reindeer (Rangifer tarandus) affects dry heath but not grass heath. J Veg Sci. 2009;20:805–13.

64. Ravolainen VT, Bråthen KA, Ims RA, Yoccoz NG, Henden JA, Killengreen ST. Rapid, landscape scale responses in riparian tundra vegetation to exclusion of small and large mammalian herbivores. Basic Appl Ecol. 2011;12:643–53.

65. Zamin TJ, Grogan P. Caribou exclusion during a population low increases deciduous and evergreen shrub species biomass and nitrogen pools in low Arctic tundra. J Ecol. 2013;101:671–83.

66. Kitti H, Forbes BC, Oksanen J. Long- and short-term effects of reindeer grazing on tundra wetland vegetation. Polar Biol. 2009;32:253–61.

67. Olofsson J. Short- and long-term effects of changes in reindeer grazing pressure on tundra heath vegetation. J Ecol. 2006;94:431–40.

68. Cahoon SMP, Sullivan PF, Post E, Welker JM. Large herbivores limit CO_2 uptake and suppress carbon cycle responses to warming in West Greenland. Global Change Biol. 2012;18:469–79.

69. Gonzalez VT, Bråthen KA, Ravolainen VT, Iversen M, Hagen SB. Large-scale grazing history effects on Arctic-alpine germinable seed banks. Plant Ecol. 2010;207:321–31.

70. Post E, Pedersen C. Opposing plant community responses to warming with and without herbivores. PNAS. 2008;105:12353–8.

71. Proulx M, Mazumder A. Reversal of grazing impact on plant species richness in nutrient-poor vs. nutrient-rich ecosystems. Ecology. 1998;79:2581–92.

72. Illius AW, O'Connor TG. Resource heterogeneity and ungulate population dynamics. Oikos. 2000;89:283–94.

73. Bråthen KA, Oksanen J. Reindeer reduce biomass of preferred plant species. J Veg Sci. 2001;12:473–80.

74. den Herder M, Virtanen R, Roininen H. Effects of reindeer browsing on tundra willow and its associated insect herbivores. J Appl Ecol. 2004;41:870–9.

75. Dormann CF, Skarpe C. Flowering, growth and defence in the two sexes: Consequences of herbivore exclusion for Salix polaris. Funct Ecol. 2002;16:649–56.

76. Eskelinen A, Oksanen J. Changes in the abundance, composition and species richness of mountain vegetation in relation to summer grazing by reindeer. J Veg Sci. 2006;17:245–54.

77. Gough L, Shrestha K, Johnson DR, Moon B. Long-term mammalian herbivory and nutrient addition alter lichen community structure in Alaskan dry heath tundra. Arct Antarct Alp Res. 2008;40:65–73.

78. Grellmann D. Plant responses to fertilization and exclusion of grazers on an arctic tundra heath. Oikos. 2002;98:190–204.

79. Lehtonen J, Heikkinen RK. On the recovery of mountain birch after Epirrita damage in Finnish Lapland, with a particular emphasis on reindeer grazing. Ecoscience. 1995;2:349–56.

80. Manseau M, Huot J, Crête M. Effects of summer grazing by caribou on composition and productivity of vegetation: Community and landscape level. J Ecol. 1996;84:503–13.

81. Nellemann C, Jordhøy P, Støen OG, Strand O. Cumulative impacts of tourist resorts on wild reindeer (Rangifer tarandus tarandus) during winter. Arctic. 2000;53:9–17.

82. Nellemann C, Vistnes I, Jordhøy P, Strand O. Winter distribution of wild reindeer in relation to power lines, roads and resorts. Biol Conservation. 2001;101:351–60.

83. Olofsson J, Strengbom J. Response of galling invertebrates on Salix lanata to reindeer herbivory. Oikos. 2000;91:493–8.

84. Olofsson J, Hulme PE, Oksanen L, Suominen O. Importance of large and small mammalian herbivores for the plant community structure in the forest tundra ecotone. Oikos. 2004;106:324–34.

85. Olofsson J, te Beest M, Ericson L. Complex biotic interactions drive longterm vegetation dynamics in a subarctic ecosystem. Phil Trans Roy Soc B. 2013;368:20120486.

86. Pajunen AM. Environmental and biotic determinants of growth and height of arctic willow shrubs along a latitudinal gradient. Arct Antarct Alp Res. 2009;41:478–85.

87. Pedersen C, Post E. Interactions between herbivory and warming in aboveground biomass production of arctic vegetation. BMC Ecol. 2008;8:17.

88. Post E. Erosion of community diversity and stability by herbivore removal under warming. Proc Roy Soc B. 2013;280:20122722.

89. Tømmervik H, Johansen B, Riseth JÅ, Karlsen SR, Solberg B, Høgda KA. Above ground biomass changes in the mountain birch forests and mountain heaths of Finnmarksvidda, northern Norway, in the period 1957–2006. Forest Ecol Manage. 2009;257:244–57.

90. van der Wal R, Brooker R, Cooper E, Langvatn R. Differential effects of reindeer on high Arctic lichens. J Veg Sci. 2001;12:705–10.

91. Vistnes I, Nellemann C, Jordhøy P, Strand O. Effects of infrastructure on migration and range use of wild reindeer. J Wildlife Manage. 2004;68:101–8.

92. Zamin TJ, Grogan P. Birch shrub growth in the low Arctic: The relative importance of experimental warming, enhanced nutrient availability, snow depth and caribou exclusion. Environ Res Lett. 2012;7:034027.

Permissions

List of Contributors

Bo Söderström
Mistra Council for Evidence-Based Environmental Management, Royal Swedish
Academy of Sciences, P.O. Box 50005, SE-104 05 Stockholm, Sweden

Katarina Hedlund
Department of Biology, Lund University, SE-223 62 Lund, Sweden

Louise E Jackson
Department of Land, Air and Water Resources, University of California Davis, One Shields Avenue, Davis, CA 95616, USA

Thomas Kätterer
Department Ecology, SLU, P.O.Box 7044, 750 07 Uppsala, Sweden

Emanuele Lugato
Joint Research Centre, Land Resource Management – Soil action, Institute for Environment & Sustainability (IES), European Commission, Ispra, VA, Italy

Ingrid K Thomsen
Department of Agroecology, Aarhus University, P.O. Box 50, DK-8830 Tjele, Denmark

Helene Bracht Jørgensen
Department of Biology, Lund University, SE-223 62 Lund, Sweden

Neal R Haddaway
Centre for Evidence-Based Conservation, School of Environment, Natural Resources and Geography, Bangor University, Bangor, Gwynedd LL57 2UW, UK

Annette Burden
NERC Centre for Ecology and Hydrology, Bangor, UK

Chris D Evans
NERC Centre for Ecology and Hydrology, Bangor, UK

John R Healey
School of environment, Natural Resources and Geography, Bangor University, Bangor, Gwynedd LL57 2UW, UK

Davey L Jones
School of environment, Natural Resources and Geography, Bangor University, Bangor, Gwynedd LL57 2UW, UK

Sarah E Dalrymple
School of Natural Sciences and Psychology, Liverpool John Moores University, James Parsons Building, Byrom Street, Liverpool L3 3AF, UK

Andrew S Pullin
Centre for Evidence-Based Conservation, School of Environment, Natural Resources and Geography, Bangor University, Bangor, Gwynedd LL57 2UW, UK

Jeremy Sweet
Sweet Environmental Consultants, 6 Green Street, Willingham, CB24 5JA Cambridge, UK

Kaloyan Kostov
Agrobioinstitute, Dragan Tzankov 8, 1164 Sofia, Bulgaria

Sini Savilaakso
Center for International Forestry Research, P.O. Box 0113 BOCBD, Bogor 16000, Indonesia

Claude Garcia
Department of Environmental Systems Science, ETH Zurich, Universitaetstrasse 16, 8092 Zurich, Switzerland
CIRAD, Research Unit Goods and Services of Tropical Forest Ecosystems, Avenue Agropolis, 34398 Montpellier,Cedex 5, France

John Garcia-Ullo
Department of Environmental Systems Science, ETH Zurich, Universitaetstrasse 16, 8092 Zurich, Switzerland

Jaboury Ghazoul
Department of Environmental Systems Science, ETH Zurich, Universitaetstrasse 16, 8092 Zurich, Switzerland

Martha Groom
Interdisciplinary Arts & Sciences, University of Washington, Bothell, 18115 Campus Way NE, Bothell, WA 98011-8246, USA
Program on the Environment, University of Washington, Seattle, WA 98195-1800, USA

Manuel R Guariguata
Center for International Forestry Research, P.O. Box 0113 BOCBD, Bogor 16000, Indonesia

Yves Laumonier
Center for International Forestry Research, P.O. Box 0113 BOCBD, Bogor 16000, Indonesia
CIRAD, Research Unit Goods and Services of Tropical Forest Ecosystems, Avenue Agropolis, 34398 Montpellier,Cedex 5, France

Robert Nasi
Center for International Forestry Research, P.O. Box 0113 BOCBD, Bogor 16000, Indonesia

Gillian Petrokofsky
Biodiversity Institute, Department of Zoology, University of oxford, South Parks Road, Oxford OX1 3PS, UK

Jake Snaddon
Biodiversity Institute, Department of Zoology, University of oxford, South Parks Road, Oxford OX1 3PS, UK

Michal Zrust
Zoological Society of London, Indonesia Programme, Jl. Gunung Gede 1 No.11A, Bogor 16151, Indonesia

Jaqueline Garcia-Yi
Technische Universitaet Muenchen, Chair of Agricultural and Food Economics, Alte Akademie 12, 85354 Freising, Germany

Tiptunya Lapikanonth
Program on Sustainable Resource Management, Hans-Carl-von- Carlowitz-Platz 2, 85354 Freising, Germany

Hanum Vionita
Program on Sustainable Resource Management, Hans-Carl-von- Carlowitz-Platz 2, 85354 Freising, Germany

Hanh Vu
Program on Sustainable Resource Management, Hans-Carl-von- Carlowitz-Platz 2, 85354 Freising, Germany

Shuang Yang
Program on Sustainable Resource Management, Hans-Carl-von- Carlowitz-Platz 2, 85354 Freising, Germany

Yating Zhong
Program on Sustainable Resource Management, Hans-Carl-von- Carlowitz-Platz 2, 85354 Freising, Germany

Yifei Li
Program on Sustainable Resource Management, Hans-Carl-von- Carlowitz-Platz 2, 85354 Freising, Germany

Veronika Nagelschneider
Program on Sustainable Resource Management, Hans-Carl-von- Carlowitz-Platz 2, 85354 Freising, Germany

Birgid Schlindwein
Library of the Technical University of Munich, Branch Weihenstephan, Maximus-von-Imhof-Forum 1-3, 85354 Freising, Germany

Justus Wesseler
Technische Universitaet Muenchen, Chair of Agricultural and Food Economics, Alte Akademie 12, 85354 Freising, Germany
Chair of Agricultural Economics and Rural Policy, Wageningen University, Hollandseweg 1, 6706KN Wageningen, The Netherlands

Maria Ojanen
Center for International Forestry Research (CIFOR), Jalan CIFOR, Situ Gede, Sindang Barang, Bogor (Barat) 16115, Indonesia

Daniel C Miller
School of Natural Resources and Environment, University of Michigan, 4024 Dana Building,40 Church Street, Ann Arbor, MI 48109, USA

Wen Zhou
Center for International Forestry Research (CIFOR), Jalan CIFOR, Situ Gede, Sindang Barang, Bogor (Barat) 16115, Indonesia

Baruani Mshale
School of Natural Resources and Environment, University of Michigan, 4024 Dana Building,40 Church Street, Ann Arbor, MI 48109, USA

Esther Mwangi
Center for International Forestry Research (CIFOR), Jalan CIFOR, Situ Gede, Sindang Barang, Bogor (Barat) 16115, Indonesia

Gillian Petrokofsky
Department of Zoology, The Tinbergen Building, South Parks Road, Oxford OX1 3PS, UK

Neal R Haddaway
Centre for Evidence-Based Conservation, School of the Environment and Natural Resources and Geography, Bangor University, Bangor LL57 2UW, UK

David Styles
School of the Environment and Natural Resources and Geography, Bangor University, Bangor LL57 2UW, UK

Andrew S Pullin
Centre for Evidence-Based Conservation, School of the Environment and Natural Resources and Geography, Bangor University, Bangor LL57 2UW, UK

Kaloyan Kostov
Agrobioinstitute, 8 Dragan Tzankov Blvd, BG-1164 Sofia, Bulgaria

Christian Frølund Damgaard
Department of Bioscience, Aarhus University, P.O. Box 314, Vejlsøvej 25, 8600 Silkeborg, Denmark

Niels Bohse Hendriksen
Department of Environmental Science, Aarhus University, Frederiksborgvej 399, DK-4000 Roskilde, Denmark

Jeremy B Sweet
Sweet Environmental Consultants, 6 Green Street, Willingham, Cambridge CB245 JA, UK

Paul Henning Krogh
Department of Bioscience, Aarhus University, P.O. Box 314, Vejlsøvej 25, 8600 Silkeborg, Denmark

Magnus Land
Mistra EviEM, The Royal Swedish Academy of Sciences, Box 50005, SE-104 05 Stockholm, Sweden

Cynthia A de Wit
Department of Environmental Science and Analytical Chemistry (ACES), Stockholm University, SE-106 91 Stockholm, Sweden

Ian T Cousins
Department of Environmental Science and Analytical Chemistry (ACES), Stockholm University, SE-106 91 Stockholm, Sweden

Dorte Herzke
Norwegian Inst Air Res, FRAM High North Res Ctr Climate & Environm, Tromso, Norway

Jana Johansson
Department of Environmental Science and Analytical Chemistry (ACES), Stockholm University, SE-106 91 Stockholm, Sweden

Jonathan W Martin
Division of Analytical and Environmental Toxicology, 10-102C Clinical Sciences, Edmonton, Alberta T6G 2G3, Canada

Marcel Korth
CEE Johannesburg, Centre for Anthropological Research, University of Johannesburg, Johannesburg, South Africa

Ruth Stewart
CEE Johannesburg, Centre for Anthropological Research, University of Johannesburg, Johannesburg, South Africa EPPI-Centre, Social Science Research Unit, Institute of Education, University of London, London, UK

Laurenz Langer
CEE Johannesburg, Centre for Anthropological Research, University of Johannesburg, Johannesburg, South Africa

Nolizwe Madinga
CEE Johannesburg, Centre for Anthropological Research, University of Johannesburg, Johannesburg, South Africa

Natalie Rebelo Da Silva
CEE Johannesburg, Centre for Anthropological Research, University of Johannesburg, Johannesburg, South Africa

Hazel Zaranyika
CEE Johannesburg, Centre for Anthropological Research, University of Johannesburg, Johannesburg, South Africa

Carina van Rooyen
CEE Johannesburg, Centre for Anthropological Research, University of Johannesburg, Johannesburg, South Africa.

Thea de Wet
CEE Johannesburg, Centre for Anthropological Research, University of Johannesburg, Johannesburg, South Africa

Michael Meissle
Agroscope, Institute for Sustainability Sciences ISS, Reckenholzstrasse 191, 8046 Zurich, Switzerland

Steven E Naranj
USDA-ARS, Arid-Land Agricultural Research Center, 21881 North Cardon Lane, Maricopa 85138, AZ, USA

Christian Kohl
Julius Kühn-Institut, Institute for Biosafety in Plant Biotechnology, Erwin-Baur-Str. 27, 06484 Quedlinburg, Germany

Judith Riedel
Agroscope, Institute for Sustainability Sciences ISS, Reckenholzstrasse 191, 8046 Zurich, Switzerland

Jörg Romeis
Agroscope, Institute for Sustainability Sciences ISS, Reckenholzstrasse 191, 8046 Zurich, Switzerland

Lucas Gutiérrez Rodríguez
Center for International Forestry Research (CIFOR), Jalan CIFOR, Situ Gede, Sindang Barang, Bogor, Barat 16115, Indonesia

Nick Hogarth
Center for International Forestry Research (CIFOR), Jalan CIFOR, Situ Gede, Sindang Barang, Bogor, Barat 16115, Indonesia

Wen Zhou
Center for International Forestry Research (CIFOR), Jalan CIFOR, Situ Gede, Sindang Barang, Bogor, Barat 16115, Indonesia

Louis Putzel
Center for International Forestry Research (CIFOR), Jalan CIFOR, Situ Gede, Sindang Barang, Bogor, Barat 16115, Indonesia

Chen Xie
China National Forestry Economics and Development Research Center, State Forestry Administration, Hepingli Dongjie No. 18, Beijing 100714, China

Kun Zhang
China National Forestry Economics and Development Research Center, State Forestry Administration, Hepingli Dongjie No. 18, Beijing 100714, China

Claes Bernes
Mistra Council for Evidence-Based Environmental Management, Royal Swedish
Academy of Sciences, P.O. Box 50005, SE-104 05 Stockholm, Sweden

Kari Anne Bråthen
Department of Arctic and Marine Biology, University of Tromsø, NO-9037 Tromsø, Norway

Bruce C Forbes
Arctic Centre, University of Lapland, P.O. Box 122, FIN-96101 Rovaniemi, Finland

James DM Speed
University Museum, Norwegian University of Science and Technology, NO-7491 Trondheim, Norway

Jon Moen
Department of Ecology and Environmental Science, Umeå University, SE-901 87 Umeå, Sweden